SpringerWienNewYork

Cristina Prisacariu

Polyurethane Elastomers

From Morphology to Mechanical Aspects

SpringerWienNewYork

Dr. Cristina Prisacariu
The Romanian Academy's Institute of Macromolecular Chemistry "Petru Poni"
Aleea Grigore Ghica Voda 41 A
700487 Iasi
Romania
crispris@icmpp.ro

SpringerWienNewYork is a part of Springer Science + Business Media
springer.at

Typesetting: Camera ready by the authors

Printed on acid-free paper
SPIN 80012835

With 145 Figures

Library of Congress Control Number: 2011932685

ISBN 978-3-7091-0513-9 e-ISBN 978-3-7091-0514-6
DOI 10.1007/978-3-7091-0514-6
SpringerWienNewYork

This book is dedicated to Professor Adrian A. Caraculacu, the "father" of dibenzyl based polyurethanes, and not only.

Foreword

The World Wide Web offers wonderful resources. This is a comfortable medium for today, especially for those who have grown up in a virtual world of information. But printed books remain unique vehicles of information and ideas. They are convenient and organized. They can be thought provoking. Many current textbooks are excellent encyclopedias of information. But too many are overbearing and stately, often mannered in presentation. So they go unread or they have a small audience. And that is a shame. I therefore try here with a writing style that I hope will people to read, to pause and consider, and then to read further.

I also think a scientific book needs to be part of a scientific conversation with the readers. A book must not look like a "*cemetery of dead ideas*", (*Miguel de Unamuno*, 1913). It has to be filled with ideas and evidence that go beyond what is found in the journal articles. A scientific book has to act as an argument, as a commentary, not merely a summary of other people ideas and works. It needs to allow the author to enter into dialogue with the readers and must be able to generate discussion with audiences. A scientific book has to give the readers a relevant description of the topic as well as its overall perspective, argument, or purpose. Authors should always immerse themselves in their books, writing a book they would like to read.

Not doing so, there is the risk that "*the books have the same enemies as people: fire, humidity... weather, and their own content.*" (*Paul Valery, 1921*)

* * *

This book is designed as a review of the state of the art but also brings new contributions in highly specialized subjects, in a still rapidly moving field: the fundamentals of polyurethane elastomers: their chemistry and properties and materials choice for formulation development.

Why choosing to write a book on the fundamental science of polyurethanes? These materials have played a pivotal role in achieving the improvement of the human condition. They have long become an inherent part of our everyday lives. In fact polyurethane products are almost everywhere we look. Just think of a Parisian boulevard or elsewhere, sitting outside a café, think—people watching. Watch them wherever you happen to be—at a bus stop, walking down the high street, sitting in a meeting... Think of what they are wearing, how they are walking (their clothes, their shoes...), what their job might be. Or chose an object in the room. You will find polyurethanes everywhere we look.

Things are moving. If only thinking of the footwear industry—how much it has changed in a century in so many significant ways: machinery, footwear testing, location of manufacturing plants, and especially alternatives to leather where polyurethanes play a leading role. Inexpensively made, polyurethanes do have consistent properties.

Year 1894. The T. & A. Bata Shoe Company in Zlín, The Czech Republic by the siblings Tomáš, Anna and Antonín Bat'a – the eighth generation of shoc-making Batas. (http://www.batova-vila.cz/EN/Thomas-Bata-Foundation-History.html)

Or think of the developments brought by polyurethanes in the clothing industry... Here are example pictures of the benefits of progress by doing polyurethane clothes... isn't it remarkable? You can judge yourselves ☺...

Year 2010 new design — 100% PU jacket.

Year 1910. Dress designed by Paul Poiret. (http://upload.wikimedia.org/wikipedia//commons/4/4b/Poiretdress.jpg)

Polyurethanes are unique materials where the elasticity of rubber is combined with the toughness and durability of metals which allows the engineer to replace rubber, plastic and metal with the ultimate in abrasion resistance and physical properties. Polyurethanes have a large variety of applications due to their unique properties, such as high strength, high hardness, high modulus, and high elongation at break. Therefore there are so many problems these materials can solve: from automotive parts to building and construction, from medicine to electronics, from textiles to furniture. And even so, this is not all.

PUs applications

PUs Applications	Amount of polyurethane used (millions of pounds)	Percentage of total
Building & Construction	1,459	26.8%
Transportation	1,298	23.8%
Furniture & Bedding	1,127	20.7%
Appliances	278	5.1%
Packaging	251	4.6%
Textiles, Fibers & Apparel	181	3.3%
Machinery & Foundry	178	3.3%
Electronics	75	1.4%
Footwear	39	0.7%
Other uses	558	10.2%
Total	5,444	100.0%

Source: http://en.wikipedia.org/wiki/Polyurethane#History

Picture of a polyurethane clepsydra counting history and present of PUs development in the areas of fundamental research and applications

Polyurethanes can be manufactured in an extremely wide range of grades, in densities from 6 to 1 220 kg/m^3 and from flexible elastomers to rigid, hard plastics.

Polyurethane materials exist in a variety of forms including flexible or rigid foams, chemical resistant coatings, specialty adhesives and sealants, and elastomers. Most polyurethanes are *thermoset materials*; they cannot be melted and reshaped as thermoplastic materials. Once the reactions have ceased the thermoset polyurethanes are cured and cannot be heat shaped without degradation. The thermal stability results from the croslinking degree of polymer chains (the crosslink density) and from the nature and frequency of repeating units within the polymer chains.

Thermoplastic polyurethane elastomers are polymers that bridge the gap between rubbers and plastics. They can be used in a wide range of properties, from hard rubbers to soft engineering thermoplastics as they are elastic and melt-processable. They can be processed on extrusion as well as injection, blow and compression molding equipment. They can be vacuum-formed or solution-coated and are suited for a wide variety of fabrication methodologies. They provide a considerable number of physical property combinations: high resilience, good compression set, resistance to abrasions, tears, impacts, weather, and even hydrocarbons. Such materials offer flexibility without the use of plasticizers as well as a broad range of hardness and high elasticity.

The world's consumption of polyurethanes has been growing continuously because as materials they offer so many different properties and unique ecological and economic advantages. In fact, its consumption rate is increasing by up to 5% a year. (source: www.inntecsrl.eu/cms). And there seems to be no end to this trend, quite the opposite actually: polyurethane systems are finding new fields of application and market segments: in Europe or the U.S., in Central or South America, in Asia, Africa or Australia.

Over 2 million people are employed in industries involving polyurethanes, of which more than 800 000 workers are in the EU. The market value of the polyurethanes industry in the EU today is over €150 billion (source: www.polyurethane.org).

Cristina Prisacariu
April 2011

Preface

Everything connects to everything else. (Leonardo da Vinci)

Things are always changing, while everything connects together to everything else, to form a balance of our chances, experiences and things, so to maintain *a structure*. From letters that build the word to language, from the relationship between our actions to their consequences, everything is connected. From social networking to digit communication, or from neuron networks in the brain to chemical neuromediators generating thoughts and viewpoints of people, everything is connected.

They all work together, while each relationship is a discrete thing in itself.

The overwhelming majority of research studies rely on the opinions of individuals, but all of them are linked.

The connections go beyond human relationship. Various systems of interacting things act together to perform a common goal. The details make the product. In this picture, polyurethanes architecture is a small world where everything is linked to everything too by a two — phase microstructure consisting of hard and soft domains. The overall properties of polyurethanes depend on the intrinsic properties of each of the phases, which in turn depend on details of molecular packing of the constituents within the phases, including the density of hydrogen bonds.

There are no hard and fast rules for obtaining the optimum polyurethane elastomers end products, success depends on good formulation selection with well chosen and appropriate processing parameters. Polyurethane elastomers are polymers with remarkable versatility. Through suitable choice of the diisocyanates and diols combined in their synthesis, a huge range of physical properties are achievable. For example, these may be varied from those typical of soft elastomers to those of hard plastics, simply by varying the dominant diol from a flexible long-chain molecule to a small molecule such as ethylene glycol.

This versatility is combined with the processing advantages of a thermoplastic. It is therefore understandable that the polyurethane elastomers are employed in an exceptionally wide range of manufactured products.

An important feature of polyurethane elastomers is that their elastomeric behaviour is highly sensitive to chemical and physical structure of the material, that are potentially under the control of the synthesist. For this advantage to be exploited more effectively, however, there is a need for deeper understanding of the mechanisms by which structure determines the properties of importance.

The present book aims to advance this understanding, by means of a systematic study of the effects of varying chemical composition of model polyurethane elastomers on: (a) their physical structure at the important nanometre length-scale, and (b) the resulting mechanical properties of interest.

The book reviews aspects from the up-to-date literature focused on these topics including our research. In addition the book records selected results from an international NATO project collaboration between The Department of Engineering Science, University of Oxford and our Romanian laboratory at the Institute of Macromolecular Chemistry "Petru Poni", Iasi. This has made possible an unusually comprehensive study of the structure and important physical properties of polyurethane elastomers—a class of polymer of such great industrial importance.

A particular topic approached in this book is new insight into the physical origin of inelastic effects in reinforced elastomers, to assist with the development of physically-based constitutive models.

In the book, a study was made of how aspects of the constitutive responses of polyurethane elastomers vary with composition: the polyaddition procedure, the hard segment, soft segment and chain extender (*mostly diols*) were varied systematically in a large number of systems of model and novel materials. Results were related to: microstructural changes, on the basis of evidence from X-ray scattering, scanning and transmission electron microscopy, atomic force microscopy, and also dynamic mechanical analyses, differential scanning calorimetry and IR dichroic measurements.

The book covers aspects from the morphology to mechanical aspects focused on the elasticity and inelasticity of amorphous to crystalline polyurethane elastomers, in relation to their sensitivity to chemical and physical structure. In such polymers, resilience of the material is an important attribute. In many applications they are in commercial competition with other, relatively soft, elastomeric materials. The choice of material for any given application then hinges on a spectrum of key properties offered by relatively soft polymers—stiffness and strain recovery characterizing their elasticity, but also inelastic effects such as hysteresis and stress relaxation. In these respects the mechanical properties of polyurethane elastomers are similar to those of other elastomers.

In this book, inelastic effects correlated with systematically varying the chemical structure, were also investigated by including quantitative correlations between the magnitude of the Mullins effect and the fractional energy dissipation by hysteresis under cyclic straining, giving common relations approached by all the materials studied.

A major structural feature that was explored in this book is the relationship between the nature of the hard segment (crystallizing or not) and that of the soft segments. Crystallinity has been sometimes observed in the commercial polyurethane elastomers hard phase but this is usually limited to only a few percent for most hard segment structures when solidified from the melt.

But there is one particular diisocyanate, 4,4'-dibenzyl diisocyanate (DBDI) that, in the presence of suitable chain extenders (diols or diamines), gives rise to significant degrees of crystallinity and this is included in the present work.

Schematic of the flexible 4,4'-dibenzyl diisocyanate DBDI crystallizing

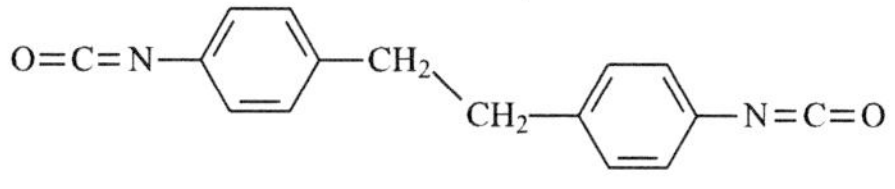

Schematic of a conventional rigid diisocyanate, 4,4'-diphenylmethane diisocyanate (MDI) non-crystallizing

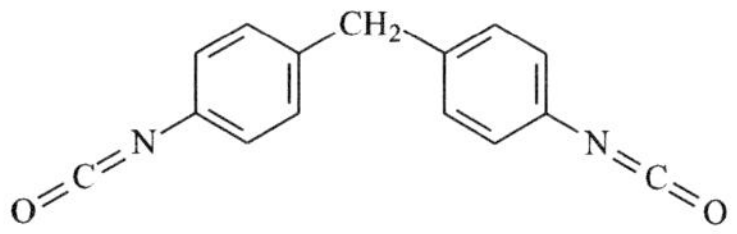

Recent developments in polyurethane materials have been made in our Romanian laboratory, by producing and studying polymers based on the diisocyanate DBDI giving dibenzyl based hard segments (only available from Romania), that allows the variation of hard domain crystallinity as a key structural variable. The conformational mobility of the novel DBDI causes an unusually wide range of mechanical, physical and chemical properties, associated with the possibility of pronounced phase separation into a domain-matrix morphology, and with a high tendency to crystallization and self-association by hydrogen bonding, which is not available with the conventional diisocyanates in traditional melt-cast polyurethanes.

Thus, new polymers were achieved by us, with a controlled ordering of copolymer hard segment blocks on the macromolecular chain.

This book concerns with the study of a large series of systems of PUs based on the aromatic DBDI diisocyanate as compared to materials derived from conventional diisocyanates, mainly 4,4'-diphenylmethane diisocyanate (MDI), as one of the most representative and commonly used aromatic diisocyanates.

The present book is organized into 6 chapters. Chapter 1 describes general aspects on the chemistry of polyurethane elastomers: their origins and development, the principles and synthesis mechanisms, as well as general considerations on the main chemical parameters that define such materials, i. e. diisocyanate, macrodiol and chain extender. Selected considerations regarding the reactivity of diisocyanates, the hydrogen bonding and its dynamic and quantum aspects are also discussed in this chapter.

In chapter 2, there are investigated aspects regarding the morphology and phase separation on two large categories of polyurethane elastomers: materials based on *single* diisocyanates, and materials derived from *mixtures* of diisocyanates. Results are discussed in terms of the structural studies: wide and small angle X-ray scattering (WAXS) and (SAXS), and transmission and scanning electron microscopy

(TEM and SEM) revealing particularities on the polyurethane elastomers morphology. Conventional materials derived from rigid diisocyanates (*mainly MDI*) are compared with the relatively recent polyurethane elastomers obtained by us, that contain crystallizing hard segments generated by DBDI which displayes a variable geometry. *Such materials have higher flow stress in the hard phase caused by stronger phase segregation.*

Thermal techniques such as differential scanning calorimetry (DSC) and dynamic mechanical analysis (DMA) used to describe the thermal behaviour of the materials are also detailed in this book. A comparison is made in chapter 3, between the thermal behaviour of conventional materials and those derived from dibenzyl structures, from the perspective of the *materials with single diisocyanates as compared to those based on mixtures of diioscyanates*. The thermal behaviour of a series of novel polyurethane blends obtained with the isocyanate DBDI has also been described in chapter 3.

The mechanical performance of polyurethane elastomers is strongly affected by the higher-ordered structure of hard segments on the macromolecular chain, and this is also discussed to a large extent in the present book. Inelastic features in the polyurethane elastomers constitutive response are investigated in chapter 4. Their constitutive responses are sensitive to microstructural detail on length scales of order nm and the hard-domain crystallinity exerts a strong effect on inelasticity of the elastomers and this is largely discussed in chapter 4. Tensile properties and the cyclic tensile responses are investigated and the correlation between the Mullins response and hysteresis is made for a large number of polyurethane elastomers with varying systematically the chemical structure: diisocyanates, macrodiols and chain extenders. A comparison has been made between the mechanical response of elastomers achieved without and with excess of isocyanate NCO groups. For the latter, a study of the kinetics and postcure reaction mechanisms has been devoted.

Special consideration was given to the investigation of polyurethane elastomers deformation induced morphological developments and crystallization. In chapter 5, a brief review has been made of selected research in this area, by including materials with hard segments crystallizing or non-crytallizing, based on diisocyanates of variable geometries.

Chapter 6 is dedicated to a detailed description of a new family of crosslinked polyurethanes, with regard to their thermorheological characterization as shape memory materials and the prediction of their shape-memory performance.

Recent progress has been done by us, in the field of PUs chain extended with diols. For this reason, in the present book I have mainly focused on this cathegory of materials.

Understanding the reaction pathways involved, in resolving the subtle morphological evolution at the nanometre level, and capturing mathematically the complex, large-deformation nonlinear viscoelastic mechanical behaviour are assumed to bring new important insights in the world basic research in polyurethanes and towards applied industrial research in this area, and this is the overall target of this book.

Which is why, this book is addressed not only to chemists from both academia and industry, but to all those who are interested in the latest developments in the science of elastomeric materials.

I started from simple to complex approaches so that hopefully this book is also useful and of interest for graduate students and in general for younger audience.

However, while doing so, undoubtedly I am aware that the road to success is always "under construction" ☺.

Acknowledgements

Gratitude is the memory of the heart. (Jean Baptiste Massieu)

Being thankful means to share, and this is what I desire to do, by bringing here my deep appreciation and thankful thoughts to all those whom I have felt close to me while writing this book.

My deepest gratitude and affection to my mentor, Professor Victor Bausic, formerly Professor of Engineering Science at Gh. Asachi Technical University Iasi, Romania, for bringing out the best in me and for opening and widening my perspectives. I am thankful for every moment I spent with him. Gratitude spurs me on to prove myself worthy of what he has have done for me.

I am very grateful to Professor Paul C. Buckley, Professor of Engineering Science at The Department of Engineering Science, University of Oxford. Our eleven year collaboration resulted in valuable research part of which is included in this book.

I wish to thank my friends and collaborators, Dr. Chris Martin (University of Manchester, UK) and Mr. Robert Olley (University of Reading, UK) for extremely helpful assistance in the structural characterization of the materials. I always enjoyed scientific discussions and work with them and wish them all the best for their career and future. I consider myself lucky I had the chance to meet Robert and work with him over the last decade. I thank him for sharing with me his exceptional knowledge in the field of investigation of polyurethane morphology. I thank him for teaching me the in and outs of electron microscopy.

I desire to express my gratitude to Professor Bogdan Simionescu, the Head of the Romanian Institute of Macromolecular Chemistry "Petru Poni", Iasi, where I currently work. He has substantially widened the perspectives of our institute so that today our institute has an excellent international recognition. I deeply appreciate this.

I wish to thank my colleagues and friends, Dr. Virgil Barboiu, Dr. Anton Airinei, Dr. Elena Scortanu, Dr. Mariana Bercea, Dr. Sergiu Coseri and Mr. Bogdan Agapie. Their loyalty, helpfulness, and encouragement have been very precious gifts to me.

I wish to thank my mother, my father and my father-in-law. My deepest gratitude to them! They have lighted and followed me patiently and continuously so I could keep "THE flame" and power within me while writing this book.

I thank my husband, my son and my loving Cicero for being with me whenever I needed them. I thank them for being patient with me; for accepting me as I am. I know it has been difficult for them. As Marcel Proust said, they really are my "charming gardeners who make my soul blossom."

I am very grateful to my publisher, Mr. Stephen Soehnlen, for his delicate and continuous encouragement while writing of this book. He is a warm person with whom I could discuss much more than just work. Thanks a lot!

The permission granted by Elsevier, John John Wiley & Sons, Taylor and Francis, Sage Publications and Nature Publishing Group, to reproduce some of our works (published in "Polymer", "Encyclopedia of Analytical Chemistry", R.A. Meyers Ed., "J.Appl. Polym.Sci", "J.Macromol.Sci.", "IJPAC", "High Perfomance Polymers", and "Polymer J"), included in this book is also gratefully acknowledged.

The permission granted by The Romanian Academy's journal "Rev. Roum. Chim." is also thankfully acknowledged.

I would like to express my gratitude to all those who provided support, offered comments, allowed me to quote their remarks and assisted in the editing, proofreading and design.

Cristina Prisacariu
April, 2011

Contents

Abbreviations and acronyms

PU	polyurethane
SS	soft segment
HS	hard segment
MD	macrodiol
CE	chain extender
DI	diisocyanate
I	isocyanate index
2,4-TDI and 2,6-TDI	toluene-2,4- and toluene-2,6-diisocyanate
MDI	4,4'-methylenebis(phenyl isocyanate)
DBDI	4,4'-dibenzyl diisocyanate
HDI	1,6-diisocyanatohexane
HMDI	hydrogenated MDI
IPDI	isophorone diisocyanate
PEG or PEO	poly(oxyethylene) glycol
PEA	poly(ethylene adipate)diol
PBA	poly(butane adipate) diol
PTMO or PTHF	poly(oxytetramethylene) diol or polytetrahydrofurane diol
PBU	poly(butadiene)diol
PCL or PCD	polycaprolactone diol
BD, BG or BDO	1,4-butanediol
EG	ethylene glycol
DEG	diethylene glycol
M_n	number-average molecular weight of polymer [g/mol]
GPC	gel permeation (size exclusion) chromatography
SEM	scanning electron microscopy
DSC	differential scanning calorimetry
DMA	dynamic mechanical analysis
TG	thermogravimetric analysis
SAXS	small angle X-ray scattering
WAXS	wide angle X-ray scattering
ASTM	American Society for Testing and Materials

T_G	glass transition temperature
T_m	melting point
E^*	complex modulus
E'	loss modulus
E''	loss factor $\tan\delta$
E_c	input energy
E_r	recoverable energy

Chapter 1
Chemistry of polyurethane elastomers

1.1 Origins and development

If you want to understand today, you have to search yesterday. (Pearl Buck)

In order to discuss about the origin and history of polyurethanes, I should start with the history of the diisocyanates. Before year 1850, Wurtz and Hoffman prepared and investigated the properties of both aromatic and aliphatic diisocyanates. Yet, the development of real interest in these raw materials was discouraged at the time because of the difficulty in the original preparatory methods. In year 1884 this difficulty was overcome by Hentschel who developed a convenient way for the preparation of diisocyanates, i.e. the phosgenation of amines. Still, little commercial interest was shown in these until year 1937. Polyurethanes chemistry started in year 1937 at the I.G. Farben Laboratories, a subdivision of Bayer Corporation, in Leverkusen, Germany, where Heinrich Rinke first prepared 1,6-hexamethylene diisocyanate (HDI) and Otto Bayer developed the diisocyanate polyaddition process. The basic idea was related to spinnable products made of HDI and hexa-1,6-diamine (HDA) followed by the publication of German Patent DRP 728981 on November 1937: *"A process for the production of polyurethanes and polyureas"*. The team of inventors consisted of Otto Bayer & al. The utilization of the polyaddition principle to produce polyurethanes from liquid diisocyanates and liquid polyether or polyester diols opened new perspectives, especially when compared to already existing plastics obtained by polymerizing olefins, or by polycondensation. Initial work before year 1940 was focused on polyurethane fibers. In 1938, Heinrich Rinke produced a polymer based on octamethylene diisocyanate and 1,4-butanediol, which he called "polyurethane". While most of the early work on polyurethanes was in the field of foams, it soon became apparent that these materials could be used as synthetic rubbers. The first polyurethane rubbers were prepared in Germany by Pinten (1940). These materials known as I-Gummi had high tensile strengths and abrasion resistance but low tear strength and poor low temperatature properties.

The development of elastic polyurethanes began as a program to find a replacement for rubber during the days of World War II when polyurethanes were utilized

C. Prisacariu, *Polyurethane Elastomers*, DOI 10.1007/978-3-7091-0514-6_1,

as a replacement for rubber, which at the time was expensive and hard to obtain. In year 1950, Bayer et al systematically studied the formulations that led to the advent of the Vulkollan rubbers. These polyurethane elastomers showed many advantages over natural rubber in that they had higher abrasion resistance and tear strength, better resistance to oxygen ageing while displaying good flexibility and elasticity. However these materials had the disadvantage of rapid breakdown in the presence of water and heat.

The first commercial polyurethanes were achieved in Germany by Bayer-Fabenfabriken and in the US by B.F. Goodrich in the 1950's (Schollenbenger et al., 1958). In subsequent decades (starting year 1952 when polyisocyanates became commercially available), there were further developments. The first commercially available polyether polyol, poly(tetramethylene ether) glycol, was introduced by DuPont in 1956 by polymerizing tetrahydrofuran.

By the early 1960's, B.F. Goodrich produced Estane, while Mobay and Upjohn marketed Texin, and Pellethane respectively in the United States. Bayer and Elastogran marketed Desmopan and Elastollan respectively, in Europe.

The polyurethane market started to develop in the 1970's. The production capacity for these products began to grow rapidly, and the industry became dominated by large chemical companies. Bayer & BASF in Germany, ICI in the UK, Dow Chemical, and Upjohn Polymer Chemicals in the USA. During the 1990's the business developed on a global basis and many of the applications became dominated by very large customers consuming increasing large volumes of products. Polyurethane products are now being increasingly made in China driven mainly by the dynamic development of the Chinese economy.

Whatever the continent, today we are surrounded by polyurethane applications in every aspect of our everyday lives. And things are moving, progress is being made continuously...

Fig. 1.1 PU roller... to keep things moving forward.

1.2 Principles and mechanisms

Inanimate objects can be classified scientifically into three major categories; those that don't work, those that break down and those that get lost. (Russell Bake)

Whether or not it is true, I do not like to think of polyurethanes in such terms, simply because these materials can have an excellent durability and lifetime expectancy; they are currently used for the most severe applications in industries such as aerospace and military applications. The inherent strength of these materials is vital in the building of other external structures which are subjected to high levels of weight, impact and weathering.

1.2.1 General considerations

Polyurethanes are a broad class of polymers produced by the polyaddition reaction of a diisocyanate or a polymeric isocyanate with a polyol, in the presence of suitable catalysts and additives. Under the name of polyurethanes, a practically unlimited number of structures can be involved. The only necessary condition is in general reduced to the presence of the urethane group, (–NHCO–O–) on the macromolecular chain with a more or less frequency. The urethane group is usually formed by reaction between isocyanate and hydroxyl groups, although alternative routes such as from bischloroformates and amines are used in special cases.

If in the case of other polymers the macromolecular structure can be displayed in general by the repetition of an only one simple structural unit, (usually well deducible even from the name of the polymer), in the case of polyurethanes the situation becomes more complicated. The polyurethane represents the first example of a polymer building by using the so called "tailoring" proceeding. In this technique which involves in general many more steps, molecular fragments of a high diversity of structures and dimensions are inserted on the same macromolecular chain.

Polyurethane elastomers (PUs) are formed typically by reacting together three chemical constituents: a diisocyanate (aromatic or aliphatic), a long-chain diol (or "macrodiol"), and a small molecule chain-extender diol or a diamine (Fig. 1.2). The resulting polymer may be considered a copolymer of the macrodiol and diisocyanate-chain extender sequences: termed the soft segment (SS) and hard segment (HS) respectively, since the SS usually has its glass transition below ambient temperature and the HS is frequently a relatively rigid aromatic molecule with the glass transition above ambient temperature.

The HS are built from alternating diisocyanate-chain extender sequences while the SS originate from the polyol. Because of rigidity and hydrogen bonding, the HS (either glassy or crystalline) are associated into hard domains acting as physical crosslinks and as filler particles within the rubbery SS matrix [1–4] (Fig. 1.4).

With regard to the SS matrix, just think of the erratic trajectories described by a swarm of butterflies flying around flowers to collect and transport pollen; the

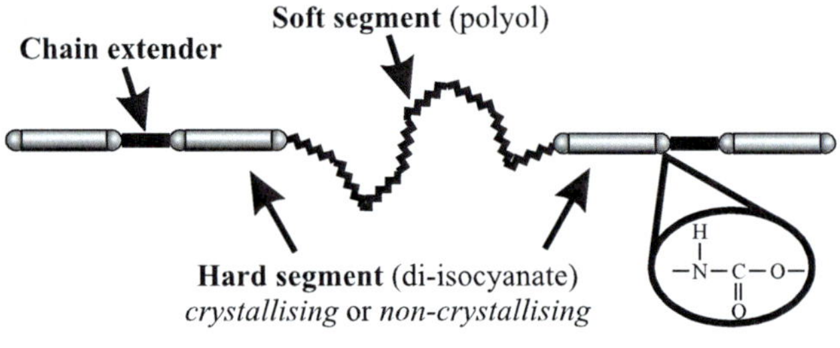

Fig. 1.2 Structure of repeat of a typical PUs. *"Realize that everything connects to everything else."* (Leonardo da Vinci)

changes in their flight directions, to explore regions rising air, switching to circling flights when they succeed in locating a patch… I am suggesting this interpretation because, why not, they could describe a SS matrix. In which case flowers would act as hard segments? ☺

Fig. 1.3 Source: http://anti-plictiseala.roiv.net/wallpaper/poze-desktop-flori-fluturi/

They like entropy, flying disordered.

Due to the incompatibility (different polarity and chemical nature) between HS and SS, phase separation occurs in most PUs. The degree of phase separation and domain formation depends on the HS and SS nature and sizes, on the type of the diisocyanate and polyol employed to produce prepolymers, on the type of the chain extender, and on the molecular weight of the SS. It is also influenced by the hydrogen bond formation between the urethane linkages, by the manufacturing process, and reaction conditions [3–6].

Usually, microphase separation is incomplete and the HS and SS phases still contain certain amounts of the other segment. The mean domain size increases from 10 to 20 nm as the HS content increases. The shapes of hard domains can be in the form of spheres 5-20 nm, or long needles 5 nm thick and 50-300 nm long. PUs phase segregation behaviour has been studied by a variety of characterization techniques, including electron microscopy, small angle X-ray and neutron scattering, infrared dichroism, dynamic mechanical analysis, and differential scanning calorimetry [6–16].

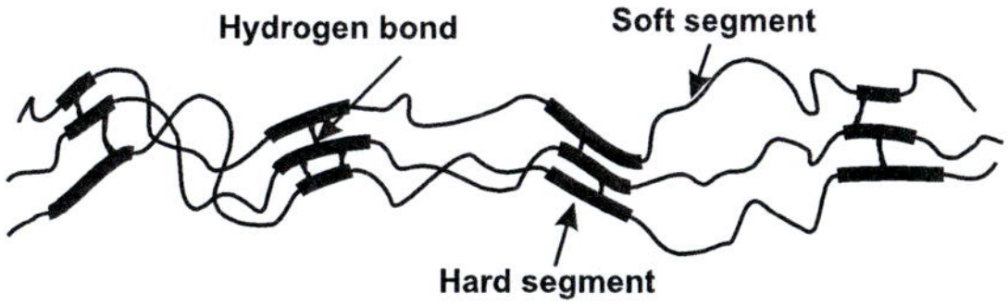

Fig. 1.4 PUs alternating hard segment (HS)–soft segment (SS) structure.

The urethane groups link the HS and the flexible SS together by means of both covalent bonds and hydrogen bonds. They are usually arranged at the borderline between domains. Urethane groups are known for their ability to self-associate via hydrogen bonding (Fig. 1.5). They form linear hydrogen bonds, in either a parallel or an anti parallel fashion. Urethane can form infinite stacks of hydrogen bonded arrays.

Hydrogen bonding has a considerable effect on the kinetics of urethane group formation. Annealing favors stronger hydrogen bonds and increases the PUs melting points as it determines a greater uniformity of the polymer network. The maximum amount of interphase H bond is in phase mixing. In pure homopolyurethanes (i.e. achieved with only HS [4]), the number of hydrogen bonds is bigger than in their corresponding PUs containing SS, where it occurs partial mixing between HS and SS.

PUs elastomeric behaviour requires highly flexible chains, i.e. a low degree of intermolecular interaction, and the presence of crosslinks which prevent sliding of the chains against their neighbors causing plastic flow. The crosslinks can be of a chemical or physical nature. Physical crosslinking can be achieved through hydrogen bonding and hard domain formation while chemical crosslinking is introduced via tri- or multifunctional constituents. Once introduced, chemical crosslinks cannot be easily destroyed by thermal treatment as it is the case with physical crosslinks, except in some special cases of labile chemical groups, producing an irreversible network. Therefore, physically crosslinked PUs allow multiple melting or dissolution of the material which is of great practical importance [1–4].

The PUs properties depend strongly on their macromolecular structure, i.e. the nature and functionality of their constituting (macro)monomers. In general the

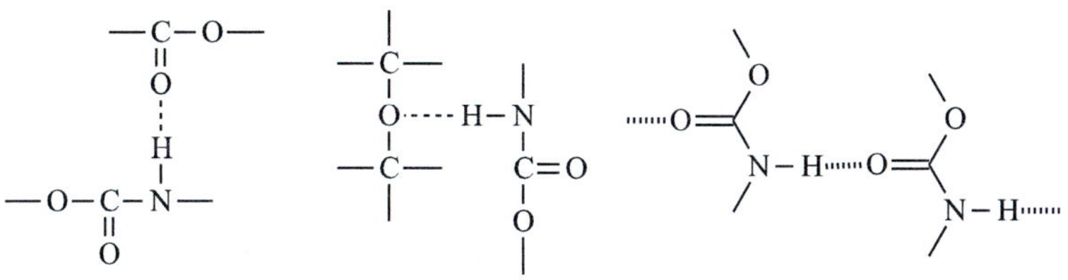

Fig. 1.5 Hydrogen bonding interaction in polyurethanes.

types of PUs that are commercially available are either thermoplastic or thermoset. Thermoplastic PUs can be thermally processed while thermoset elastomers are crosslinked and cannot be thermally processed.

For a polymer to be genuinely thermoplastic, adventitious crosslinking by the action of absorbed moisture must be avoided [1, 3, 4, 17] and this means the chemical composition expressed in numbers N of moles ($N_{diisocyanate} : N_{macrodiol} : N_{chain\ extender}$), is constrained by stoichiometry $N_{diisocyanate} = N_{macrodiol} + N_{chain\ extender}$. But dramatic variations in properties may be obtained by varying the ratio $N_{diisocyanate} : N_{macrodiol}$, thereby changing the fraction of the HS in the copolymer. However, the PUs mechanical properties are influenced not only by the HS fraction. In particular, since the diisocyanate is terminated at each end by urethane –NH–CO–O– linkages that are potentially able to hydrogen bond to corresponding groups on neighbouring molecules, mechanical properties will also depend on the extent to which this potential is realized.

By changing the formulation, materials can be produced with properties ranging from soft elastomers to relatively hard reinforced rubbers However, in contrast to the covalent crosslinks in conventional rubbers, the physical crosslinks provided by the HS microdomains in PUs can be melted, allowing the materials to be moulded or extruded.

The advantages of PUs are related to their high hardness for a given modulus, high abrasion and chemical resistance, excellent mechanical and elastic properties [1–3]. Another important factor is the possibility for the synthesist to tailor the structure by varying the type and the ratio of starting components during processing. Their ability to be fabricated by almost any of the conventional technologies has also been important in leading to their present universal applications [1–4, 18].

1.2.2 Synthesis mechanism

PUs microstructure and mechanical behaviour are strongly dependent on the synthesis method employed. The various methods for producing PUs can be differentiated according to the medium of preparation (bulk, solution, water) and the addition sequence of reactants (*one-step or prepolymer synthesis routes*).

The main difference between the materials prepared by the prepolymer and the one-step methods involves the chain build-up. The PUs obtained via the prepolymer method are statistically more regular in the chain sequence of polyester-diisocyanate-glycol-diisocyanate-polyester, whereas PUs obtained by using the one-step method, (assuming the polyester and the glycol are of equal activity), have a more random sequence. A higher order of crystallinity is obtained in the one-step polymers. The one-step route begins with the slightly favored reaction between glycol and diisocyanate which produces highly crystalline, mobile chain elements. These areas of crystallinity, acting as crosslinks, increase the tensile strength of the one-step PUs.

In the *one-shot polyaddition procedure* all reagents (SS macrodiol, diisocyanate and chain extender) are added at once during the initial reaction. Although this is a commonly used industrial technique, this procedure does not have the control required to yield regular block sequences [2–4]. However the process is faster, easier and more reproducible and can be used to best advantage where the reaction rates of the diol components with the diisocyanate are comparable [1–4, 19, 20]. If the reactivity of both groups in an isocyanate are the same, then the ratio of relative reaction rates will be 2, since the concentration of the second NCO group is half that of the original. Since the diisocyanate and macrodiol components are usually incompatible at lower temperatures, the reaction will take place at the phase interface and the stoichiometric balance could be changed. Incompatibility results in structural heterogeneity, in a change of the average HS length, in segregation during processing and in lower solubility in solvents [21–23]. Incompatibility has the same effect on the prepolymer process too, but a greater regularity.

A higher order of crystallinity in the one-step polymers is also important. The one-step polymers begin with the slightly favored reaction of glycol and diisocyanate which produces highly crystalline, mobile chain elements. Therefore, order can be established before extended polymer growth has occurred. These areas of crystallinity, acting as crosslinks, increase the tensile strength of the one-step elastomers. Although materials are molded at high temperatures, the melting point of the one-step elastomer is still higher [24]. Complete random disorder is not attained during short exposure to temperatures below the melting point and therefore the crystalline order persists in the one-step polymer.

In the one-shot method, factors like the different reactivity of isocyanate groups and the different hydroxyl groups of a polyol and a chain extender may affect the distribution of HS in the chain or even produce a mixture (oligomer) of diisocyanate–chain extender and diisocyanate–macrodiol homopolymers [4, 24, 25].

In contrast in the *two-step polymerization technique* (Fig. 1.6), in the first step of the reaction a prepolymer is produced through the reaction of a SS oligomer with an excess of diisocyanate, followed by chain extension with a short diol(urethane or ester) or a diamine (urea, urethaneurea, amide, or ester-amide) to form the HS and also to increase the overall molecular weight of the polymer.

As the two-step method is more controlled, it produces linear PU chains, fewer side reactions and polydispersities near 2, which is an expected result of step-growth polymers [2, 4, 24].

PU structures obtained by the two step polyaddition procedure tend to be more regular than the corresponding polymers obtained via the one step route. This is because the prepolymer route caps the polyol with the diisocyanate and then connects these oligomers with the chain extenders. This provides a more regular hard–soft–hard–soft sequence than in the one-step synthesis route where the HS size distribution is narrower. The structural regularity may result in better mechanical properties since HS aggregate or crystallize much easier to form physical cross-linking points [26, 27]. As shown, a two-stage polymerization gives a product of a narrower HS molecular weight distribution on the macromolecular chain than does a one-stage polymerization. This tends to reduce the residual stresses and to improve the

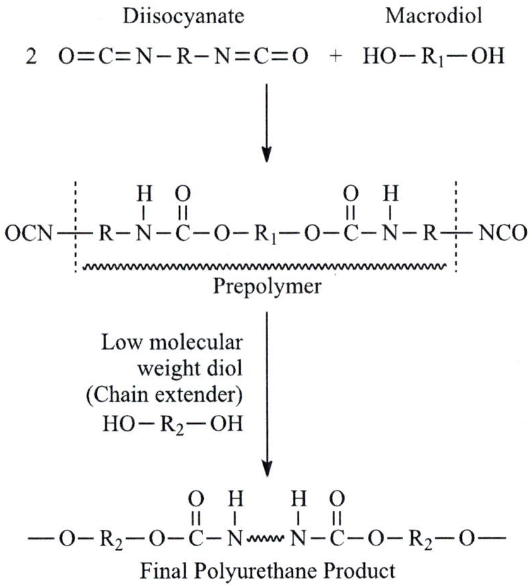

Fig. 1.6 Schematic of the PUs two-step polymerization synthesis route based on diol chain extenders.

mechanical properties in the material especially in the higher deformation region apparently due to the formation of more organized domains under stress.

An ideal PU primary structure assumes a perfect alternating block copolymer. However, under practical conditions, the SS structure, as well as the urethane reaction, follows a statistical Flory distribution, which is why the HS formed is not perfectly alternating and the block length may vary [28].

The bonding of SS and HS microphases in the PU copolymer structure (either by homoselection or by heteroselection) is possible only by joining the isocyanic (NCO) rest of one of the soft (S) or hard (H) elements to the free hydroxyl (OH) group of another S or H element (Fig. 1.7). The total number of isocyanic groups taken into reaction equals the total number of OH groups so to ensure theoretically the formation of infinite macromolecule lengths. In principle there may appear regular structures where the order of S and H segmented repeats periodically (type 1), or irregular structures (types 2, 3 etc).

If the reaction rates of the SS and HS inclusion on the increasing macromolecular chain are equal, it results a HS–SS statistical sequential distribution. If the two

Fig. 1.7 Hard and soft segment distribution on the PU macromolecular chain.

reaction rates are unequal, accumulations appear initially on the increasing chain of the sequences characteristic to the quickest process, followed by the bonding of the sequences which form with a slower rate.

To assess the effect of sequential distribution on PUs with the same chemical structure but with different orders, in the Romanian laboratory we synthesized three series of analogous materials by using different synthesis routes: (a) a sub-set of materials with a complete uncontrolled ordering obtained by the one-step route; (b) similar structures as in case (a), but with a normal quasi ordered structure obtained by the diisocyanate prepolymer synthesis followed by extension with a low molecular diol chain extender, by using the conventional two-step polymerization technique; (c) similar polymers as in the cases (a) and (b) but achieved with an increased sequential ordering of the type –S–HH–S–HH–S–HH– made up by a special inverse step by step synthesis that started with the construction of a hard ordered reactive intermediate OCN–HH–NCO, or more detailed an isocyanate (I)–chain extender (CE) structure of the type I–CE–I–CE–I, which was only then reacted with the macrodiol. As shown [29], such materials with a more regular structure have a more rigid network which requires higher initial strain energies. However this results in lower tensile strengths along with higher values of elongation and hysteresis. A high sequential ordering of the SS and HS domains of the type (c) resulted in a higher tendency of phase separation and crystallization and in a moderate decrease in the material elastomeric performance.

1.2.3 Diisocyanates

The first essential component of a PU is the isocyanate. Molecules that contain two isocyanate groups are called diisocyanates. They are compounds containing the isocyanate group (–NCO). They react with compounds containing alcohol (hydroxyl) groups to produce PUs.

The PUs hard segments can be either aromatic or aliphatic. The aromatic isocyanates are more reactive than the aliphatic diisocyanates, which can only be utilized if their reactivities match the specific polymer reaction and special properties desired in the final product. For example, PUs made from aliphatic isocyanates are light stable [30–33], while materials made from aromatic isocyanates undergo photo degradation [34–36]. Furthermore, the reactivity of an isocyanate group may vary dramatically even within the same class of a diisocyanate.

The structure, substituents, and steric effect can influence reactivity. For example in 2,4-toluene diisocyanate (TDI), the isocyanate group *para* to the methyl group is 25 time more reactive than the other NCO group at the *ortho* position [28]. Moreover, the reactivity of the second NCO group can change as a result of the initial reaction.

Two of the most important aromatic isocyanates are 2,4-tolylene diisocyanate (TDI) *and 4,4-diphenylmethane diisocyanate (MDI)*. TDI consists of a mixture of the 2,4-and 2,6-toluene diisocyanate isomers [37–39].

Other example isocyanates commonly used in PUs applications are shown in Fig. 1.8. A series of model conventional aromatic diisocyanates is depicted: 4,4-diphenylmethane diisocyanate (MDI), 1,5-naphthalene diisocyanate (NDI) and 2,4-tolylene diisocyanate (TDI), most common commercially as an 80:20 mixture of the 2,4-and 2,6-toluene diisocyanate isomers (TDI).

Also shown in Fig. 1.8 are two of the commonly used aliphatic diisocyanates, 2,2,4-trimethyl-1,6-hexamethylene diisocyanate (TMDI) and 1,6-hexamethylene diisocyanate (HDI). These all are isocyanates of rigid geometries [39.40].

As an example, HDI is a highly crystalline diisocyanate, which forms spherulitic structures under moderate annealing conditions (room temperature, on the order of minutes) [41, 42]. Unlike HS based for example on the chain extender 1,4-butanediol and diisocyanate MDI, the couple between 1,4 butanediol and the diisocyanate HDI imparts flexibility to the hard domain, which promotes the rate and extent of microphase segregation and may moderate the PUs plastic deformation behaviour. The use of an aliphatic diisocyanate, such as HDI, also circumvents issues related to thermal degradation and light instability usually associated with the aromatic hard domain [28, 41, 42]. In addition, the high degree of crystallinity in HDI lowers its susceptibility to hydrolytic attack.

The PUs degree of phase separation or domain formation is also strongly influenced by the *symmetry of the diisocyanate* and by the selection of the chain extender, influencing the degree of interaction between the HS. A diisocyanate and a chain extender with a more symmetrical structure increases the formation of organized structures, leading to more complete phase segregation. For example, as shown by Seefried et al. [43], an increased symmetry and bulkiness in the isocyanate MDI as compared to TDI leads to better phase separation and better elastomeric properties. As also shown by us [44–46] the symmetry of the diisocyanate may lead to HS crystallization and better tensile strength, hardness and modulus.

The aromatic or aliphatic diisocyanates have profound effects on the PUs properties [47–58]. The structural rigidity of aromatic HS generally produce elastomers with high tensile strength and modulus and enhanced thermal stability [55]. However, these aromatic units are susceptible to attack by ultraviolet radiation, causing degradation and yellowing without the use of stabilizers.

Aliphatic diisocyanates, such as HDI, offer ultraviolet stability and resistance to hydrolytic degradation, but are less reactive than the aromatic HS [idem 21, 26]. These aliphatic HS may impart more flexibility to the microstructure compared to aromatic HS, as they determine a higher degree of phase segregation [56].

There are also other inexpensive diisocyanates like NDI, produced on a smaller scale. In early works, Bayer et al [57] showed that NDI exhibited excellent physical properties in water extended poly(ethylene adipate) (PEA) based elastomers, due to the rigid naphatalene structure. However, this polymer has poorer stability at high temperature (it decomposes above 150°C).

For comparison reasons and not only—as you will see, in the present book, considerable attention has been given to the aromatic isocyanate of variable geometry ***4,4'-dibenzyl diisocyanate (DBDI)*** [59–63], which (as shown in the preface of this book), is a relatively novel diisocyanate, now available commercially,

Diisocyanates

4,4-methylene-bis-(phenylisocyanate) (MDI)

1,5-naphtalene diisocyanate (NDI)

2,4-toluene diisocyanate (TDI)

2,2,4-trimethyl-1,6-hexamethylene diisocyanate (TMDI)

1,6-hexamethylene diisocyanate (HDI)

4,4'-dibenzyl diisocyanate (DBDI)

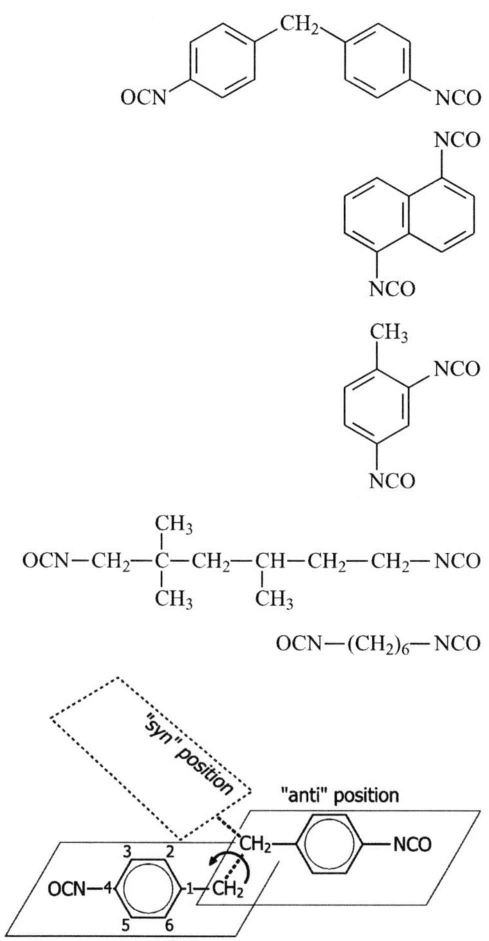

Fig. 1.8 Examples of commonly used aromatic and aliphatic diisocyanates; all of them are rigid; also included is the flexible 4,4'-dibenzyl diisocyanate (DBDI) (produced in our Romanian laboratory), in the extended linear "anti" and contorted "syn" forms.

first synthesized in our Romanian laboratory. It differs from the conventional diisocyanate MDI only in that DBDI contains two methylene groups between the aromatic rings, while the conventional MDI has only one. This apparently small difference nevertheless produces substantial changes in properties of the resulting polymer, because of the possibility of rotation around the central –C–C– bond, allowing more compact packing and crystallization between DBDI hard segments blocks [64–69].

Our research reported in this book is mainly focused on materials obtained with aromatic diisocyanates MDI and DBDI. To compare the DBDI materials, I have chosen MDI as one of the most representative and commonly used aromatic diisocyanates.

1.2.4 Macrodiols

Long-chain diols or "macrodiols", generating the SS, are typically polyester or polyether diols, with molecular weights between 500 to 5 000, although in practice, molecular weights of 1 000 and 2 000 are primarily used.

Polyols are formed by base-catalyzed addition of propylene oxide (PO), ethylene oxide (EO) onto a hydroxyl or amine containing initiator, or by polyesterification of a di-acid, such as adipic acid, with glycols, such as ethylene glycol or dipropylene glycol. *Polyols extended with PO or EO are polyether polyols. Polyols formed by polyesterification are polyester polyols.* The choice of extender, and molecular weight of the polyol greatly affect its physical state, and the PUs elastomeric properties. Important characteristics of polyols are their molecular weight, percentage of primary hydroxyl groups, functionality, and viscosity.

Common SS include polyethers, polyesters and polyalkyl glycols with glass transition temperatures in the range of –70°to –30°C. Commonly used macrodiols in the PUs synthesis are: polyalkyl-diols, such as: polyisobutylene diol [70], polybutadiene (PBU) [20, 71], or oligo-butadiene diols [72] as well as hydrogenated polybutadiene diol [20]; polyether diols: polytetrahydrofuran (PTHF or PTMO) [50–52], polyethylene glycol (PEG) or (PEO) [73], polypropyleneoxide (PPO) [73] or mixed blocks of them: PEO–PPO–PEO [74] and PPO–THF [54]; polyester diols: poly(ethylene adipate) (PEA) [4, 20], poly(butylene adipate) (PBA) [20, 73], and latterly polycaprolactone diol (PCL or PCD) [75], polyalkylcarbonate polyol [20] or mixed blocks of them, for example: poly(carbonate-co-ester)diol [76], poly(hexamethylene-carbonate)diol [77], as well as poly(hexamethylene-carbonate-co-caprolactone)diol [78] and a mixed block copolymer of polyether and polyester blocks: PCL–b–PTHF–b–PCL [79]. Examples schemes of macrodiols are shown in Fig. 1.9.

The ether bond (C–O) bond in the polyether SS and the ester (CO–O) bond in the polyester are capable of hydrogen bonding with the urethane linkages (NH–CO) in the hard domain, influencing the degree of microphase segregation; polyesters are generally stronger hydrogen bond acceptors than polyethers. These general observations are dependent upon the crystallinity or ordering of the SS, which is a function of molecular weight.

Hydroxy terminated polybutadiene (PBU) [2] has also been used for special types of PUs elastomers where products of low water absorption and sensitivity are required. In general the mechanical strength of this class of PUs is lower than that of their polyester and polyether analogues with the advantage of giving PUs of very low glass transition temperatures and hence temperature flexibility.

The degree of phase separation is strongly influenced by the type of the SS [2–4, 80]. Polyester polyols differ from the polyether polyols as they virtually have no

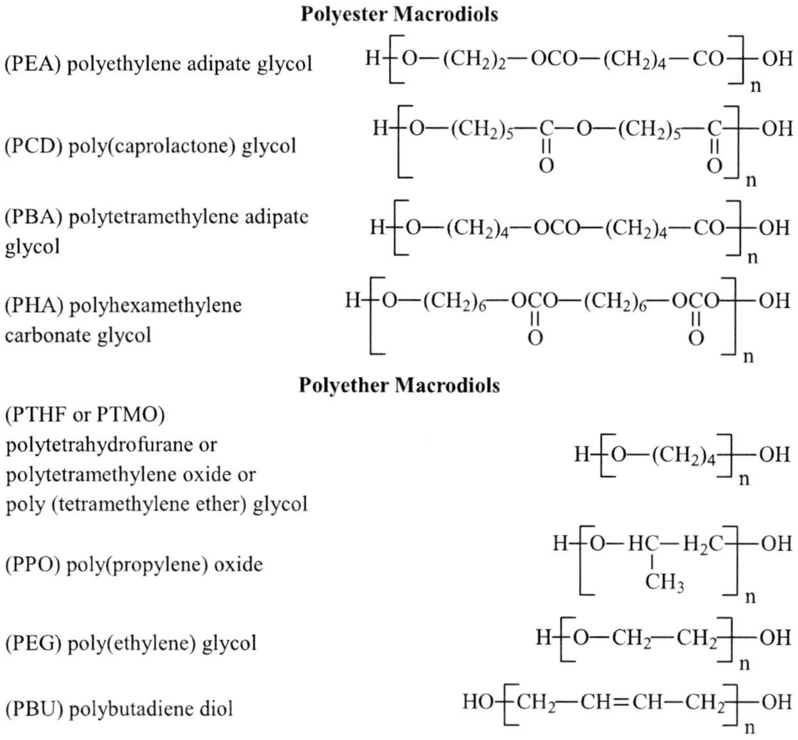

Fig. 1.9 Examples of commonly used macrodiols in the PUs synthesis.

unreactive end groups, a factor which contributes substantially to a higher strength of polyester based PUs. The increased polarity of the ester carbonyl group leads to stronger hydrogen bonding between HS and SS. Thus polyester PUs are stiffer and have a higher strength than the polyetheric PUs with a better resistance to high temperature, solvents and oxidation. Polyether polyols give higher resilience with a good hydrolysis resistance.

As shown by Sung and Schneider [81], Clough and Schneider [82], Shibayama [83] and also by ourselves [80], phase mixing is enhanced in polyester based PUs due to the increased number of ester groups capable to form hydrogen bonds with the urethane group from the HS. Higher molecular weight polyethers provide better phase separation in polyether than in polyester PUs.

Stronger phase segregation has been observed in polyether based PUs especially when employed with crystallizable HS. [3, 4, 80]. Polyesters have stronger intermolecular interactions between the N–H and the ester carbonyl, which leads to higher interfacial mixing and higher modulus. However these interactions lead to increased temperature dependence of the PU properties including more hysteresis or stress softening.

The selection of the SS type also has a great influence of the PUs structure and properties. The polyether polyols have a good hydrolisis resistance. The increased polarity of the ester carbonyl group leads to stronger hydrogen bonding between the HS and SS, which increase the material strength. In contrast, polyester based PUs are stiffer and stronger, with a better ressitance to solvents and high temperatures. Polyester polyols differ from the poyether polyols as they virtually have no unreactive end groups, which contributes significantly to the increase of the strength of the material.

Polyethers usually have a lower glass transisiton temperature and a weaker interchain force than polyesters, therefore rendering PUs with more reduced mechanical properties. This is often attributed to the stronger hydrogen bonding between the NH and the ester carbonyl group, rather than urethane NH–ether oxygen bond. Among polyethers, the PTMO based PUs have the best physical properties, which reflect the regularity of its chain structure and its ability to crystallize upon stretching.

Polyester PUs have good material properties, but they are susceptible to hydrolytic cleavage of the ester linkage while polyether based urethanes have relatively high resistance to hydrolytic chain scission. Polyethylene oxide (PEO) based materials exhibit poor water resistance due to the hydrophilic nature of the ethylene oxide. Although the PUs mechanical properties obtained with polypropylene oxide (PPO) are not as good as those made from PTMO, PPO has also been widely used because of its low cost and reasonable hydrolytic stability [2].

1.2.5 Chain extenders

Typical chain extenders are low molecular weight (Mw) < 400 difunctional intermediates designed to react with the isocyanate groups to become part of the HS. These materials typically produce urethane or urea based HS which are highly regular and polar, leading to significant local phase ordering and possibly crystallinity.

The chain extenders play a very important role. Without a chain extender, a PU formed by directly reacting diisocyanate and polyol generally has very poor physical properties and often does not exhibit microphase separation.

The introduction of a chain extender may increase the HS length to allow HS segregation, which results in good mechanical properties, such as an increase in the modulus and an increase in the HS glass transition temperature of the polymer.

PUs chain extenders can be categorized into two classes: aromatic diols and diamines and the corresponding aliphatic diols and diamines. In general PUs chain extended with an aliphatic diol produce a softer material than do their aromatic chain extended counterparts [84]. Also, diamine chain extenders are much more reactive than diol chain extenders and give properties superior to those of similar polymers prepared with the equivalent diol chain extenders. This is due to the HS (urea linkage) which has a higher density of hydrogen bonding, which results in a higher glass transition temperature and higher thermal stability. However, for the same reason, PU ureas made from diamine chain extenders tend to be less soluble

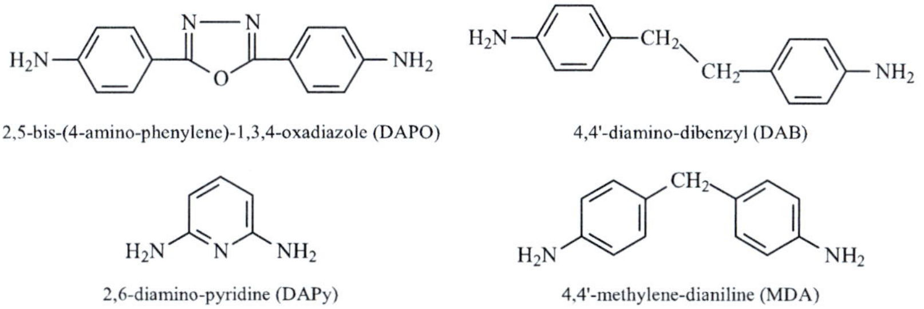

Fig. 1.10 Example diamine chain extenders.

in common solvents and therefore they are more difficult to melt process. Moreover, due to electron delocalization, the aromatic chain extenders have less reactivity than aliphatic chain extenders, which can be favourable in reactions that need to be highly controlled.

Examples of diol chain extenders (Table 1.1) include 1,6-hexanediol (HG) [20], ethylene glycol (EG) [58, 60, 61], 1,4-butanediol (BDO, BD or BG) [53, 75, 77, 78], diethylene glycol (DEG) [61], diacetylene diols [50] such as: 2,4-hexadiyne-1,6-diol or 5,7-dodecadiyne-1,12-diol and some aromatic diols: triazine diols [72, 85] or 4,4'-(ethane-1,2-diyl)bis(benzenethiohexanol) [86].

Table 1.1 Example diol chain extenders

(EG) (ethylene glycol)	HO–$(CH_2)_2$–OH
(BG or BDO) (butylene glycol)	HO–$(CH_2)_4$–OH
(DEG) (diethylene glycol)	HO–$(CH_2)_2$–O–$(CH_2)_2$–O

Examples of diamine chain extenders (Fig. 1.10) include 2,5-bis-(4-amino-phenylene-1,3,4-oxadiazole) (DAPO), 4,4'-diamino-dibenzyl (DAB), 2,6-diamino-pyridine (DAPy) or 4,4'-methylene-diamine (MDA).

The chain extender structure strongly influences the PUs mechanical performance. Modifying the ratio between the polyol and chain extender, PUs may result in a change from a hard, brittle material to a rubbery elastomer, as a result of the variation of the HS concentration (defined as the ratio of the mass of the non-polyol components to the total mass of the polymer) [2–4].

Also, the study of the effect of the chain extender length on the structure of MDI/diol hard segments has shown that BDO and longer diol chain extenders (up to octandiol) produce PUs structures with properties that depend on whether the diol has an *even* or *odd* number of methylene (CH_2) groups. The even diol polymers adopt the lowest energy fully extended conformations that allow for hydrogen bonding in both directions perpendicular to the chain axis. Such a hydrogen bonding network would not be possible for the odd diol polymers in the extended conformation because they adopt contracted, higher energy conformations. Both the *odd* and *even* diol polymers adopt staggered chain structures with triclinic unit cells, but the even diol polymers have a higher crystalline order. The first two members of the homologous series, prepared by using ethylene glycol and propandiol, are exceptions to the above behaviour and adopt contracted unstaggered structures. These two diols are too short to allow packing of the diisocyanate units in the same way as for the longer chain extenders [3, 4].

When a diamine is used as a chain extender, better physical properties may result than using a diol, due to stronger interchain interaction from the urea linkage. However, the use of an aliphatic glycol containing aromatic structure may result in materials with higher modulus and hardness [87].

In the present book we will focus mainly on PUs derived from diol chain extenders. This is due to recent progress we have done in the research area of these materials.

1.2.6 General considerations regarding the isocyanate structure

The isocyanate group represents a strained structure able to perform addition reactions when interacting with various compounds that can be generally considered as phyloisocyanaic reaction partners. As a rule, the addition performs at the N=C double bond and only exceptionally at the C=O group [88, 89].

The isocyanate group reactivity is based on the polarisation of these groups represented by the following resonance limit structures [1] (equation 1.1):

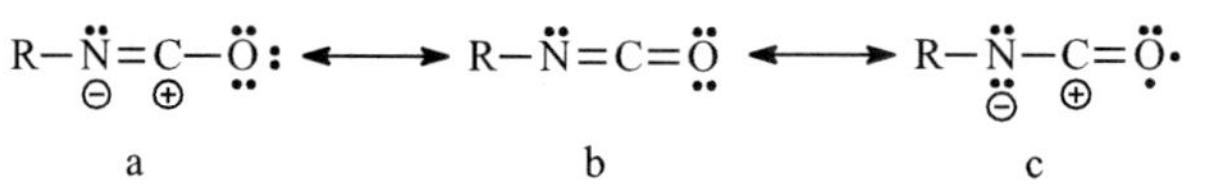

(1.1)

According to the molecular orbital theory, the isocyanic group has a linear structure, the two cumulated double bound N=C and C=O lying in principle on the same axis. The π electrons from the respective double bond are situated in two different perpendicular planes. As shown by Caraculacu et al. [89], due to the higher electronegativity of the nitrogen and oxygen atoms the electron density is shifted toward the ends of this system. Theoretically the π electrons of the N=C and of C=O groups should not interact with each another, and the both double bonds should be considered as isolated groups.

The problem of the electron density distribution in the case of the isocyanate group has been investigated. There still is large disagreement between researchers. This element is very important as it determines the reaction mechanism by specifying which of the double bonds, C=N or C=O is the first affected in the initial stage of the reaction [89].

With regard to the urethane group formation, Kozak et al [90] concluded that the electronic deficit at the NCO carbon is not a sufficient condition to promote the reaction.

The NCO group deviation from the linearity although apparently of a smaller importance, induces the possibility of the existence of two planar isomeric forms corresponding to the *cis* or *trans* structure [90]. The possible effects of this isomerism on the energetic states belonging to some supposed structures which could appear as a result of the interactions between the NCO methyl isocyanate group and the C=C methyl metacrylate double bond during their copolymerisation, were evaluated by quanto-chemical methods by Nizelsky [91].

The complexity of the fundamental understanding of isocyanate chemistry increases even more if considering other particular isocyanate structures. There are known very large series of isocyanate types, unexplored until now in the polymer chemistry, in which the radical bonded to the –NCO group is not an aromatic or aliphatic hydrocarbonate rest. Consequently the electronic structure and therefore the reactivity of these isocyanates can be strongly influenced by the nature of the alternative radical attached to the nitrogen atom of the NCO group. These substituents can display more or less electron withdrawing or donating capacities. The isocyanates with an electron-withdrawing group, such as sulfonyl [92, 93], chlorosulfonyl [94], phosphorous [95] and carbonyl [96, 97], attached to the isocyanic group have been largely studied. The same consideration was given in the opposite case when the presence of an electron-donating atom such as oxygen, nitrogen or sulphur, modifies the chemical behaviour of NCO system. Reichen [98] presented an interesting review which illustrates the multiple possibilities of these new classes as of compounds.

Even though the area of isocyanates chemistry becomes more and more wide, the largest majority of applications still remains based on the classic addition reactions leading in general to the appearance of urethane or urea groups. Numerous works were dedicated to the elucidation of the mechanism of these fundamental reactions. However, many fundamental problems are yet unclear so far.

1.2.7 Reactivity of diisocyanates

In our previous papers [45], a series of PUs based on dibenzylic monomers were presented. These monomers belong to a large category of diaromatic compounds with variable geometry. This is due to rotation around the A–B bridge as shown in Fig. 1.11, where X and Y = functional groups, and A and B = –CH–, –CR, R–, –CO–, –NH–, –NR–, –S–, –SO–, and $–SO_2–$. As previously shown by us [59–63] the presence of the dibenzylic structure induces significant effects on the macromolecular chains, especially those in connection with the rearranging capacity of the molecular fragments and thus to the increase in the degree of crystallinity. This is reflected in the PUs mechanical properties. During the synthesis of urethanic polymers through homogeneous polyaddition, disproportionation accompanies the reaction, leading to the simultaneous synthesis of PUs with different relative ratios of the components both in solution and suspension media.

In order to clarify the mechanism of this process, the dibenzyl diisocyanates reactivities were studied by us. For comparison reasons, the reactivity of one of the most common diisocyanates, 4,4-MDI was also investigated.

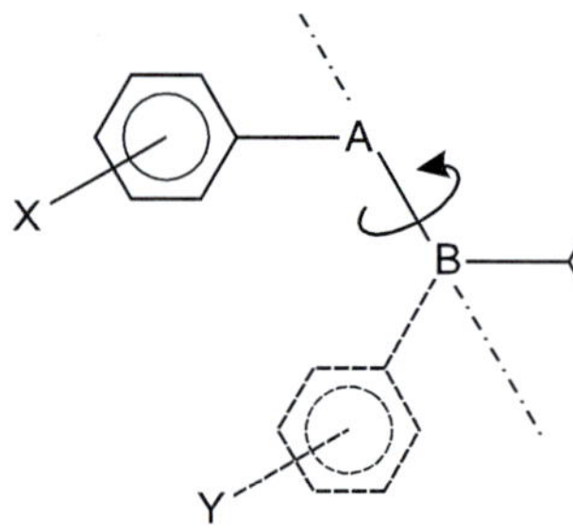

Fig. 1.11 Schematic of dibenzyl monomers.

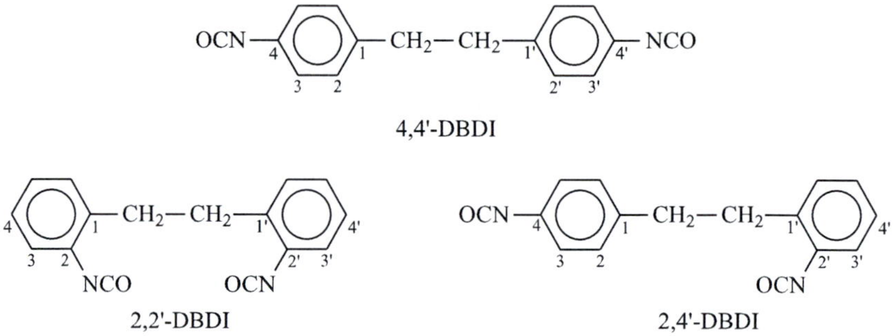

Fig. 1.12 Schematic of DBDI intermediates.

A series of intermediates (Fig. 1.12) obtained by us was considered. They were synthesized with two identical functional groups situated in different rings: 2,2'-DBDI, 2,4'-DBDI and 4,4'-DBDI as described elsewhere [45]. A study was made by us on the reactivity of 2,2'-, 2,4'-, and 4,4'-DBDI with n-butanol in benzene.

The concentrations of all reaction species were monitored by using high performance liquid chromatography (HPLC). The NCO/OH group ratio had a value of 2 as it has in the case of PUs prepolymer synthesis conditions. Benzene was selected as the reaction solvent because of its low polarity, thus permitting the monitoring of the catalytic effects of the alcohol and the resulting urethanes. The effect of various functional groups and solvents on polyaddition processes was also studied, e.g., the influence of alcohol concentration, the addition of urethane, ester, and ether groups, and polar solvents.

The molar fraction (–FN) of unreacted diisocyanate was determined as a function of the reduced time τ (Fig. 1.13) [44]. Due to the changes in the rate constants during reaction, both the initial and average rate-constant values were calculated. The rate constants of the initial and average stages of the process and the values of the rate-constant ratios were determined. These values corresponded to the reactions of the first and second NCO groups of the symmetrical DBDI isocyanates (2,2' and 4,4') and to those of the 4,4'-MDI reaction. Note that the reactivity values of 4,4'-DBDI

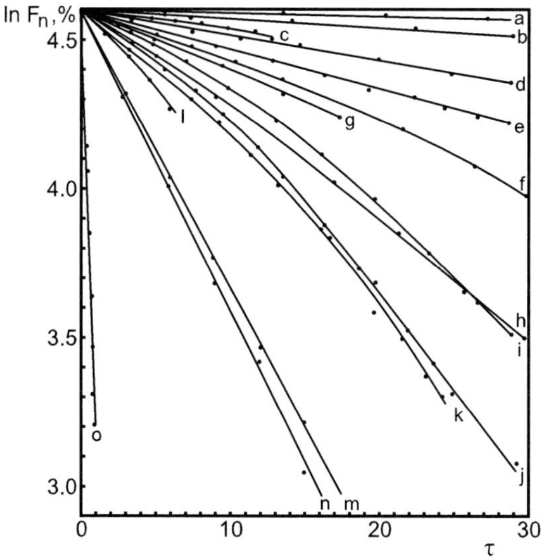

Fig. 1.13 Dependence of the ln F_N (%) function on τ. Number of kinetic runs: a = 10, b = 6, c = 4, d = 12, e = 7, f = 14, g = 3, h = 15, i = 11, j = 1, k = 2, l = 5, m = 9, n = 8, o = 13 [45].

were very close to those of 4,4'-MDI, and these values barely changed during the reactions.

In the case of 2,2'-DBDI, geometric effects were clearly evident. Reactivity was influenced by the steric effects wich lead to significant intramolecular catalytic activity. These effects were responsible for the whole reaction pattern.

From the reactivity standpoint, 2,4'-DBDI was similar to 2,4'-TDI, although the selectivity of the NCO groups at the ortho and para positions was found to be lower than that of 2,4'-TDI. The former material has an important practical implication since it is a liquid with a low vapour pressure and therefore reduces the risk of inhalation.

In a second series of investigations [44-46], Caraculacu et al., studied the reactivity of the functional groups 4,4'-DBDI. The rate constants of the reactions of diisocyanates with n-butanol were determined by using the IR spectophotometric method of Bailey [99]. The rate constants of the consecutive reactions were determined by the time ratio method developed by Frost and Pearson [100].

The kinetic data indicated that the overall reactivity of a diisocyanate and its decrease as the reaction progresses, were determined by two factors: (a) the electron attracting power (electophilivity) of the radical attached to the isocyanate groups and (b) the interaction between the two functional groups.

In the case of 4,4'-MDI and of 4,4'-DBDI, the activating influence of the isocyanate and urethane groups, respectively, was transmitted by an inductive mechanism through the methylene and ethylene groups. Simultaneously, the methylene and ethylene groups had a deactivating influence. The balance among these opposing effects during the process was illustrated by the decrease of the rate constants [59].

By studying the reactivity of 4,4'-DBDI and comparing it with other conventional diisocyanates, it observed that the reactivity of 4,4'-DBDI was very similar to that of the diisocyanate 4,4'-MDI.

This is one of the reasons for which in the present book we refered only to the 4,4'-DBDI isomer. Another thing is that from our earlier studies, we could onclude that the utilization of this isomer resulted in much better mechanical propeties than in the case in which 2,2'-DBDI or 2,4'-DBDI were employed in the PUs synthesis.

For simplification reasons, in the present book, we denoted 4,4'-DBDI as DBDI, as long as this was the only isomer which we used in the synthesis of the materials that make the subject of this book. Also for simplification reasons we denoted 4,4'-MDI as MDI.

1.2.8 The hydrogen bonding

The number of earlier publications can be a measure of the importance of a scientific topic. Numerous authors tried to predict more accurately the influence of hydrogen bonding on the PUs reactants reactivity and on the urethane reaction mechanism [89, 101–103]. The complexity of the hydrogen bonding systems determined by their high instability and the lack of proper information on the nature and number of associated alcohol species, maintained however so far a high degree of incertitude.

From the perspective of macromolecular science and especially for the isocyanate chemistry, the main problem resides in the elucidation of the hydrogen bond associates structure especially in the case of diols and polyols as in fact they finally represent the products involved directly in the macromolecular synthesis. Certainly a similar question is also risen in the case of diamines leading to the urea groups formation.

1.2.8.1 Dynamic and quantum aspects of hydrogen bonding

Along with the physical methods which will be discussed in the following chapters, the problem of the hydrogen bonding should also be considered from the quantum point of view. Different orbital molecular calculus were approached [104–107]. However, unfortunately at present the *ab initio* calculations are still insufficiently developed. There is still sensible discrepancy between some of the calculated and experimentally determined spectral data [89, 108].

With regard to the hydrogen bonding more or less recent developments, the main accent has been shifted from the old global equilibrium characterisation achieved in the case of some particular association transient structures forms, towards the dynamics of the process [109, 110–113].

Panayiotou et al [109] developed an unified statistical thermodynamic model for hydrogen-bonded fluids and presented an interesting equation of state approach, according to which the basic assumption was the division of intermolecular forces into physical (van der Waals) and chemical (hydrogen bonding) forces. For multicomponent systems of molecules, a general formalism is presented regardless the number of hydrogen (proton) donor and acceptor groups [89]. Also, the model applies to self-associated substances and may be used to describe the thermodynamic behaviour of these systems. The equation of state approach was compared to the experimental values for the vapour pressure and heat of mixing and there were obtained good results [89].

It concluded that the main aspect which should be considered is the equilibrium between the formation and rupture of every hydrogen bonds and not the ensemble equilibrium between various associates [89]. Veytsman [114] considered systems of molecules with a one type of proton-donor group and a one type of proton-acceptor group and proposed an approximate combinatorial expression for the number or routes to form hydrogen bonds that does not invoke the existence of associates [89].

1.2.8.2 The hydrogen bonding in glycols

In the PUs macromolecular synthesis glycols acts as one of the most important raw materials. The glycols have a high tendency to self-associate but in this case it is possible to appear more or less intramolecular hydrogen bonding as well, depending on the molecular structure facilities [89].

In earlies studies (1964), Robertson et Stutchbury [115] evidenced the importance of this kind of association and inferred it for kinetical misunderstandings. By studying the IR glycol spectra, Kuhn [116] showed that in diluted unpolar solvents a series of diols presented two ν_{OH} bands, while others showed only one band. The higher frequency band at 3 630-3 644 cm^{-1} was attributed to the free ν_{OH} and the lower 3 478-3 590 cm^{-1} to intramolecular band. At higher glycol concentrations it appeared a third band at 3 185-3 415 cm^{-1} attributed to intermolecular associations [89]. The difference $\Delta\nu_{OH}$ between the free ν_{OH} and that of intramolecular bands represented a measure of the hydrogen bond force. Kuhn et al [117] suggested that in 1,4-diols there exist a higher chain flexibility and the intramolecular hydrogen bonding becomes very strong. This fact is characterised by a high ν_{OH} value of 100-160 cm^{-1} [89]. Steric factors which determine the intramolecular hydrogen bonding force are the reciprocal approaching of the two glycol OH groups, below the limit of 3.3 Å, and the posibility of the O–H···O structure to adopt a linear form.

The hydrogen bond strength in diols varies in the following order [89, 118]:

1,4-diols > 1,3-diols > 1,2- and 2,3-diols

A correlation between this situation and the reactivity towards phenylisocyanate was made by Caraculacu et al [89, 119].

The high reactivity of 1,4-buthanediol in the first step of the reaction, when it reacts with only a single OH group was explained in terms of the existence of an increased nucleophilicity of the oxygen from the OH proton donor group, involved in the intramolecular hydrogen bonding formation [89, 120].

The intramolecular hydrogen bonding in diols [121] with a longer *life time* existence was examined by A.J. Benigno et al [122] using fluorescence transient spectra (540 nm excitation) generated with an amplified synchronously-pumped dye laser. The authors used an indirect method by which it was measured the life time of the flurescent hydrogen bond mixt associate for a series of hydroxilix compounds with resorufin considered as a reference compound. The ethanol-resofurin hydrogen bond associate presented the smallest life time, whereas the life time for the 1,3-butanediol-resofurin associate was found to be 4.1 times longer as compared to that of ethanol. These results were attributed to the presence of the internal hydrogen bonding in 1,3-butanediol (1,3-BDO) [89]. As 1,4-butanediol has a higher tendency to form internal hydrogen bonding than 1,3-butanediol [118, 123], the lifetime of internal hydrogen bonding in 1,4-butanediol should be longer. Basing on these observations, and on the kinetical measurements, a good concordance was reported by Caraculacu et al., between the hydrogen bond life time and the reaction rate of different OH type in the urethane formation [89, 124].

Chapter 2
Structural studies on polyurethane elastomers

2.1 General considerations on the phase separation and morphological features

The whole is more than the sum of its parts (Aristotle)

In addition to chemical interactions, the physical arrangement of the molecules plays an important role.

Remember the widely accepted paradigm that *"the whole is more than the sum of its parts"*? Cumulative effects and interactions in a PU material system reveal more than the parts hidden properties. The reverse is effective as well.

Morphology is defined as *the science of structural forms*. PUs properties originate in their nanoscale phase separation by the competition between the hard and soft phases. PUs morphology can be studied at different levels of structure according to their domain sizes [4]. At an intermediate level a two phase structure is observed, where the dimensions of domains are of the order of 5–100 nm.

It is well-known that the PUs tend to exhibit phase separation where the SS units confer elastomeric behaviour while the microphase rich in HS provides physical cross-linking—see [4, 74, 78, 80, 125–127], for example the review by Dietrich and Hespe [126]. In addition to the microphase separation of segments, the macrophase separation of molecules with different composition has to be considered in discussing the PUs morphology and their properties. As this phenomenon is supposed to be common for PUs, it is also important to study the control and characterization of the heterogeneity of segmented macromolecules and also the relationship between properties and the molecular heterogeneity.

Because of the basic thermodynamic incompatibility of the segments, localized microphase separation occurs, leading to the well recognized domain structure, and the properties of the final bulk material are strongly influenced by the extent of microphase separation and by the morphological characteristics of the domains (Fig. 2.1 and 2.2).

C. Prisacariu, *Polyurethane Elastomers*, DOI 10.1007/978-3-7091-0514-6_2,

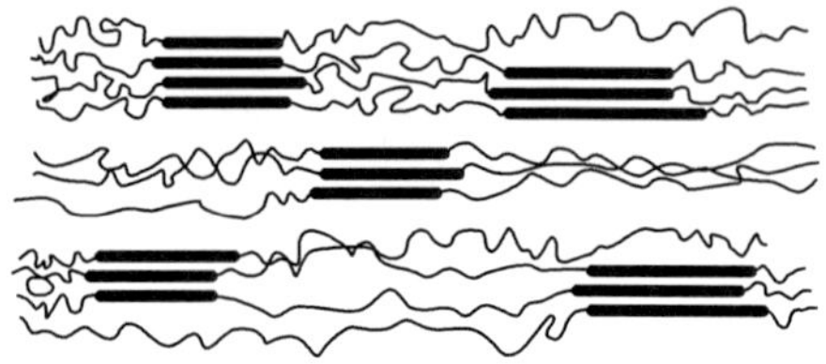

Fig. 2.1 Schematic representation of a well phase segregated PU system

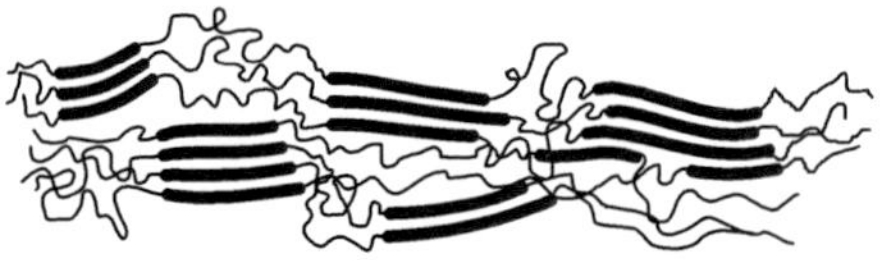

Fig. 2.2 Schematic representation of a phase mixed PU system

It is energetically favourable for the SS and HS not to mix. Thus during cooling from above a critical order-disorder temperature, spontaneous segregation of SS and HS into separate soft (SS-rich) and hard (HS-rich) phases occurs by the process of spinodal decomposition. To achieve elastomeric performance, the SS must be the majority constituent by mass, and the phase structure then takes the form of discrete hard domains dispersed within a soft matrix. Such a phase structure impacts on mechanical properties, and a further structural parameter of importance, therefore, is the *degree of phase separation*.

PUs phase separation strongly depends on the hydrogen bond formation between the urethane linkages, as well as on the manufacturing process, and reaction conditions [4, 128–132]. It also depends on the SS type and molecular weight. As the SS molecular weight decreases, the phase mixing increases [77]. Furthermore, the molecular weight of the SS is a dominant factor since higher molecular weights easily form rich phases on account of entropic contribution. As long as the size of the microphase-separated structure is of a nanometer scale, PUs can be classified as nanostructured materials [131].

The extent of the hydrogen bonding which is related to the phase separation is affected by the material structure, composition and temperature. In linear PUs, Tiffan and Terenzi found that almost all NH groups at room temperature are hydrogen bonded [132]. Similar observations were made by Boyarchuk et al [133] both for polyether and polyester PUs.

The overall mechanical properties of the material depend upon the relative volume fractions of soft and hard phases, and on the intrinsic properties of each of the phases. These in turn depend on details of molecular packing of the constituents within the phases, including the density of hydrogen bonds.

There is a wide variety of compositional variables which can affect the degree of phase segregation, the organization of the HS and consequently the PUs mechanical properties. The *symmetry of the diisocyanate* is an important factor. For example, Sung et al. studied a series of polymers based on the diisocyanates 2,4-TDI or 2,6-TDI, extended with BDO, and polyether or polyester SS of molar masses 1 000 or 2 000 g/mol. In the 2,4-TDI polymers, the *asymmetric* placement of the isocyanate residued with respect to the methyl group resulted in some head-to-tail isomerization in the HS. In contrast, in the 2,6-TDI based materials, this problem was absent since the molecule is *symmetrical* [134]. Other compositional variables were undertaken in this study: the macrodiol type (polyether or polyester SS macrodiol), the SS molecular weight and the length of the HS.

All of the 2,4-TDI polymers had amorphous domain structures with HS glass transitions below 100°C, whereas all of the 2,6-TDI based polymers exhibited semicrystalline HS structures with melting temperatures in the range of 130° to 200°C [134]. The interurethane hydrogen bonding was proved to be insufficient in order to stabilize the HS domains and to assure extensive phase segregation. As shown by Sung et al, there were other conditions that also contributed to PUs phase incompatibility such as the increase in the SS molecular weight, and the use of a less polar SS like hydroxy terminated polybutadienes (PBU).

For segmented PUs synthesized with PBU in bulk, Xu et al [21] showed that such materials contain two fractions of segmented macromolecules with quite different composition and average HS length. The morphological complexity of such PUs was interpreted by taking into account the domain formation due to the segregation phenomena on different levels, namely, macrophase separation and microphase separation. The general morphological features observed in the PBU based PUs were useful to understand the morphology-property relationships of other segmented copolymers.

Comparing the effects of polyether and polyester SS, the polyester macrodiol PBA introduced greater phase mixing than the PTMO soft segments, thereby raising the glass transition temperature of the SS [4, 134, 135]. These results confirmed that the hydrogen bonding to ester is stronger than to ether.

There was also evidence that the hydrogen bonding interaction can interfere with the SS crystallization, as long as the SS crystallization did not occur in polymers like those based on 2,4-TDI-PTMO (molar mass 2 000 g/mol), whereas a slight improvement in phase segregation occured in the polymers with 2,6-TDI-PTMO of molar mass 2 000 g/mol [134]. The SS segment crystallized freely as indicated by a further decrease in the glass transition temperature.

Similar comparisons were made for materials obtained with the diioscyanate MDI by using the polyester PBA of analogous molar mass. The MDI materials showed that hydrogen bond dissociation can occur within an ordered domain structure at temperatures well below the domain transition temperature and with little regard for improvement in the degree of order [134]. The thermal behaviour of hydrogen bonding appeared to be virtually independent of the PUs structural organization.

Previous publications [136, 137] reviewed models proposed for HS achieved with the diisocyanate MDI and extended with BDO. A detailed morphological analysis [137] of a series of such materials showed that the hard microdomain structure was in qualitative agreement with the model proposed by Koberstein and Stein [136]. This model was based on the partial solubilization of short HS into the soft microphase. HS shorter than the critical length for microphase separation were presumed to remain within the soft microphase while longer segments aggregated into lamellar hard microdomains of thickness proportional to the critical sequence length. Models of this nature provided a means for relating chemical structure and thermodynamics to the PUs properties.

Mc Kiernan [138, 139] studied the influence of the hydrogen bonding on the crystallization behaviour of a series of linear, *aliphatic* m,n-polyurethanes $[O{-}(CH_2)_m{-}O{-}C(O){-}NH(CH_2)_n{-}NHC(O)]_x$ derived from long chain aliphatic diols. The studies showed that, hydrogen bonding still influenced the crystallization process of the long-chain, aliphatic PUs. The X-ray and electron diffraction, and infrared spectroscopy indicated that long alkane PUs segments have interchain and intersheet distances similar to that observed for PUs of higher hydrogen bonding densities. The hydrogen bonding controls the crystallization, packing, and morphology of such materials. In addition, high-temperature infrared studies showed the existence and high concentration (~75%) of hydrogen bonding in such PUs even in the melt [138, 139].

The three-dimensional hydrogen-bonding ability to act as an additional driving force for more complete phase segregation determines stronger interaction between the HS.

Hydrogen bonding can be detected and studied by means *of IR spectroscopy*. The changing distribution of hydrogen-bonding properties in the hard and soft domains can be assessed by this technique. The hydrogen bonded and free N–H and urethane carbonyl C=O are the peaks of interest [140–148]. PUs phase separation in PUs can be characterized by measuring the intensity and position of the hydrogen-bonded N–H stretching vibration. HS–HS hydrogen bonding (N–H···O=C) in the hard domains and HS–SS (N–H···O–) hydrogen bonding in phase-mixed states were identified. In the case of well segregated PUs systems, there is significant N–H···O=C hydrogen bonding, since both units are associated with the HS. N–H can form strong hydrogen bonding with the oxygen of the ether groups associated with the SS, where observation of the hydrogen bonds indicates the existence of a dispersed phase consisting of HS mixed with SS [149]. In general, the N–H groups free from hydrogen bonding have a stretching vibration at 3 450 cm^{-1}. In contrast, the groups involved in hydrogen bonding have much lower frequencies, ranging from 3 300 to as low as 3 200 cm^{-1}. The exact position depends on the strength of the hydrogen bonding. The spectroscopic data can then be correlated to the macroscopic structural transformation, mainly the degree of phase mixing or the separation kinetics.

The extent of hydrogen bonding is affected by the nature of the SS. Eisenbach and Gronski [150] found that 40% of carbonyls in polyether urethanes are hydrogen bonded, which is much less than with polyester urethanes on the same basis but at a higher HS content [151].Their results are in agreement with the findings of other authors [152–155] who also found that the ratio of band intensities of the bonded C=O to the free carbonyl groups decreased as the SS length decreased, i.e. when the volume fraction of the soft phase increased.

In some of our recent papers we also made evidence that polyetheric PUs phase separate to a much greater extent than the corresponding polyester-based polymers [135]. This is consistent with previous studies, that showed polyether macrodiols to give greater phase separation than polyester macrodiols [4, 14]. The reason is believed to be the availability of a >C=O group on each monomer in a polyester for possible hydrogen bonding with the >N–H groups on the HS. This lowers the free energy of mixing of HS and SS that drives phase separation, relative to those macrodiols where this is absent. The polyethers PUs are extensively hydrogen bonded which involves the N–H group as the proton donor and the urethane carbonyl, the ester carbonyl (in polyester materials) or the ether oxygen (in polyether PUs) as proton acceptor. As shown [4, 135], although nearly 100% of the N–H groups are involved in the hydrogen bonding, only 60% of the urethane carbonyls are hydrogen bonded, while the remaining N–H groups hydrogen bond with the SS. This was explained in terms of the phase intermixing of a large hard domain surface area, due to tiny domains. Srichatrupimul and Cooper [153] observed that the hydrogen bonding between the SS and HS is stronger and dissociates at higher temperatures in the interurethane hydrogen bonding.

Numerous studies of the PUs morphological features and phase segregation behaviour have been carried out by means of a variety of characterization techniques, including transmission and scanning electron microscopy (TEM) and (SEM), small angle X-ray scattering (SAXS) and atomic force microscopy (AFM), and by infrared dichroism, or thermal methods like dynamic mechanical analysis (DMA), and differential scanning calorimetry (DSC). If one could say that the variety of the techniques used to investigate the PUs phase separation suffer from emotional disorder, this would be called *phase separation anxiety...* ☺

These techniques allow visions of the microphase separation in polymers. As one can say that eye is the mirror of the soul, microphase is the mirror of the polymer structure. Remember the well known Alexander Pope's quote a few hundreds of years ago where he was asking *"why has not man a microscopic eye?"* Certainly if he had been a scientist nowadays and if he had had access to these techniques, he woudn't have stated that man has not a microscopic eye *"for the plain reason he is not a fly"*.

The first direct evidence for the formation of a two-phase structure was obtained from the SAXS studies by Bonart [156]. However detailed information could not be obtained from the single peak in their SAXS profiles without some ambiguity, as long as PUs have a disordered two-phase morphology. The two-phase microstructure was observed by Koutsky et al, by using *transmission electron microscopy (TEM)* both on polyether and on polyester PUs [157]. They found that the sizes of the hard domains varied between 30 Å to 100 Å. Later TEM works were carried out by Xu et al., Chen-Tsai et al., and Serrano et al., on the morphology of a series of TDI based PUs extended with BDO, with varying the HS fraction. The two-phase morphology was evidenced [158–160]. Yet, on the basis of the obtained evidence it still could not be assumed that complete phase separation occurred. In fact, there was evidence that appreciable hydrogen bonding exist between the HS and SS, which involved incomplete phase separation. HS domains were observed as grains of 3-10 nm dimensions by Koutsky et al [157] in solvent etched and iodine stained PUs samples by means of TEM. Finer fibrils (6 nm width) were observed by Fridman and Thomas [161]. Fibrillar structures were reported by Schneider et al. where the smallest fibrils observed were 20-30 nm in width [162]. For PUs structures of the type MDI:PBU:BG but of variable HS concentrations, Li et al. evidenced a rod-like

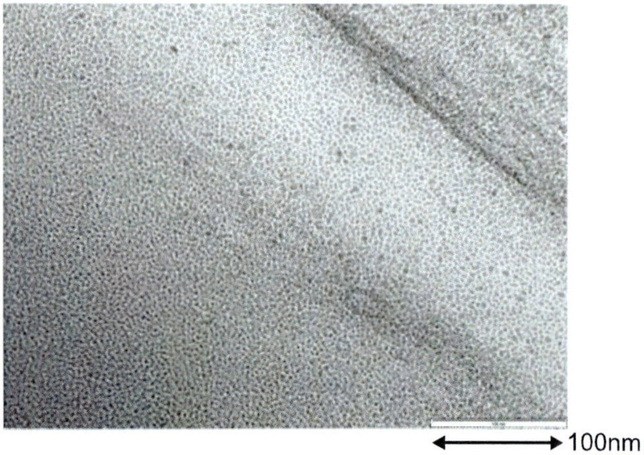

By courtesy of Dr Kayleen Campbell, University of Queensland

Fig. 2.3 The two-phase structure of a PU material based on MDI:PTHF:BDO. Soft domains are represented by white areas.

or lamellar structure for materials with a HS percentage of 42%-67%. At lower percentages, (<31wt %), the HS phase was dispersed in the SS matrix in the form of short cylinders or spheroids [163, 164].

An example TEM micrograph of a MDI-BDO based PU based on PTHF (Mw=2 000 g/mol), is shown in Fig. 2.3.

The use of TEM to image the nanoscale morphologies of PUs is a very useful technique that has the ability to image at high magnifications provided there is sufficient contrast between the electron density of the two phases. The electron density of one of the phases is sometimes enhanced using staining agents that make this technique dependent on the efficacy of the staining procedure. Phase separation can be detected by TEM if there is sufficient contrast between the phases. Serrano et al [165] were able to observe two phases in MDI/PBU/BDO polyurethanes even without staining, in samples with 55% and 75 wt% hard segments. Yet, TEM experiments are limited by the possibility of beam damage, and are also tedious and time-consuming due to the microtomy involved in cutting samples into few tens of nanometers thin sections. Imaging at high magnifications under TEM can also lead to misleading phase-contrast artifacts observed at a scale length of approximately 100 Å under slight defocus conditions.

Li et al used various physical methods to study the morphology of PBU based materials with variable of HS content [166]. Segmented PUs with low HS contents were found to have a morphology of dispersed short HS cylinders embedded in the matrix of the PBU based SS. An alternating HS-SS rodlike or lamellar microdomain structure was characteristic of materials with higher HS contents. At very high HS percentages, a morphology with a dispersed SS phase was observed.

HS and SS domains can organize to form crystalline superstructures, especially in the case of solution cast samples. Spherulites with diameters from several thousands of nm up to about 20 μm were observed by *small-angle light scattering (SALS)* [167, 168], and by using *optical and electron microscopic methods* [169–171].

More recently, compositional heterogeneity has also been detected for various PUs, such as those based on poly(caprolactone) glycol PCL/PET polyesters [172] or MDI:PCL:BDO based PUs [162], by means of *gel permeation chromatography and thin layer chromatography.*

It is necessary to consider both the microphase separation of HS and SS of the same chain into different domains, and also macrophase separation of macromolecules as a whole (segregation on the molecular level) in order to explain the morphology and properties of these materials.

Small-angle X-ray scattering (SAXS) is another important tool to make evidence of SS and HS phase separation. SAXS offers several important advantages and although interpretation is less direct than are the TEM or AFM techniques, the sample preparation for SAXS avoids the potential for staining artifacts that may arise with TEM. X-rays are also less damaging to the samples than are the electron beams used in TEM, allowing measurements to be made continuously over many minutes. SAXS is used to characterize the domain structure in the size range of tens of nm. The scattering is related to morphology on the nanometer scale, but it originates from a relatively large volume (mm^3) of the material under investigation although microbeam techniques can sample smaller volumes. Therefore the results give a more statistically significant view of the morphology [173]. The X-ray are scattered by the electrons of a sample and the amplitude of the scattering radiation is given by the three-dimensional Fourier transformation of the electron density distribution. Analysis of the intensity as a function of angle reveals information regarding the morphology of the material. SAXS results for patterns of phase separated PUs show a distinct peak. This is indicative of a quasi-periodic fluctuation in electron density within these materials.

Bragg equation is used to estimate the characteristic morphological periodicity from the scattering vector corresponding to the SAXS peak position, which is related to the scattering angle and the wavelength of the X-rays. The periodicity given by the Bragg spacing corresponding to the scattering angle at peak intensity usually ranges from about 10 to 25 nm and it increases with increasing the SS molecular weight [4, 135, 173–175].

Numerous structural SAXS studies were done on PUs. For example, materials achieved with MDI-diol based HS were studied by Bonart et al [14, 15] and by Wilkes and Yusek [174]. The X-ray patterns obtained by them showed a single Bragg reflection in the diffraction pattern of the poly(MDI-BDO) hard segments at $d \approx 7.9$ Å, azimuthally inclined at ~30° to the meridian. In addition, an intense amorphous halo at $d \approx 4$ Å was observed. Bonart assigned the Bragg reflection to planes inclined at ~60° to the fibre axis, and suggested that these planes arise from a staggering of adjacent chains so that intermolecular C=O···H–N hydrogen bonds are formed between the urethane groups. This lead to the conclusion that the PU chain structure is analogous to that for the α-form of nylon 6.6.

Koberstein and Stein [136] also studied the PUs phase separation by using SAXS. The maximum in the small angle region was interpreted as the mean separation distance between the HS. The intensity profile contained information about the interfacial thickness and about the purity of the domains.

SAXS was also used by Li et al [163] to investigate the structure of a series of segmented PUs of variable HS percentage, for materials based on the couple diisocyanate-chain extender, HDI/BDO. Thet found that the phase structure was insensitive to the increasing HS content and thermal treatment. Structure property behaviour of the MDI/BDO based PUs was studied through the effect of composition ratio of the HS by Abouzahr et al [176]. They observed that morphological changes occur as the HS fraction is increased. The texture changed from that in which little HS domain content exists al low HS levels (15%), to that in which the polymer has an interlocking domain morphology at higher HS content (35% and 40%). Low hysteresis and high extension was obtained when isolated HS exist (25% HS). Thermal treatment of the samples lead to domain disruption and to HS-SS mixing.

In general crystalline domains are disrupted the least while the fastest recovery is displayed in samples with a non-crystalline domain texture, but this is going to be discussed to a great extent in Chapter 4 of this book.

For PU structures of a high HS content (above ~30%), continuous HS domains could be clearly identified and traced in structural transitions [177–182]. On the other hand, the PUs of a low HS composition (less than 30%) were believed to have only discontinuous HS domains of a micellar nature [180] in the SS solvent-like matrix, especially when the sequence length of the HS is short [179]. Y.S. Sun et al [183], and L.Y. Chiang et al [184], investigated the micellar structural characteristics of segmented PUs of a low HS content and a short HS sequence length, by using the contrast variation provided by *SANS and SAXS*. The polyether MDI/PTMO based materials were prepared from the reaction of the chain extender BDO with diisocyanated prepolymers [184]. PUs were synthesized so that they contained 19.3 wt% hard segments. These two SAXS and SANS profiles were very similar in a wide Q range, except the sharp raising peak for SAXS in the Q region lower than 0.02 $Å^{-1}$. The results indicated a general two-phase structure with low-Q sharp peak observed in the SAXS profile, which were attributed to the scattering from the partially crystalline PTMO domains of few hundreds angstroms estimated from the broad WAXS peak widths of the PTMO chains. The PTMO crystalline domains were virtually invisible to neutrons due to the very low scattering contrast [183]. The high value for the critical aggregation concentration of the HS in the PTMO matrix indicated that the HS of a short MDI sequence can dissolve in elastomer well up to ~30%.

The increase in the HS content produces a decrease in the PUs melting enthalpy. As the melting enthalpy of is related to the interactions between the polymeric chains (so-called virtual crosslinking), the trend followed by the melting enthalpy indicates that the virtual crosslinking becomes less important as the HS content increases. PUs such as those containing TDI have a high HS content and show a melting enthalpy lower than other materials. TDI usually contains a mixture of 2,4- and 2,6-isomers and two kinds of urethane groups are possible.

Abouzahr and Wilkes also studied the effect of HS content on a series of MDI-based PUs [176]. Depending on the HS content, the properties of the elastomers were divided into four regimes. At very low HS percentages, they found that the polymer was poorly phase separated and exhibited poor elastomeric properties. At slightly higher HS contents the polymer was better phase separated, but the HS were small and isolated. This lead to higher modulus, toughness and extensibility and consequently to low hysteresis. At higher HS percentages, the HS became interpenetrating and the material had a much higher modulus and hysteresis. At percentages of 50% HS and higher, the phases inverted from HS domains in a soft matrix to soft domains in a HS matrix and the polymer became brittle, high modulus plastic. Similar observations were made by us on PUs based on crystallizable HS [185].

PUs with MDI:BDO hard segments and oligomeric ethyleneoxide end-capped poly(propyleneoxide) SS containing 5-100 wt% hard blocks contents were investigated by Saiani et al [186, 187]. In their works, the PU without oligomeric diol (100% HS) exhibited one glass transition temperature and no scattering peak in the SAXS experiments, indicative of a single-phase structure. Adding just 10 wt% SS resulted in a microphase separated structure [187].

By using wide-angle X-ray diffraction (WAXD) and SAXS, Bonart et al [14] proposed a model according to which the HS were considered as extended chains.

Later, on the basis of results from SAXS and thermal analysis, Koberstein and Stein [136] and Leung and Koberstein [137] proposed a model according to which the HS domains thickness domains was controlled by the shortest HS chain insoluble in the SS phase. The HS longer than the critical length were either coiled or even folded so to be included in the HS domains. HS shorter than the critical length were dissolved in the SS phases. They also observed that the HS mobility, and the strength of the HS interaction between themselves were influenced by *the annealing temperature*. For example, in materials derived from *aromatic* diisocyanates like MDI/BDO-based segmented PUs, the HS mobility was low, and the HS interactions between them were strong, which resulted in slower phase separation after the melt was quenched to lower annealing temperatures.

Koberstein et al [180] investigated a series of thermoplastic PUs based on poly -ε-caprolactone SS and *aromatic* MDI:BDO based HS. To assess the hard block size on the PUs morphology, Koberstein et al. used poly -ε-caprolactone SS because of its miscibility with the HS. A range of block lengths showed that the compatibility level combined in polymers with smaller block lengths showed no phase segregation as inidcated by SAXS measurements. The short hard block lengths were assumed to dissolve into the soft phase of such PUs systems. Increasing the block lengths, the position of the SAXS peak indicated phase separation but only of short range orders. This is consistent with a morphology with a short term order in the form of lamellar HS domains being dispersed in the SS continuous phase.

Replacing aromatic with *aliphatic* diisocyanates like 1,6-hexamethylene diisocyanate (HDI) while keeping the same structure with BDO as chain extender, resulted in HS of greater flexibility. Due to the higher HS mobility, the phase separation rate increased owing to the relatively higher HS mobility and consequently lower PU system viscosity.

In materials obtatined with aliphatic diisoyanates, the insensitivity of the hard domain length to increasing HS content has been reported for a number of HDI-BDO polyurethanes; this behaviour has been attributed to chain-folding of the lamellar morphology.

Chu et al (1992) studied the phase structure of segmented PUs as a function of the HS flexibility, by using synchrotron SAXS analysis in order to assess *the effects of thermodinamics vs. kinetics* during the phase separation [188]. Using the value of the relative invariant from the SAXS analysis and the glass transition temperature of the SS determined by DSC, they found that an increased HS content did not lead to the enhance of the phase separation. This finding does not match with the theory according to which higher HS contents lead to lower compatibility between HS and SS, which should promote phase segregation. This behaviour was explained by Chu et al., as a function of kinetics.

Other methods *associated with solid-state NMR* have been also used in numerous works to investigate the PUs phase separation behaviour [189-193]. Dumais et al., detected phase separation in NMR line broadening experiments using labeled chain extenders [191]. The two-dimensional wide-line-separation nuclear magnetic resonance spectroscopy (WISE-NMR) provides a particularly interesting alternative tool for determination of the PUs phase composition. With this experiment, a correlation has been established between chemical structure and segmental mobility as reflected in ^{13}C chemical shifts and ^{1}H line shapes, respectively. Segments dispersed in HS rich domains are immobilized, resulting in broad components in the ^{1}H spectra. The fraction of HS and SS distributed in each phase can therefore be calculated from the resolved ^{13}C spectrum. Using this technique, the degree of phase separation in piperazine-based PUs has been established [194].

The PUs microstructure can be also investigated by means of *atomic force microscopy (AFM)*. Phase images obtained via AFM, enable visual representation of the PUs microphase separated morphology. AFM records the surface topography of materials by measuring attractive or repulsive forces between the probe and the sample. Vertical deflections caused by surface variations are monitored as a raster scan drugs a fine tip over the sample. A detailed description of different modes in AFM technology has been described in [195].

The use of AFM to examine the surface morphology of polymers is now well established While different variations of AFM are available, tapping-mode AFM has gained popularity due to the lower forces involved, and because there is only intermittent contact between the sample and the tip, unlike for example in contact-mode AFM. This technique allows simultaneous detection of height and degree of phase segregation. It provides information on the variations in topography and local stiffness. *'Height'* images are obtained by storing the vertical *'z'* position of the AFM scanner-head as it scans an *'x-y'* surface, while maintaining a constant *'set point'* amplitude. Simultaneously, *'phase'* images are obtained by detecting the phase shift between the actual oscillation of a tip and its drive oscillation. In addition, since AFM images are essentially three dimensional plots of data points, this technique also enables semi-quantitative analysis of the images by means of surface roughness and power spectral density calculations.

The AFM technique has been increasingly used as an alternative to transmission electron microscopy (TEM) for the visual interpretation of the PUs nanostructure where the surface morphology was imaged by using the tapping mode AFM. The high modulus hard domains and the low modulus soft blocks appeared as light and dark regions respectively.

The development of a PUs hard domain continuous morphology upon increasing the HS size has been investigated [196]. Using AFM, the hard domains were found to preferentially orient with their long axis along the radial direction of the spherulites. The hard domain connectivity was found to increase with increasing the percentage of HS content. For example, materials such as those based on the SS macrodiol polyethylene glycol (PEO) displayed extremely rough surfaces, limiting their resolution via AFM [176, 196]. At lower HS contents (22 wt%), the deformation mechanism involved extension of the SS. The HS were randomly dispersed in a continuous soft domain without significant aggregation which is why the alignment and break-up of the hard domain structure did not occur. At higher HS content (33 wt%), the hard domains were observed to be still randomly dispersed in a continuous soft domain, which was reflected in higher interdomain spacings. With increasing the HS content, an inter-lockong HS morphology developed, leading to the decrease of the average distance between the hard domains, which was confirmed by the SAXS studies. The development of a hard domain continuous morphology upon increasing the HS size has been observed in numerous other segmented PUs [176, 196].

PUs usually possess the microphase separated structure with interdomain spacing of 10-20 nm. But what happens if the film *thickness* is smaller than the spacing? As a rule, PUs block copolymers loose the microphase separated structure with the thickness which is smaller than the periodic structure, or forms an island structure, in which the periodic structure exists [197]. Decreasing the PUs film thickness, the size of the microphase separated domains decrease as evidenced by AFM. Decreasing the film thickness the aggregation of the HS chains becomes weaker [131].

PUs phase separation is also strongly influenced by *temperature*. The changes in the hydrogen bonding characteristics occurring in soft and hard domains are correlated with the macroscopic phase transformation as a function of temperature. Increasing the temperature above the 'melting' transition of the hard domains, results in a homogeneous structure accompanied by the disassociation of most of the HS–HS hydrogen bonding. If rapidly cooled (quenched), the high-temperature structure becomes "frozen", which results in a significant number of interactions between the HS and SS. Quenching to lower temperatures retains the degree of phase mixing at the elevated temperature. If the quenched temperature is much lower than the SS glass transition temperature, the slow rate of segmental separation at low temperature prevents the reformation of the HS–HS hydrogen bonding. Raising the temperature, the low temperature structure undergoes phase separation to become a room temperature structure. In this way, direct correlation between microscopic evidence for hydrogen bonding characteristics in soft and hard domains and macroscopic phase transformation can be obtained [149]. While the degree of microphase separation becomes weaker with increasing the polymerization temperatures, thermal treatments usually determine the increase of the degree of microphase separation [198, 199]. When the PUs are annealed at certain temperatures, the HS domains with melting points above the annealing temperature are progressed.

2.2 Phase separation kinetics

Every process which is not forbidden must occur. (totalitarian principle)

Remember the totalitarian principle according to which *"every process which is not forbidden must occur"*? Not only the characterization of the PUs phase separated structure is important, but, from the perspective of PUs fundamental point of view, the mechanism associated with the phase separation is also extremely important and equally interesting. Occurrences in the phase separation process are strongly affected by the variety of parameters in operation.

Phase separation kinetics for PUs block copolymers has been extensively studied. Various techniques have been applied to study the aspects of PUs phase separation kinetic behaviour. For example, scattering techniques, mainly SAXS can give the average size of the hard domains evolving from a homogeneous single phase. By monitoring mechanical properties as a function of phase separation time, the phase separation kinetics can also be deduced, but the direct relationship between the macroscopic mechanical properties and the degree of phase separation is generally hard to establish.

The rate and degree of phase separation depends on structural parameters such as interaction parameter between two phases, presence of hydrogen bonding and crystallization of one or both phases, segment size and segment size distribution, and also experimental conditions, i.e. annealing and storing temperature and time) [4]. The rate of phase separation depends on the regime of cooling, or isothermal temperature if the process is studied isothermally. In general the experiments consist of heating the PUs sample at high temperature (annealing) and then quenching to a storage temperature, at different time intervals. On quenching, HS ordering needs pulling of the SS ends, while stretched SS retractive forces act during reheating of the sample, which may cause HS disruption. This leads to partial mixing of phases, at elevated temperatures, above the glass transition temperature of the HS. Heating above the melting point of the hard phase causes complete mixing with SS [4]. At high temperatures part of short HS are soluble in the soft phase. Decreasing temperature, shorter and shorter segments become excluded.

A series spectroscopic analysis of the phase separation kinetics in model PUs was made by Lee et al [200]. Isothermal phase separation kinetics was studied by raising the PUs sample temperature quickly to the phase separation temperature and measure the changes in the infrared spectrum at constant temperature. They assigned spectroscopic features characteristic of urethane linkages dispersed in the

SS as compared to interurethane hydrogen bonding confined to the HS domains. With a specially designed sample cell [200], they trapped a phase-mixed structure at a temperature 60°C below the glass transition temperature of the SS. When the quenched PU sample was brought up to a higher temperature, the increase of one spectroscopic component and the corresponding decrease of the other provided a direct measurement of phase-separation kinetics.

The results of the study clearly indicated that the phase separation process involves a gradual transfer of the HS dissolved in the soft matrix into the hard domain. As shown [200], frequency and intensity differences at different temperatures for both the hydrogen bonded and free C=O vibrations can be significant and difficult to define. However the measurement of the relative intensity of the two components at constant temperature removes this difficulty.

2.3 Structural studies on polyurethane elastomers with crystallizable hard segments

Crystallization is the action of the mind that discovers fresh perfections ... at every turn of events. (Stendhal)

"Crystallization" is a concept developed in 1822 by the French writer Stendhal, to describe mental metamorphosis. In the summer of 1818 Stendhal went to the salt mines of Hallein near Salzburg with his friend Madame Gherardi, where they discovered the phenomenon of salt "crystallization". He defined the term "crystallization" and used it as a metaphor for human relationships. *Remember Stendhal's* classic (1882) "De l'Amour" where he explains crystallization as a natural "mental process" drawing from proofs of the perfection. Imagine a tree branch covered with crystal deposits left in Salzburg's salt mines for a few months.

The notion of "crystallization" was pivotal for Stendhal who extended this metaphor into his theories on art and on literature: the process of making and perfecting art and the evocative power of the artist. He defined them as "crystallization processes".

Obviously the author of "Le Rouge et le Noir" wouldn't have imagined that the "crystallization" concept defined by him, will be extended to describe things like polymers. Moreover, in terms of crystallization, the morphology of polymers is even much more complicated—although I am sure artists or writers would dislike this.☺ But here, while writing this, I feel on a safe territory, far from people of art, so I can dare to make such an assertion.

It is axiomatic that an individual polymer molecule which possess a high degree of chemical and structural regularity among its chain elements is capable to undergo crystallization. For a polymer with a high regular ordered, structure, conditions must be found that are kinetically favourable for crystallization.

The understanding of the structure and properties of semi-crystalline polymers involves numerous variable techniques and theoretical approaches. Essentially, all

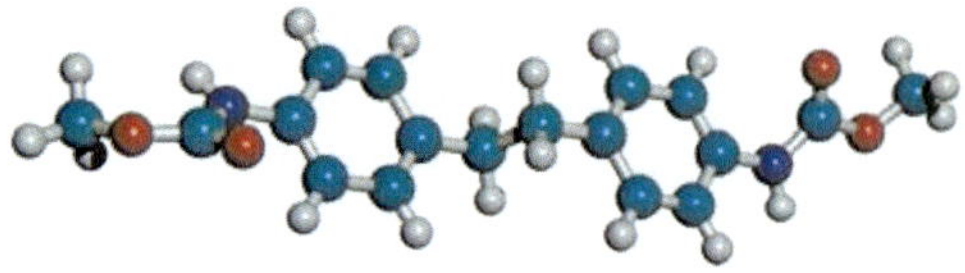

Fig. 2.4 Extended linear "anti" DBDI position

Fig. 2.5 Contorted "syn" DBDI position

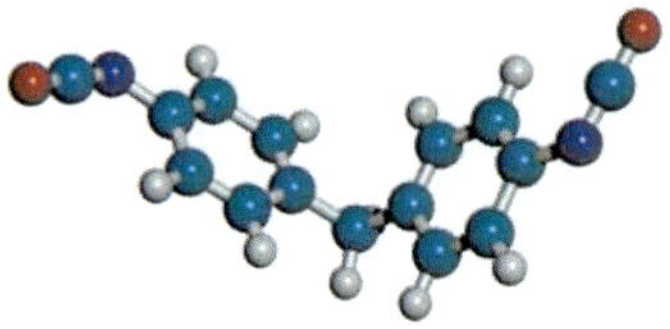

Fig. 2.6 Conventional rigid 4,4-diphenylmethane diisocyanate (MDI) non-crystallizing

properties are controlled by morphology, which in turn is determined by the crystallization mechanisms. From the thermodynamic point of view, the nature of a crystallization process is governed by both thermodynamic and kinetic factors, which can make it highly variable and difficult to control.

PUs morphology is very complicated not only because of the two-phase structure, but also because of other physical phenomena such as *crystallization* and hydrogen bonding in such systems. The crystallinity of phases and the size of domains along with the molecular composition, control the PUs morphology and macroscopic properties. Crystallization of the HS is an important determinant of the extent of phase separation, and the development of crystallinity is dependent on the hydrogen bonding and other interactions between the chains in the hard domains [201].

Crystallinity has been observed in the soft phase when the macrodiol chain is long enough, and it is also sometimes present in the hard phase. The latter is usually limited to only a few percent for most HS structures when solidified from the melt, but there is one particular diisocyanate, *4,4'-dibenzyl diisocyanate (DBDI)* that, in the presence of suitable chain extender, gives rise to significant degrees of crystallinity [60, 61, 135], and this is included and detailed in the present book.

As reported by ourselves [60, 61, 135] and by Lyman and Gowerr [58, 202], the MDI molecule introduces the rigid –Ph–CH_2–Ph– moiety in the elastomeric PU hard segments. In contrast when using DBDI, the specific –Ph–CH_2–CH_2–Ph– moiety introduces a variable geometry into the hard segments due to the possibility of internal rotation of this isocyanate around the –CH_2–CH_2– ethylene bridge.

This leads to the appearance of both "anti" and "syn" rotational conformations, which coexist in the DBDI based PU macromolecules, (Fig. 2.4–2.6). As a result, in this latter case the PU macromolecules can adopt a more compact packing which enhances significantly the ability to order in crystalline structures involving predominantly the "anti" form [60]. Shown in Fig. 2.4 and 2.5 are the extended linear "anti" and contorted "syn" DBDI positions as compared to the conventional rigid 4,4-diphenylmethane diisocyanate (MDI) non-crystallizing (Fig. 2.6).

The rotational isomerism of the dibenzyl biaromatic (DBDI) systems allows multiple spatial arrangements, with different reciprocal influences of the functional groups situated in the two rings. As shown, the angle between the plane and the $-CH_2-CH_2-$ linkage is 70.5°. As earlier shown in this book, by studying the reactivity of DBDI and comparing it with other conventional diisocyanates, we concluded that the reactivity of DBDI was very similar to that of the conventional rigid diisocyanate MDI [45].

The PUs with DBDI based HS investigated by Caraculacu et al [59] showed a fiber repeat distance which is that expected for a fully extended chain. In this conformation, chains readily pack to form fully hydrogen-bonded sheets. The X-ray diffraction spectrum showed a periodicity of 19.7 ± 0.2 Å and was similar to that of the triclinic form of nylon 66, nylon 6 and nylon 8 [59]. This was in contrast with the analogous materials based on the conventional rigid diisocyanates like MDI, where the extended chain conformation has the appearance of a large zig-zag structure where, although chains with this conformation can be readily packed together, only one half of the potential hydrogen bonds are formed [59]. In the case of the materials with MDI, X-ray determination showed a periodicity of 15.7 ± 0.1 Å with qualitatively similar arrangement but belonging to the hexagonal lattice crystalline system [59]. As previously shown by us, in polymers with MDI, the HS are associated but not so ordered as they are in the materials with DBDI where HS are both ordered and associated, which determines a crystalline PU structure [59, 60].

A conformational mobility of the diisocyanate like DBDI (Fig. 2.7) may cause an unusually wide range of mechanical, physical and chemical properties, associated with the possibility of pronounced phase separation into a domain-matrix morphology, and with a higher tendency to crystallization and self-association by hydrogen bonding [61, 127]. Hard domains having a higher flow stress in the presence of the diisocyanate DBDI, have been associated with increased hydrogen bonding which was enhanced in numerous cases by HS crystallinity [61]. The primary effect of employing flexible HS in the PUs synthesis, was shown to be a closer self-association of HS by hydrogen bonding.

Shown in Fig. 2.8 (a) and (b) are example microphotos of the two-phase structure for two analogous materials based on PTHF of molar mass 2 000 g/mol and chain extended with BDO. They differ only in the type of diisocyanate, the rigid MDI or DBDI displaying a variable geometry. Transmission optical microscopy standard method was used to investigate the microstructures.

Both images reveal clear heterogeneity of the samples. This is consistent with previous observations made by ourselves and by others on polyether macrodiols based PUs to give greater phase separation than polyester macrodiols [4, 14],

Fig. 2.7 2-D network of hydrogen bonds in a DBDI based material

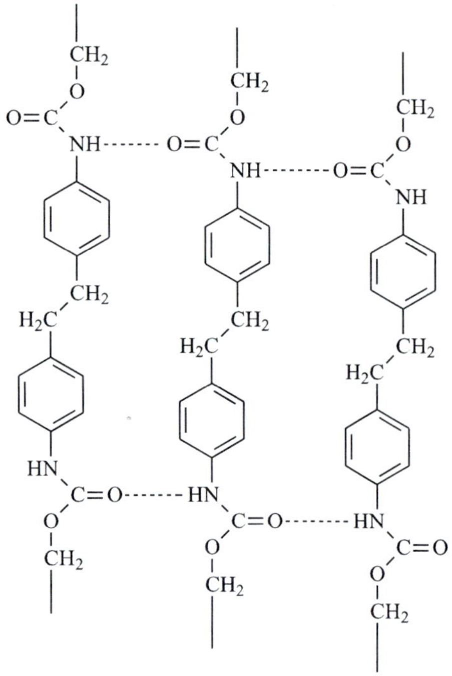

regardless the type of the diisocyanate used in the PU syntesis. Moreover, the inclusions (2 μm diameter or less) observed in the DBDI based structure were associating with a higher tendency to HS crystallize characteristic for materials based on the couples DBDI-EG or DBDI-BDO [61].

In the DBDI based polymers there was a clear indication from data that the physical origin of the flow stress must be relative displacement of the hydrogen-bonded HS [61, 135, 175]. In deformed PUs, the anisotropy of the structures relative to the strain direction was clearly visible in the 2D images (Fig. 2.9). Samples were

Fig. 2.8 Example two-phase structures: (a) MDI:BDO:PTHF; (b) DBDI:BDO:PTHF at 1 000 μm magnification

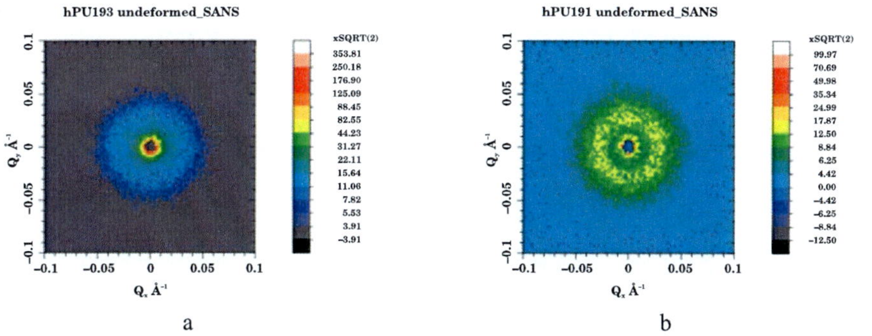

Fig. 2.9 SANS patterns from two bulk PU elastomers; they differ only in the type of diisocyanate, the flexible DBDI and rigid MDI: (a) DBDI:BDO:PTHF; (b) MDI:BDO:PTHF; by courtesy of Dr. D. Bucknall, Virginia Tech University.

strained to up to 300% levels of extension. Depending on the type of diisocyanate (crystallizing or not), they showed various degrees of anisotropy, with the highest values corresponding to the PTHF/DBDI based materials. The macroscopic deformation was present in the molecular levels that were probed with neutrons.

There were tendencies to phase separation, with a characteristic length of ca 20 nm, and, when DBDI was employed with BG or EG, to crystallization of the hard phase through its effect of increasing the flow stress.

In our previous beam time (RB510315) we observed a strong scattering peak at low Q in the various PUs (see Fig. 2.10) which is consistent with the microphase separation in block copolymers predicted by Leibler [203]. The phase domain dimensions varied between 10-20 nm, depending on the PUs specific composition. After straining to 300% (Fig. 2.10) the morphology is clearly disrupted with a reduction or in some PUs a total loss of domain structure.

The AFM method was also used by ourselves to examine and compare DBDI with MDI based films surfaces and to measure their surface topography. Topographical and phase contrast AFM images were taken from different zones of the materials. The replacement of DBDI with MDI determined morphological variations and a decrease of the average height texture parameter, both for the molten and for freeze fracture surfaces. The separation between crystalline and amorphous phases was better observed in the AFM phase contrast images. The crystalline domains were observed to have higher hardness while the amorphous domains have higher elasticity. The separation between crystalline and amorphous phases was more pronounced in the DBDI polymer. Example topographical and phase contrast AFM images for freeze fracture surfaces are shown in Fig. 2.11 for the same two materials (analogous structures) as in Fig. 2.8 and 2.9. They differ only in the type of diisocyanate, the model rigid MDI or the flexible DBDI.

The SEM investigation made by us clealy revealed the strong influence of the variable geometry of the diisocyanate DBDI on the formation of hard domains with a pronounced crystallinity, especially when EG or BDO were employed as chain

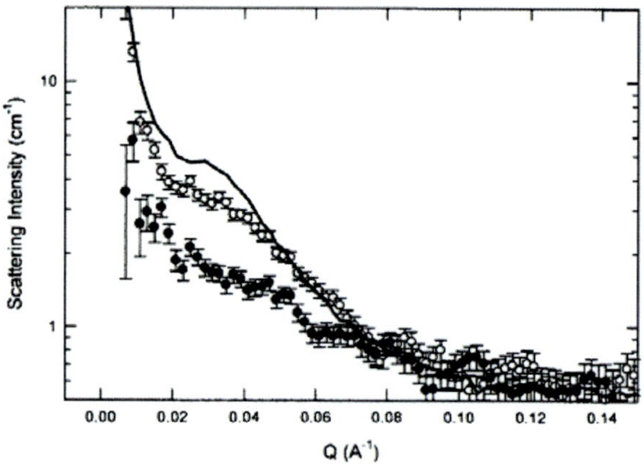

Fig. 2.10 SANS scattering from a fully hydrogenous PU before (solid line) and after straining to 300%, with data in the parallel (open circles) and perpendicular (solid circles) to the draw direction.

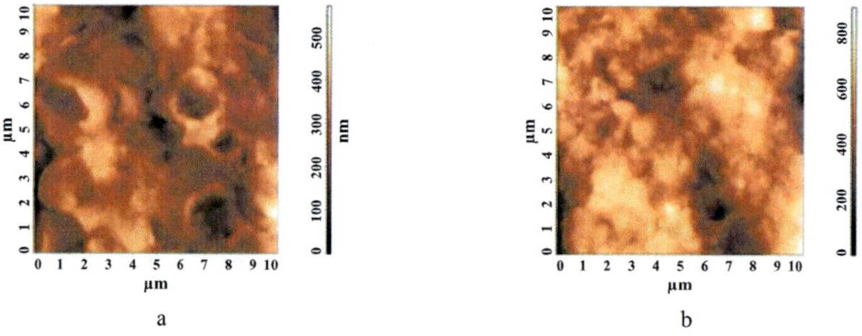

Fig. 2.11 Topographical and phase contrast AFM images for freeze fracture surfaces for two analogous materials; they differ only in the type of diisocyanate. (a) (DBDI:BDO:PTHF); (b) MDI:BDO:PTHF. Courtesy by Mrs. I. Stoica, Institute of Macromolecular Chemistry, Petru Poni, Iasi, Romania

extenders. Shown in Fig. 2.12 and 2.13 are the SEM pictures of two PUs with similar structures, where only the type of isocyanate differs: MDI non-crystallizing, DBDI crystallizing. With DBDI the polymer matrix becomes more mobile and allows the chain extender-diisocyanate segments to move and crystallize in self-associations [204]. Investigation of PUs by means of SEM revealed the fact that the nature of the hard phase and the degree of phase separation in PUs can be controlled via the chemical structure.

The materials displayed a relatively coarse structure on 10 μm scale but which varied between the polymers. Fig. 2.12 displays the features of a DBDI based original sheet. At 0.1 mm magnification the arrow points to a particularly prominent area of the banding which is typical of this specimen. The most prominent feature is what is referred to as the *coarse structure* of about 10 μm in scale: this is due

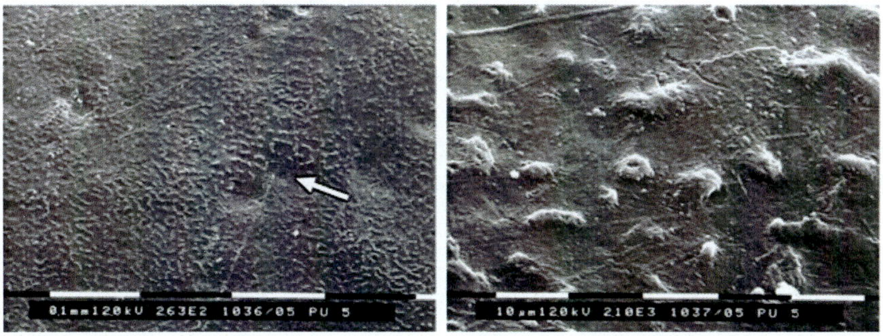

Fig. 2.12 SEM of a PU based on DBDI (original sheet) at two magnifications [204].

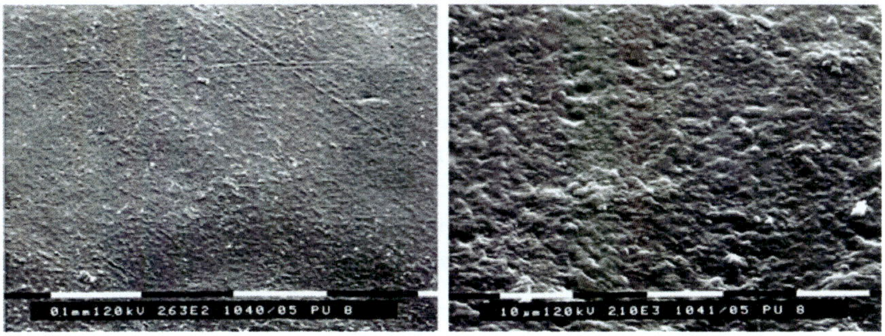

Fig. 2.13 SEM of a PU based on MDI (original sheet) at two magnifications [204].

to large extent of PU phases separation in the dibenzyl-based material. There are only the faintest hints of a texture on this scale in the equivalent material with MDI. This banding is made up of phase-segregated lumpy regions, which are either finer or not present at all in the material with MDI. Fig. 2.13 displays the features of the analogous MDI based PU original sheet. The banding also appears, but much less pronounced. PU with DBDI which displays X-ray crystallinity [60, 127] tends to give a rougher surface morphology than PU based on MDI.

The most prominent feature is what is referred to as the coarse structure which is due to large extent of phases separation in the polymers derived from DBDI. As previously shown, the dibenzyl based polymers did not retract after rupture [204]. After step stretching of the materials, the presence of the dibenzyl structures gave a much rougher surface morphology than that observed in the MDI based polymers [204].

Several chemical etching techniques were tried on these materials. In Fig. 2.14, etched surfaces of MDI and DBDI based materials are shown.

In the material based on MDI (Fig. 2.14(a)), the relatively resistant material is in majority, with non-resistant pockets being consumed by the etchant. As observed elsewhere [60], such pockets represent materials precipitated during polymerization.

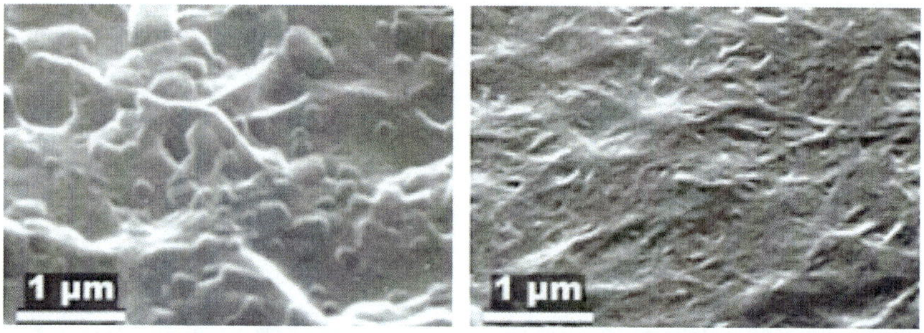

Fig. 2.14 SEM of etched PU specimens: (a) MDI; (b) DBDI [204].

In the materials with DBDI (Fig. 2.14(b)) the overall etched texture is much flatter. The regions of chemical segregation are more or less equally attacked. The material with DBDI does not show the precipitated crystals. A more detailed SEM description on the morphology of MDI and DBDI based PUs and mixtures of them, is made in section 2.3.2.2. where the materials are characterized and compared from two perspectives: (a) effect of the SS macrodiol nature (polyester or polyether); (b) effect of type, and number of diisocyanates (crystallizing or not) and their order of introduction in the reaction synthesis.

2.3.1 Structural studies on polyurethane elastomers obtained with single diisocyanates

The degree of phase segregation between HS and SS depends both on the degree of interaction between the HS to each other and towards the SS and on the mobility of the polyurethane chains. Hard domains having a higher flow stress in the presence of the diisocyanate DBDI, have been associated with increased hydrogen bonding which was enhanced in numerous cases by HS crystallinity [60, 61, 135]. The primary effect of employing flexible HS in the PUs synthesis, was shown to be a closer self-association of hard segments by hydrogen bonding. In such polymers there is a clear indication from data that the physical origin of the flow stress must be relative displacement of the hydrogen-bonded HS.

A systematic study of PUs with HS of variable crystallinity was made by us [60, 61, 127, 135]. Two diisocyanates were considered: the frequently employed MDI, and its close relation DBDI, that is of special interest because of its tendency to crystallize on cooling from the melt in the presence of some chain extenders [135]. The family of model PUs was synthesized for this work in the authors' Romanian laboratory. They were all three-component systems combined in stoichiometric proportions, and consisting of: (1) a diisocyanate—either MDI or DBDI; (2) a SS macrodiol—PEA, PTHF, or PBA; and (3) a small molecule diol as chain extender—anhydrous EG, DEG, or BDO. The macrodiols were all of molar mass 2000 ± 50 g/mol.

The three components were always mixed in the molar proportions hard segment:macrodiol:chain extender = 4:1:3, giving hard segment mass fractions in the region of 40%, and isocyanic index I = 100 which means that they are truly thermoplastic. They do not have the potential for further reaction with ambient humidity to produce chain lengthening and allophanate cross-linking, seen in similar polymers but with excess isocyanate groups (e.g. I = 110). Synthesis was carried out by the prepolymer route described previously by Prisacariu et al [135]. Information on the nanometre-scale physical structures of the materials was gained by X-ray scattering, using synchrotron radiation at the UK Daresbury Laboratory. Wide-angle X-ray scattering (WAXS) studies were carried out using Station 16.2 SMX, with X-ray wavelength 82 pm. The 2D patterns were radially averaged to produce 1D intensity profiles, as illustrated in Fig. 2.15. In some of the materials, predominantly some of those based on DBDI, sharp peaks were observed in the WAXS intensity versus 2θ scans, indicating some crystallization of the hard segments [135]. There was no indication of crystallinity arising from the soft segments in these materials. The scattering intensity was separated into amorphous halo (I_a) and crystal diffraction (I_c) components by fitting Gaussian peaks to those crystal diffraction peaks visible, following the procedure used in a previous paper [135], and a degree of crystallinity χ was determined from the ratio of the integrated intensities:

$$\chi = \frac{\int I_c \mathrm{d}\theta}{\int (I_a + I_c)\,\mathrm{d}\theta} \tag{2.1}$$

Values of χ are included in Table 2.1.

Small-angle X-ray scattering (SAXS) studies were carried out using Daresbury Station 16.1 with the RAPID 2-dimensonal detector, with X-ray wavelength $\lambda = 141$ pm and a camera length of 4 m. Scattering intensities were radially averaged, to obtain 1-D patterns of intensity versus $q = \left(\frac{4\pi}{\lambda}\right)\sin\theta$. Three features are of particular interest.

Firstly, wide variations were observed in the scattering intensity for different materials, indicating differing degrees of phase separation. This was quantified as follows, following Saiani et al [187]. For each scattering pattern, the measured intensity (in arbitrary units) was normalized for specimen thickness and incident beam intensity. Then the high q tail of the curve of normalized intensity $I_n(q)$ was fitted to Porod's Law for scattering by a two-phase system with sharp phase boundaries $I_n = \frac{K}{q^4} + I_b$. Excellent agreement was found in all cases except PU_1. For the other polymers the mean R^2 for fitting to Porod's Law was 0.996. These fits provided the background intensity I_b (assumed independent of q) and Porod constant K. The corrected intensity $I = I_n - I_b$ was then employed in determining the relative scattering invariant Q from the relation

Table 2.1 Chemical and physical structures of the PUs studied in this work. Q is the relative SAXS scattering invariant obtained from the intensity distribution, d is the dominant long period obtained from $q*$ the position of the peak SAXS intensity, A/V is the particle surface-to-volume ratio obtained from SAXS, and χ is the degree of crystallinity as determined from WAXS [135].

Material	DI	MD	CE	Hard segment volume fraction ϕ_H	Q	d Nm	A/V Nm^{-1}	χ
PU_1	MDI	PEA	EG	0.356	0.82	-	-	0.036
PU_2	MDI	PEA	DEG	0.380	2.46	18.7	0.722	0
PU_3	MDI	PEA	BG	0.372	2.30	14.1	0.816	0
PU_4	DBDI	PEA	EG	0.363	5.69	16.0	0.822	0.165
PU_5	DBDI	PEA	DEG	0.387	2.32	22.0	0.721	0
PU_6	DBDI	PEA	BG	0.378	2.47	16.8	0.659	0.104
PU_7	MDI	PTHF	EG	0.326	28.69	21.0	0.441	0.012
PU_8	MDI	PTHF	DEG	0.350	11.40	22.8	0.572	0
PU_9	DBDI	PTHF	EG	0.333	34.35	18.7	0.753	0.158
PU_{10}	DBDI	PTHF	DEG	0.356	22.91	20.1	0.519	0
PU_{11}	MDI	PBA	EG	0.323	5.64	18.7	0.586	0
PU_{12}	MDI	PBA	BG	0.338	5.47	16.8	0.835	0.027
PU_{13}	DBDI	PBA	DEG	0.344	4.96	-	0.536	0
PU_{14}	DBDI	PBA	BG	0.354	6.28	16.2	0.556	0.183

$$Q = \int_0^\infty q^2 I(q) dq \tag{2.2}$$

where the upper end of the range of integration, beyond where data were available, was obtained with the extrapolation $I = \frac{K}{q^4}$.

Secondly, all but two of the scattering patterns obtained showed a distinct peak, at a position $q*$ in the region of 0.3 nm^{-1}. In these cases, a "long period" d was calculated from $d = \frac{2\pi}{q*}$, representing a dominant repeat distance of the two-phase structure causing the scattering [135].

Finally, the availability of both parameters K and Q enabled another structural measure to be estimated: the particle surface area-to-volume ratio A/V. It was obtained as follows [205]:

$$A/V = \frac{\pi K}{Q}(1 - \phi_h) \tag{2.3}$$

where ϕ_h is the volume fraction of particles: hard diisocyanate-rich domains in the present materials. Note that ϕ_h could only be approximated, since the exact composition and densities of particles and matrix are unknown. In applying equation (2.2), it was approximated by its theoretical limit in the case of complete phase segregation, i.e. the HS volume fraction ϕ_H, calculated from the HS mass fraction, the density of the HS (taken to be 1.29 g/ml for DBDI and 1.27 g/ml for MDI [187]),

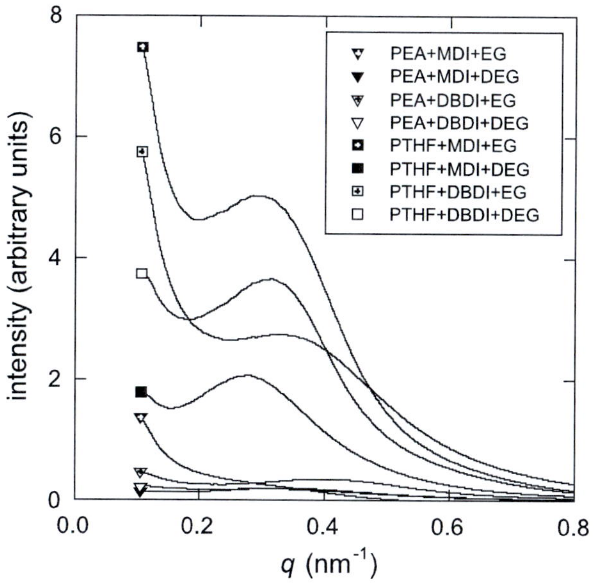

Fig. 2.15 Example 1-D SAXS patterns normalized and baseline-corrected as described in the text, for eight of the materials listed in Table 2.1, with compositions shown [135].

and the density of the SS (taken as 1.18 g/ml for PEA, 1.04 g/ml for PTHF and 1.02 g/ml for PBA). Values of ϕ_H, Q, d and A/V are included in Table 2.1.

Fig. 2.15 shows SAXS patterns for eight of the materials, providing evidence of phase separation in the present series of materials. But Fig. 2.15 also reveals that the SAXS intensity, indicating the degree of phase separation, varies greatly between the materials. In particular, it is clearly visible that the polymers with strongest scattering are those with PTHF as macrodiol. This result is quantified in Table 2.1, where the values of relative scattering invariant Q provide numerical measures of the relative degree of phase separation, since it depends on the hard phase volume fraction φ_h and the difference in electron density (ρ_e) in hard (h) and soft (s) phases as follows [135, 187]:

$$Q \propto \varphi_h(1-\varphi_h)(\rho_{e,h}-\rho_{e,s})^2 \tag{2.4}$$

In the arbitrary units of Q in Table 2.1, the PTHF-based polymers PU_7, PU_8, PU_9 and PU_{10} have Q in the range 11-34, while all the other polymers have smaller values in the range 0.8-6 [135].

The observation that PTHF-based PUs phase-separate to a greater extent than the corresponding PEA and PBA-based polymers is consistent with previous studies of polyurethanes, that showed polyether macrodiols to give greater phase separation than polyester macrodiols [4, 135]. The reason is believed to be the availability of a >C=O group on each monomer in a polyester for possible hydrogen bonding with the >N–H groups on the hard segments. This lowers the free energy of mixing of

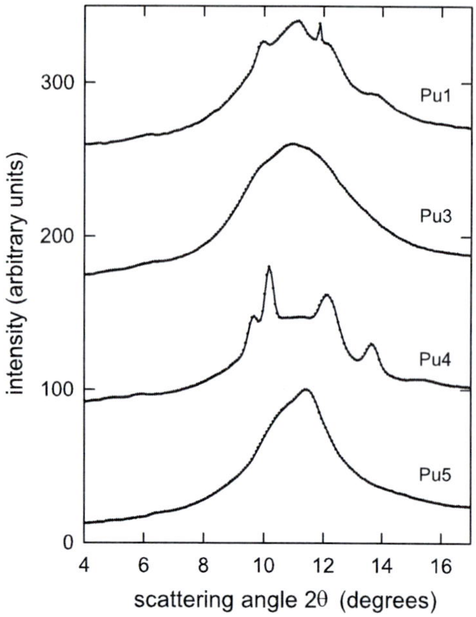

Fig. 2.16 Example 1-D WAXS patterns for four representative materials: all are based on PEA as macrodiol but differ in di-isocyanate and chain extender. PU_1 (MDI+EG) is slightly crystalline (4%), PU_3 (MDI+BG) is amorphous, PU_4 (DBDI+EG) is significantly crystalline (17%), PU_5 (DBDI+DEG) is amorphous. Consecutive patterns have been shifted vertically 80 units for clarity [135].

hard and soft segments that drives phase separation, relative to those macrodiols where this is absent, such as the polyethers.

Fig. 2.15 also shows that the SAXS patterns of most polymers had pronounced peaks in intensity, indicating a dominant repeat distance for the 2-phase structure. As quantified by the long period d, this varied from 14 to 23 nm. Such values for d and the values of hard domain area-to-volume ratio A/V listed in Table 2.1, emphasise the small sizes of the domains. An important consequence for the mechanical properties of the materials is that a large fraction of the hard segments must therefore reside at the surfaces of the domains. This fraction may be quantified in terms of $v^{1/3}$ where v is the volume of one hard segment monomer: in the present materials it varies between ca 32% (PU_7) to 62% (PU_{12}). Finally, another notable implication is that molecular mobility within the soft matrix must suffer significant constraint from the excluded volume of the hard domains. An approximate measure of the average width of the gap between domains may be obtained from $(V/A)\dfrac{1-\phi_H}{\phi_H}$. The values in Table 2.1 indicate this quantity varies between 2.1 nm (PU_3) and 4.6 nm (PU_{14}) [135]. In addition, mobility in the soft phase must be reduced by their molecular connection to the relatively immobile segments of the hard domains.

WAXS patterns gave evidence of hard phase crystallinity in some of the materials [135]. Fig. 2.16 illustrates the range of types of pattern obtained: DBDI-based polymers with intense sharp peaks indicating significant crystallinity (PU_4), DBDI-based polymers with no sharp peaks indicating no crystallinity (PU_5), MDI-based polymers with low intensity sharp peaks indicating very slight crystallinity (PU_1), and MDI-based polymers showing only an amorphous halo (PU_3).

Table 2.1 includes the degree of hard segment crystallinity χ for all the materials. It is notable that all the MDI-based polymers show no, or only slight, crystallinity (maximum 4%). The only polymers to have more significant crystallinity are those based on DBDI. This is consistent with previous reports of comparisons between melt-processed polyurethanes based on these two diisocyanates [60, 202]. But the presence of DBDI does not always lead to crystallinity: it depends on the choice of chain extender. In Table 2.1, the DBDI-based polymers with DEG as chain extender can be seen to have no detectable crystallinity, whereas those with EG and BG have degrees of crystallinity up to 18%.

The relative ease of crystallisation in DBDI as compared to MDI is readily explained in terms of greater flexibility of the DBDI molecule, arising from its –$(CH_2)_2$– bridge between the phenyl rings, compared to only –CH_2– in MDI [60, 202]. Thus DBDI hard segments can adopt a linear conformation facilitating packing and inter-chain hydrogen bonding. MDI hard segments, however, are intrinsically kinked in shape, reducing conformational mobility and thereby hindering close packing and achievement of hydrogen bonding [60, 202]. Conversely, when DEG is used as chain extender with DBDI, the central –O– atom introduces kinks into the DBDI hard segment and disrupts the chain packing that could otherwise be achieved [61].

Fig. 2.17 shows example WAXS and SAXS correlated patterns for the PEA, PBA and PTHF-based materials labeled in Table 2.1, providing evidence of phase separation in the present series of materials. The polymers with strongest scattering are those with PTHF as macrodiol, irrespective of the type of diisocyanate (crystallizing or not). The crystallinity index (χ) revealed by WAXS experiments, was higher for PUs with DBDI, extended with EG or BG at lower SAXS peak areas. These two chain extenders lead to distinct diffraction peaks. Inelasticity was greater for DBDI hard segments than for MDI [61, 175]. *Irrespective of the choice of diisocyanate (MDI or DBDI), all materials with PTHF showed much higher intensity in SAXS.* Such PUs have fewer H bonding: only 40% of the carbonyl in polyether urethanes are hydrogen bonded than polyester urethanes (PEA, PBA) with same HS content.

Materials derived from DBDI and PTHF showed both crystallinity and high phase segregation. This was in contrast with the analogous structures but achieved with MDI were the crystallinity was absent, but the materials also showed high phase segregation.There was evidence of the appearance of more or less complete phase separation associated with the formation of discrete crystalline hard domains.

The PUs excellent properties of originate mainly from their special tendency to form discrete regions in which the HS made up by diisocyanate and extender tend to self associate in separate microdomains. Sometimes these associations are ordered to such an extent that they can be evidenced by their characteristic X-ray diffraction.

A good way for us to study this process was to take into account *homopolyurethanes* as *simple model substances*, which reproduced the structure of the HS from the different complex PUs.

Simple model homopolyurethanes $[EG-MDI]_n$ or $[EG-DBDI]_n$, were achieved by us. The X-ray diffraction patterns showed a significant tendency to crystallize for the dibenzyl based PUs. Fig. 2.18 shows example WAXS diffractograms for a DBDI

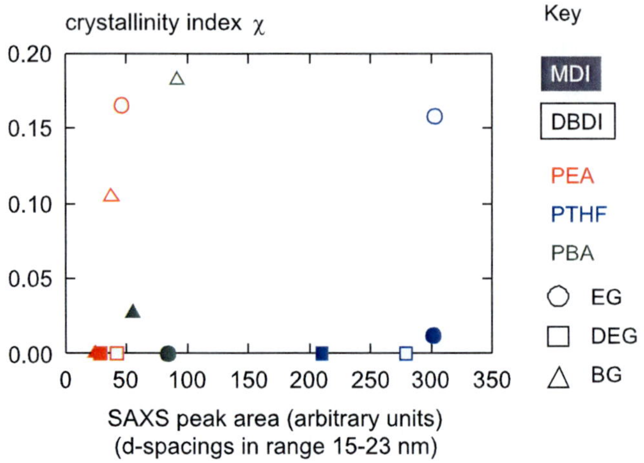

Fig. 2.17 X-ray characterization of the nano-scale structure for example PUs, plotted as crystallinity index (χ) versus SAXS peak areas; filled symbols — data for MDI-based polymers; open symbols — data for DBDI-based polymers [175]

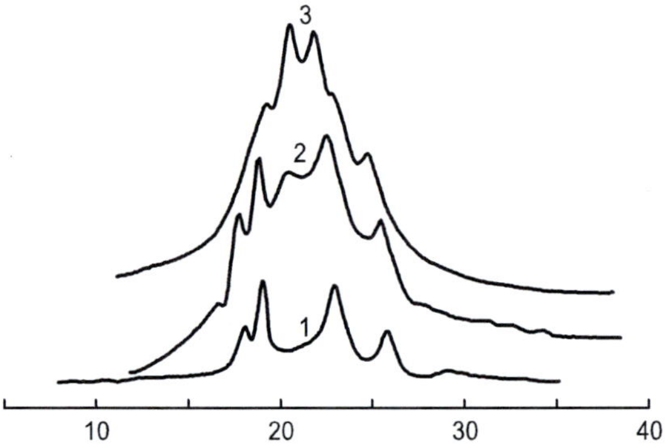

Fig. 2.18 Wide-angle X-ray scattering from: (1) homopolymer EG–DBDI; (2) – PU (DBDI:PEA:EG) with molar ratio EG/PEA = 3; (3) – PU (DBDI:PEA:EG) with molar ratio EG/PEA = 2.64 (PU_1) [206].

material obtained with EG and PEA, with the molar ratios EG/PEA = 3 (curve 2), and EG/PEA = 2.64 respectively (curve 3) [206]. Curve 1 is for the corresponding homopolyurethane [EG-DBDI] based on HS only.

For the DBDI-based material and its corresponding homopolyurethane $[EG–DBDI]_n$ (curve 2) the crystallization did not disappear with the inclusion of the HS in the block copolyurethane elastomeric matrix, indicating the appearance of more or less phase separations associated with the formation of crystalline hard domains. As seen, curve (2) for the material with molar ratio EG/PEA = 3, was very similar with curve (1) corresponding to the homopolymer, and proportional to HS

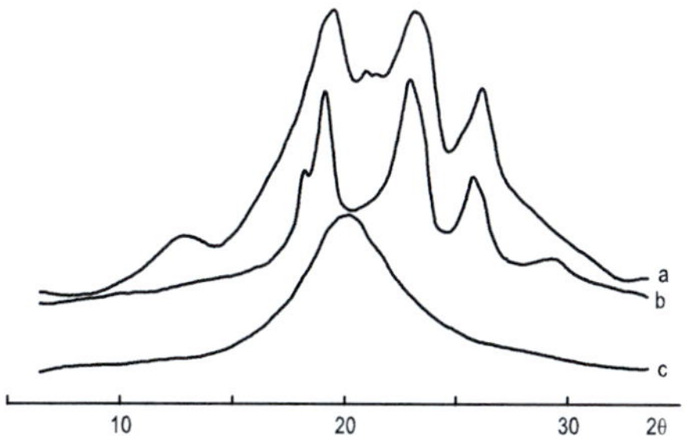

Fig. 2.19 WAXS of polyurethanes containing only hard segments: (a) $[BDO\text{–}DBDI]_n$; (b) $[EG\text{–}DBDI]_n$; (c) $[EG\text{–}MDI]_n$ [60]

content. Reducing the number of HS by lowering of molar ratio from 3 to 2.64 determined significant changes in the shapes of the diffractograms [206]. There appeared crystalline structures of other types, within which the SS were involved also.

Fig. 2.19 displays some other typical WAXS curves. As shown [60] $[EG\text{–}DBDI]_n$ represents a polymer with a high tendency of crystallization (Fig. 2.19 (b)) both in terms of quantity of crystalline material and the definition by the X-ray reflections. This behaviour is paralleled by the self-separation of crystalline polymer from its solution with time [69]. The $[EG\text{–}MDI]_n$ (Fig. 2.19 (c)) shows only a single broad peak, rather narrower than might be expected for an amorphous halo, suggesting a rather loose association of molecular chains. This material does not crystallize from solution [60].

This capacity for crystallization is characteristic not only for the $[EG\text{–}DBDI]_n$ structure, but also for similar DBDI based materials produced by changing the glycol as for example in the case of a $[BDO\text{–}DBDI]_n$ structure Fig. 2.19 (a). As shown, an extender with an even number of methylene units should allow the diisocyanate molecule to assume an almost linear "anti" conformation, so these materials should exhibit an 'odd-even' variation in their ability to crystallize as observed in similar cases [60].

The WAXS patterns showed that the crystallizability of DBDI based material largely remains even after the inclusion of the SS; this indicated the appearance of more or less complete phase separation associated with the formation of discrete crystalline hard domains. The presence of these structures induced a special mechanical behaviour in these elastomers as you will see in Chapter 4. Due to the disordered orientation of these domains in the polyether matrix the materials appear initially isotropic on the macroscale but when subjecting the polymer to stretching, some microdomains tend to orient in parallel, maintaining to some extent the so obtained anisotropy even after removing the stress. As a result some enhanced residual elongation was evident and the chain orientation was substantially confirmed by using a polarizing optical microscope. Supplementary studies on this subject were performed employing IR dichroism [60].

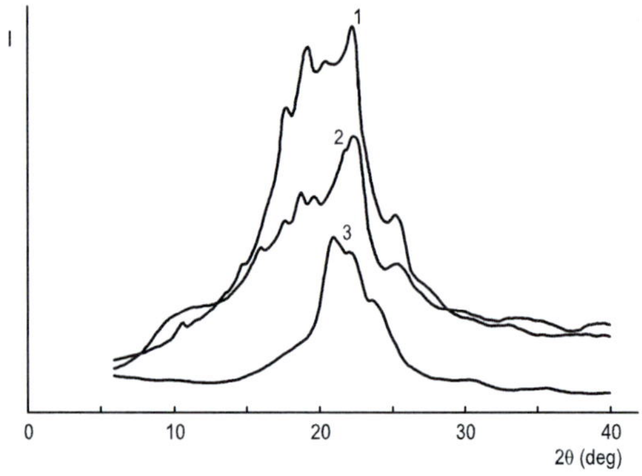

Fig. 2.20 Wide-angle X-ray diffraction from a PU of type D (DBDI-PEA-BDO, I = 110): (1) PU sheet cast directly from synthesis; (2) PU film obtained by evaporation from 10% solution in DMF; (3) PU film obtained by evaporation, then stretched to 300% elongation, and then released for 24 hours (residual elongation 175%); intensity scan perpendicular to the direction of elongation [60].

It is also noteworthy, however, that *the sample preparation technique also influences PU crystallization*, as shown for the case of a dibenzyl PU (DBDI:PEA:BDO) in Fig. 2.20. The crystallinity of the materials with DBDI is more pronounced when the polymer is obtained as a sheet cast directly as a melt following synthesis (Fig. 2.20, curve 1) than when the same PU has been obtained by evaporation from 10% solution of DMF (Fig. 2.20, curve 2). The crystallinity also changes completely if the solution-cast PU film is stretched to 300% elongation and released (Fig. 2.20, curve 3).

2.3.2 Structural studies on polyurethane elastomers obtained with mixtures of diisocyanates

As shown above one of the convenient means to reduce the DBDI based PUs crystallinity was to use the flexible chain extender diethylene glycol (DEG), which inhibits crystallization. But the crystallinity of DBDI based PUs can be also considerably reduced with the incorporation of MDI as earlier reported [61].

In the first step of our study, we have focused our work on the generally employed method of PUs synthesis in the melt, namely the prepolymer route followed by a reaction with a chain extender subsequently cured with air humidity, which consumes all of the isocyanate excess situated on the ends of the intermediary formed active PU oligomer, thereby leading to an additional degree of urea group formation [60]. *In our case, in this study, this refered to a PU synthesized by using a 1 mole of macrodiol – 4 moles diisocyanates – 2.64 moles chain extender system, so to obtain a small isocyanate excess (I = 110) situated at the ends of the above active PU oligomer.*

In the PU synthesis we started from one mole of hydroxy-terminated macrodiol which was either a polyester type—polyethylene adipate (PEA) or a polyether type—polytetrahydrofuran (PTHF).

For the preparation of the PU elastomer containing simultaneously two different diisocyanates the macrodiol was reacted in different ways with two moles of MDI and two moles of DBDI. Finally the so-obtained different prepolymers were reacted with 2.64 moles of chain extender which was in all the cases ethylene glycol (EG). There were adopted three different modes of prepolymer synthesis leading to three different active PU oligomers.

PU_{C1}: Single step prepolymer synthesis which used a melt mixture of the two isocyanates (MDI and DBDI isocyanates). In this case, the two different diisocyanates were inserted more or less randomly into the prepolymer and the resulting full polymer;

PU_{C2}: A two-step prepolymer synthesis where two moles of DBDI was first reacted with one mole of macrodiol, followed by addition of MDI;

PU_{C3}: A two-step prepolymer synthesis where two moles of MDI was firstly reacted with one mole of macrodiol, followed by addition of DBDI.

In all cases, it obtained active oligomers with final NCO groups having a calculated molecular weight of about 8 900 [60]. Finally the so-obtained different prepolymers were reacted with 2.64 moles of chain extender which was in all cases EG. There were adopted three different modes of prepolymer synthesis leading to three different active oligomers.

The composition of the PU family prepared in this study is shown in Table 2.2 and the equations are shown in Table 2.3.

Table 2.2 Compositions of the family of polyurethane elastomers prepared and studied in this work

Recipe	PU structure
PU_{C1}	CE[a]–MD[b]–(DBDI–MDI)
PU_{C2}	CE–(MD–DBDI)–MDI
PU_{C3}	CE–(MD–MDI)–DBDI
PU_{D}	CE–MD–DBDI
PU_{M}	CE–MD–MDI

[a] CE – chain extender; [b] MD – macrodiol

2.3.2.1 WAXS results for polyurethane elastomers with mixtures of diisocyanates. Crystallinity

X-Ray scattering (WAXS) were employed by us on the series of materials labeled in Tables 2.2 and 2.3, in order to follow the changes that were brought about by

Table 2.3 Synthesis routes for PU_C based on mixtures of the diisocyanates MDI and DBDI, with PEA or PTHF as soft phase [60]

$PU_{C1\ PEA}$ and $PU_{C1\ PTHF}$ type synthesis Single step prepolymer synthesis using a melt mixture of the two isocyanates (MDI + DBDI).		
MD + (2 MDI + 2 DBDI) melt mixture	⟶	U*~~~~U* + 2 (x MDI + y DBDI) prepolymer U
[U*~~~~U* + 2 (x MDI + y DBDI)] + 2,64 CE	⟶	OCN $U^*[G(UG)_{1.64}U$~~~~$U]_{2.77}$ NCO
$PU_{C2\ PEA}$ and $PU_{C2\ PTHF}$ type synthesis Two step prepolymer synthesis where DBDI is first reacted, followed by the addition of MDI.		
MD + 2 DBDI	⟶	M*~~~~M* prepolymer (M)
[M*~~~~M* + 2 MDI] + 2.64 CE	⟶	OCN $D^*[G\ (DG)_{1.64}\ M$~~~~$M]_{2.77}$NCO
$PU_{C3\ PEA}$ and $PU_{C3\ PTHF}$ type synthesis Two step prepolymer synthesis where MDI is first reacted, followed by the addition of DBDI.		
MD + 2 MDI	⟶	D*~~~~D* prepolymer (D)
[D*~~~~D* + 2 DBDI] + 2.64 CE	⟶	OCN M*[G (MG)1.64 D~~~~D] 2.77 NCO

MD - macrodiol; CE - chain extender

M* and D* = monourethane-isocyanate sequence derived from monoreacted MDI and respectively DBDI;

M and D = diurethane sequence derived from bireacted MDI and respectively DBDI;

U* and U = mono and respectively diurethane sequence derived from either randomly mono and bireacted MDI or DBDI.

changing the type and the number of isocyanates (one or two diisocyanates), and the order of their introduction in the material synthesis.

As seen in Fig 2.21, for the materials containing *two diisocyanates* of PU_{C1}, PU_{C2} and PU_{C3} types, in both of the soft segments PEA and PTHF, the X-ray diffraction patterns showed in all cases that the crystallinity decreases significantly relative to those with DBDI indicating a decrease of the HS ordering. In general when employing the PU synthesis with mixtures of DBDI and MDI, the crystallinity of the obtained PU decreased relatively the materials obtained by using the diisocyanate DBDI alone. The degree of extinction of crystallinity depended also on the used synthesis route [60].

The crystallinity was considerably reduced with the incorporation of MDI, of which the example of PU_{C1} PEA is presented in Fig. 2.21 (a). In all the PUs containing two diisocyanates, namely PU_{C1}, PU_{C2} and PU_{C3} types with either of the soft segments PEA or PTHF, the X-ray diffraction patterns showed that crystallinity decreases significantly relative to those PUs containing only DBDI (PU_D) indicating a decrease of the HS ordering [60].

To compare more critically the different routes of copolymerization PU_{C1}, PU_{C2} and PU_{C3}, and their effect on the HS ordering, a further WAXS data was collected.

These results major on the PTHF series of materials. With these also, when employing the PUs synthesis with mixtures of DBDI and MDI, the crystallinity was

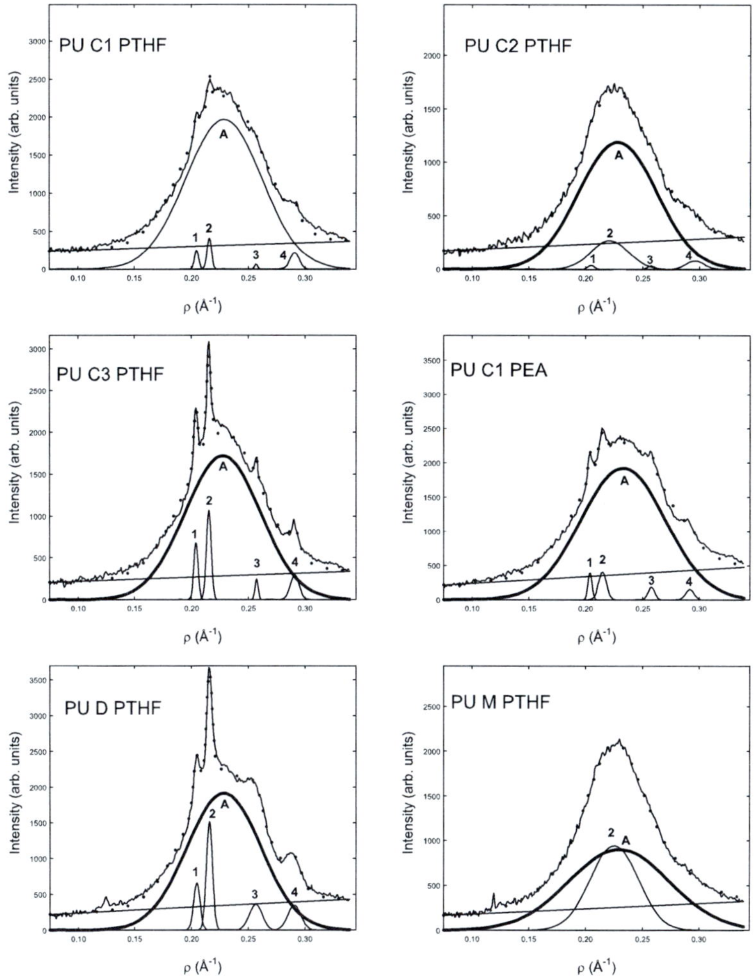

Fig. 2.21 (a) WAXS pattern collected from sample PU_{C1} with PTHF showing the separation of the scattering into crystalline and amorphous components; (b) WAXS pattern collected from sample PU_{C2} based on PTHF showing the separation of scattering components. Here, all the peaks are broad, indicating an amorphous structure; (c) WAXS pattern collected from the PTHF based sample PU_{C3} showing the separation of the scattering into crystalline and amorphous components; (d) WAXS pattern collected from a similar sample type PU_{C1} but based on the macrodiol PEA; (e) WAXS pattern collected from sample based on PTHF and derived from the single diisocianate DBDI (PU_D) showing the separation of the scattering into crystalline and amorphous components. (f) WAXS pattern collected from sample with PTHF derived from the single diisocianate MDI (PU_M) [60].

decreased relative to the unmixed PU_D materials, but it is very apparent from Fig. 2.21 that the degree to which crystallinity is depressed depends critically on the synthesis route.

Clearly, the sharp peaks associated with DBDI decreased in the order: PU_D > PU_{C3} > PU_{C1} > PU_{C2} > PU_M *corresponding to introduction of DBDI as follows: total polymer:hard segments:mixed:soft segments:none.*

The crystallization degree was higher in the case of a polyetheric material of the type PU_{C3} when the introduction in the reaction of DBDI was performed in the second stage, because the HS sequence of the EG-DBDI type becomes preponderant.

Improvement in elastomeric properties, by reducing the crystallinity in PUs based on DBDI, was obtained especially when both diisocyanates were included and reacted together in a random fashion, instead of sequentially by using a prepolymer stage [60].

It is to be noted that the substitution of the SS from PTHF to PEA did not change the above mentioned conclusions. The WAXS patterns of corresponding pairs based on PEA or PTHF were similar in all cases.

To conclude: X-ray diffraction patterns showed significant crystallisability for the PU materials based on DBDI hard segments. There was evidence of the appearance of more or less complete phase separation associated with the formation of discrete crystalline hard domains. However, when the equimolar mixture MDI+DBDI was used for hard segments (PU type C1) the crystallinity was considerably reduced [60, 61]. As mentioned elsewhere [61], in the case of PU based on the single diisocyante DBDI, the degree of crystallinity can be reduced also by employing a more flexible chain extender.

As shown [60], a general improvement in PUs properties was obtained after polymer annealing at 160°C for 30 minutes [60] when two diisocyanates (MDI+DBDI) were included in the synthesis at once in a random fashion, i.e. for the type of material PU_{C1}.

By doing this study we believe that new insight has been gained into structure-property relations in polyurethane elastomers by preparing them using mixtures of isocyanates.

2.3.2.2 Morphology of polyurethane elastomers with mixtures of diisocyanates as revealed by SEM

The materials with mixtures of diioscyanates depicted in Tables 2.2 and 2.3 were also investigated by means of SEM [60]. As before, the materials were based on PTHF or PEA.

The structure of these materials contained features at two levels. All the materials displayed a relatively structure on the 10μm scale, but which varied considerably from material to material. This is shown in Fig. 2.22 (a)–(f), for the PTHF materials. Of the materials studied in Fig. 2.22, the polymer $PU_{C3\ PTHF}$ from Fig. 2.22(c) has the coarsest structure while $PU_{M\ PTHF}$ has the finest (Fig. 2.22(e)). The materials which display X-ray crystallinity (a, c and d of Fig. 2.22) tend to give a rougher surface morphology than those which do not (b and e of Fig. 2.22). The specimens also show variations in topography such as long scratch marks which derive from

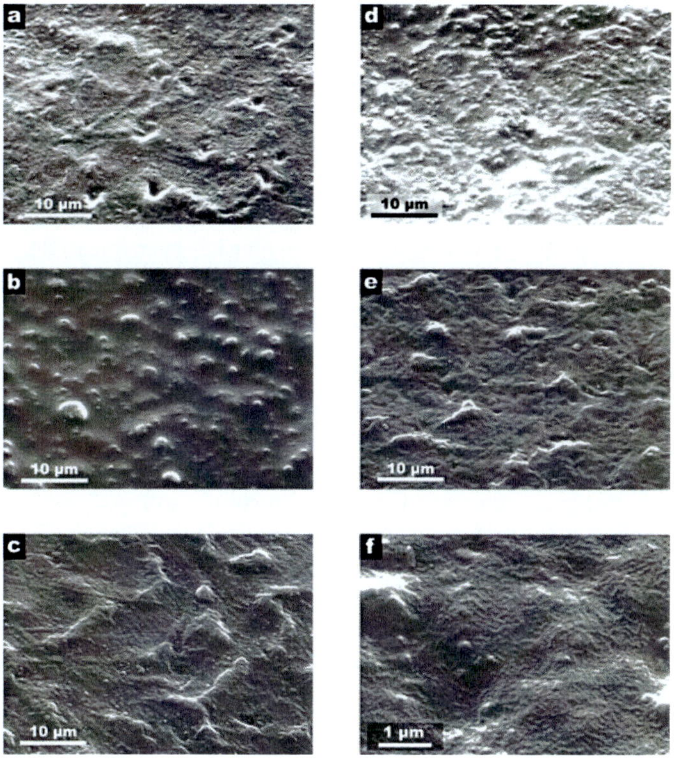

Fig. 2.22 SEM at medium magnification of: (a) $PU_{C1\ PTHF}$ (b)$PU_{C2\ PTHF}$ (c)$PU_{C3\ PTHF}$ (d)$PU_{D\ PTHF}$ (e) $PU_{M\ PTHF}$ (f) SEM at higher magnification showing fine texture of $PU_{M\ PTHF}$ [60]

the PTFE plates against which the sheets were moulded, although we have avoided showing regions where these obscure the characteristic texture of the materials.

There is also a finer texture of scale less than 100 nm, which may correspond to segregation of the two different chemical parts, namely the HS and SS. This is only observed by operating the SEM at higher accelerating voltage and smaller spot size, as in Fig. 2.22 f. Shown here is $PU_{M\ PTHF}$, but finer structure is present in the other materials also. This, however, was not intensively studied, as it would require transmission electron microscopy to give appropriate resolution, and such examination would be better if a suitable chemical etching technique were available to reveal the bulk of the material and remove features associated with the surface.

Several chemical etching techniques were tried by us on these materials [60]. Those based on alkaline reagents (whether potassium hydroxide in alkaline solution, or sodium ethoxide in ethanol) did not react enougheven to completely remove the smallest spurious features associated with the moulded surfaces. On the other hand, the permanganic reagent with phosphoric acid gave smooth and uniform attack over all the materials. In Fig. 2.23, etched surfaces of all five materials are shown at three magnification for easy comparison. $PU_{C1\ PTHF}$ (Fig. 2.23 a)) shows roundish regions which are raised and therefore more resistant to the etchant, surrounded by a flatish matrix. The scale of these regions is much larger than the length of the HS and SS, so it appears as if some precipitations of one region has occurred during the reaction: in this case the precipitated part is more resistant to etching. The whole surface appears to be strewn with small objects between 0.2 and 0.8 μm in size, which look like crystals whose edges have been rounded off by partial dissolution [60]. It may be that these are a particular chemical species, which has been released by the attack. $PU_{C2\ PTHF}$ (Fig. 2.23 b)) shows a similar structure, except that the resistant regions are much smaller, although they occupy about the same percentage of the total area. In $PU_{C3\ PTHF}$ and $PU_{M\ PTHF}$ (Fig. 2.23 c and e), it appears that the relatively resistant material is in the majority, with non-resistant pockets being eaten out by the etchant. It may be that these pockets represent material precipitated during polymerization. In $PU_{D\ PTHF}$ (Fig. 2.23 d)) the overall etched texture is much flatter, so it would seem that the regions of chemical segregation are more or less equally attacked, although what structure there is seems to show a pattern similar to Fig. 2.23 c and e. *Only $PU_{D\ PTHF}$ in Fig. 2.23 d does not show the precipitated crystals: this is the only one of the five examined samples which does not contain MDI, because the crystals are derived from units containing this monomer.* At the highest magnification, there is only a faint suggestion of fine-scale structure such as seen in Fig. 2.23 f, which is most prominent in Fig. 2.23 b (bottom). This particular enchant is therefore not ideally suited to distinguishing between hard and soft segments.

2.3.2.3 Morphology of polyurethane elastomers with mixtures of diisocyanates as revealed by AFM

The morphology of a series of polyester or polyether PUs based on mixtures of diisocyanates has been also examined by ourselves by using the AFM technique. Example PUs that were studied, based on the macrodiol PEA and chain extended with EG are depicted in Table 2.4. The molar composition of the materials was diisocyanate:macrodiol:chain extender = 4:1:3 giving an isocynate index I = 100. The preparation procedures in order to obtain PUs with single or mixtures of the diisocyanates MDI and/or DBDI has been described elsewhere [127, 207].

Normally, the surface of a PU sample fracture shows a relief composed of peaks of HS and crystal aggregates domains. The rupture occurrs most probably in the amorphous regions. Therefore, it is expected that such a surface contains more information about the sample morphology than the surface of free solidification or drying. AFM images were taken from different zones for each sample. Shown in Fig. 2.24 are images of freeze fracture surfaces for the whole series of the materials (PU_1 to PU_5).

The values depicted in Table 2.5 for the parameters H_a and S_q show some correlations between the PUs morphology and their chemical structure. Thus, the corresponding roughness order $PU_2 > PU_3 > PU_4 > PU_1 \approx PU_5$ shows that the most uneven relief appears when the materials is obtained by the two stage polyaddition

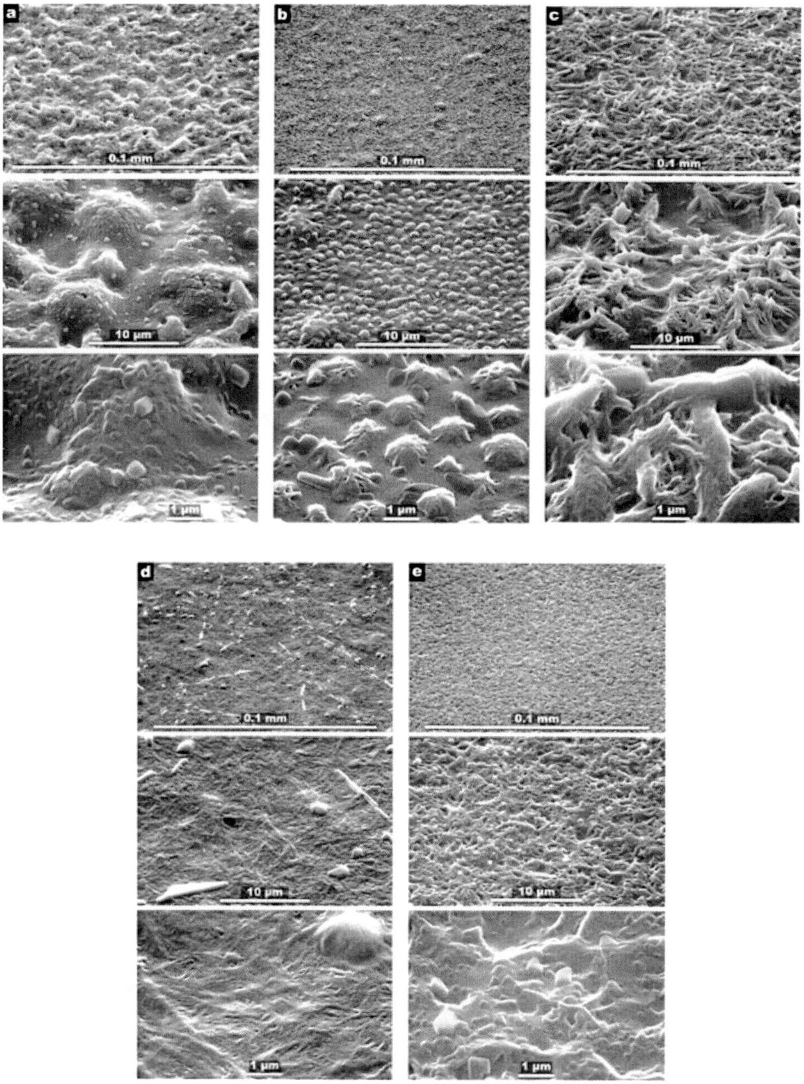

Fig. 2.23 SEM of etched polyurethane specimens at three magnifications: (a) $PU_{C1\ PTHF}$; (b) $PU_{C2\ PTHF}$; (c) $PU_{C3\ PTHF}$; (d) $PU_{D\ PTHF}$; (e) $PU_{M\ PTHF}$ [60]

procedure. A plausible explanation would be the following: when the macrodiol is first and completely reacted with one of the two diisocyanates (here denoted as DI_1), the excess of 2 moles of DI_1 to 1 mole of macrodiol in the first reaction step results in a prepolymer that mainly consists of macrodiol molecules with monoreacted DI_1 ends. At the end of the process, when the rest of the reactants is added, (i.e. 3 moles of chain extender and 2 moles of the second diisocyanate (here denoted as

Table 2.4 Compositions of the family of PUs [207]

Recipe	PU structure	PUs composition Moles of PEA	Moles of MDI	Moles of DBDI	Moles of EG
PU_1	EG–PEA–(DBDI–MDI)	1	2	2	3
PU_2	EG–(PEA–DBDI)–MDI	1	2	2	3
PU_3	EG–(PEA–MDI)–DBDI	1	2	2	3
PU_4	EG–PEA–DBDI	1	0	4	3
PU_5	EG–PEA–MDI	1	4	0	3

DI_2)), a polymer with –(DI_2–CE–DI_1–MD–DI_1–CE–DI_2–CE)– units should result (where CE is the chain extender). This is the case with the polymers PU_2 and PU_3.

Table 2.5 Surface roughness parameters from AFM height images [207]

Sample identification		Crystallinity index [135]	Surface roughness parameters		
Code	Composition	CI (%)	H_a [nm]	S_q [nm]	NSH (H_a/S_q)
PU_1	EG–PEA–(DBDI–MDI)	<20	399.4	116.1	3.44
PU_2	EG–(PEA–DBDI)–MDI	>20	1617.7	368.9	4.38
PU_3	EG–(PEA–MDI)–DBDI	>30	717.4	191.8	3.74
PU_4	EG–PEA–DBDI	~50	669.3	153.9	4.35
PU_5	EG–PEA–MDI	>25	320.1	101.6	3.15

The fact that the roughness is significantly higher for PU_2 than for PU_3 may be associated with the crystallinity, which is promoted by the presence of DBDI and is absent in the materials based on the single diisocyanate MDI (Table 2.5). The difference in crystallinity associated with DBDI and MDI was previously evidenced and ascribed to the flexibility of the hard segments, which is characteristic to the DBDI segments and absent in the MDI hard segments [127].

Regarding the roughness order, it is evident that the cross sections of the materials PU_4, PU_5 and PU_1 (where the two diisocyanates are mixed together at once in a random fashion via the one stage prepolymer synthesis route) are smoother than those obtained by the two stage prepolymer polyaddition procedures. This is not only due to crystallinity as long as this parameter is the largest for the materials PU_4, and not for the polymer PU_2. Taking into account the morphology model of the HS phase dispersed into the SS phase proposed by Christenson et al [208], our assumption was that the domains of the SS crystals joined by the HS matrices are small and dense in the PUs obtained by the one step prepolymer route, while they are bulky and rare in the PUs obtained by the two step prepolymer techniques. Such a morphology is supported by the topographic 3D and 2D images shown in Fig. 2.24

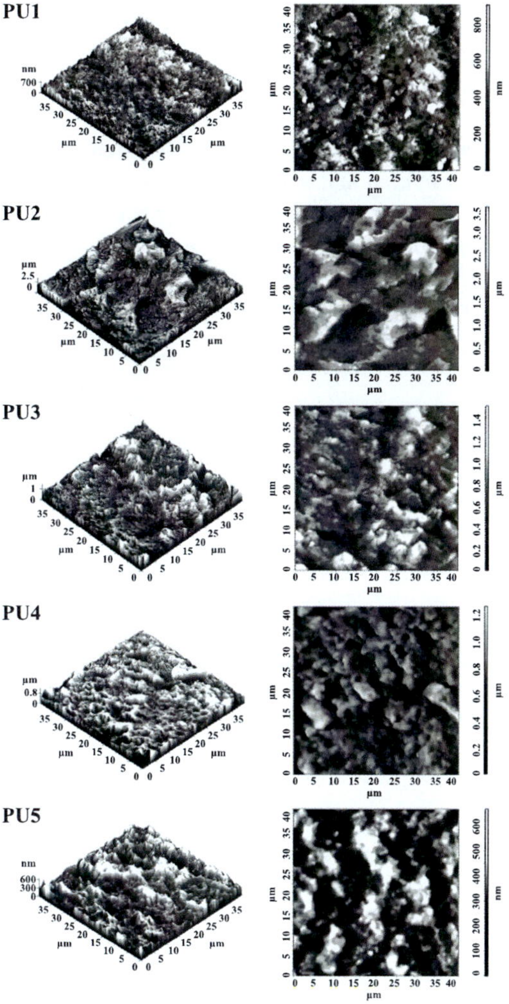

Fig. 2.24 Tapping mode AFM topographic 3D and 2D images for freeze fracture surfaces for materials listed in Tables 2.4 and 2.5 [207]

if, for example, the images of the samples PU_2 and PU_5 are compared with those of the samples PU_2 and PU_3, respectively.

It results that the different morphologies are related to the differences in the chain microstructure. Therefore, while the chains resulted from two step prepolymer syntheses, are predominantly built with –(DI_2–CE–DI_1–MD–DI_1–CE–DI_2–CE)– units, the chains obtained from one step prepolymer polyadditon processes are more heterogeneous due to the presence of amounts of other sequences like –CE–DI–MD–DI–MD–DI– and –CE–DI–CE–DI–CE–DI–. Obviously, the polymer chain heterogeneity does not promote the orientation of the chains.

It is of interest to consider the NSH parameter given by the ratio of the average height, H_a, to the RMS roughness, S_q. Discussion about this parameter is absent in the literature. As shown in Table 2.5, the NSH follow the order of the values of H_a

and S_q, but the differences between the materials are smaller. By correlating NSH with the characteristics of the surface relief it observed that NSH increased not only with the roughness, but also with the decrease of the relief uniformity. Comparing the materials PU_4 and PU_2, which showed much different values for H_a and S_q, but almost similar NSH values, it resulted that the relief of the material PU_2 is more uniform than that of the polymer PU_4.

The AFM images of fracture surfaces revealed that the PUs obtained by the two stage prepolymer poliaddition provedure, have a more uneven relief than the those of the material obtained by the one step prepolymer technique. In addition, DBDI increases this difference, but leads to more uniform relief forms than does MDI.

Chapter 3
Thermal behaviour of polyurethane elastomers

3.1 DMA experiments

If an experiment works, something has gone wrong. (Murphy's quote)

While I agree it is always good to be realistic, I never like to be pessimistic. Certainly the wellknown Murphy's quote does not apply to the experiments that are described in this book. There is no study described here that would count on the success of a single experiment ☺

Along with differential scanning calorimetry (DSC), dynamic mechanical analysis (DMA) provides methods for probing the microphase separation characteristics through changes in the glass transitions of the components. DMA experiments measure the ability of a viscoelastic material to store and dissipate mechanical energy. They are based on the differences in the load response of the viscous and elastic components to a small sinusoidally applied strain. The resulting stress that is measured using an appropriate sensor, lags behind the applied strain by a phase angle δ. For a perfectly elastic material that stores all the input energy, the phase lag is 0° and for a perfectly viscous material is 90° out of phase. Thus the stress and strain are in phase for an elastic material.

The complex Young modulus (E^*) consisting of contributions of a storage modulus (E'), and of a loss modulus (E''), is measured (eq. 3.1). The complex modulus reflects the inherent viscoelastic nature of the polymer, where stress and strains will be out of phase with one another.

$$E^* = E' + iE'' \tag{3.1}$$

The E' storage modulus quantifies the energy stored elastically by the material upon deformation, while the loss modulus E'' is a measure of the energy which is dissipated as heat during deformation. From the ratio between the loss and storage moduli, a third parameter, the loss factor ($\tan\delta$) can be calculated. This translates to a measurement of the ratio of the energy absorbed by the sample as heat to the energy used by the sample to return to its original shape. The loss modulus and loss

C. Prisacariu, *Polyurethane Elastomers*, DOI 10.1007/978-3-7091-0514-6_3,

factor are indicators of the polymer ability to internally dissipate the energy under the test conditions, which is why peaks in the $\tan\delta$ plots are often referred to as damping peaks. The DMA technique is sensitive to the various transitions which a polymer undergoes as a function of changing the temperature.

Information on the PUs different flexible and rigid blocks can be obtained by means of DMA spectra in which the various segments are respectively observed to melt in sequence as the temperature is raised. The storage modulus E', provides information regarding the stiffness of the material, while the loss factor $\tan\delta$ measures the degree of molecular motion. The glass transition temperature (T_G) of the SS is defined as the α transition in the $\tan\delta$ curve.

Significant features of a $\tan\delta$ plot include the location of the low temperature maximum, where smaller areas are associated with better phase separation of PUs hard and soft phases, as well as less of the hard phase being involved in the glass transition. The magnitude of $\tan\delta$ across the application temperature, which is proportional to the ratio of energy absorbed as heat by sample to energy returns as movement of the sample. Also significant is the location of the high temperature upturn, which is associated with the PUs softening. Very low temperature γ and β peaks are associated with reordering of HS domains.

The degree of phase mixing can be estimated from the shift of the glass transition temperature (as with DSC), and also from the storage modulus E', which has an intermediate value between the moduli of the two phases and depends on the relative amount of each phase.

3.1.1 DMA behaviour of polyurethane elastomers based on single diisocyanates

Phase separation in PUs observed by using the DMA technique was first proposed by Cooper and Tobolsky [209] based on the presence of two thermal transitions in the dynamical mechanical data that were assigned for the glass transition temperature (T_G) of the SS and HS softening point. Two transitions are a sign of incompatibility whereas only one transition is expected for compatible polymers. Hartmann et al. [210] measured the thermal and PUs dynamic mechanical properties by varying the diol molecular weight, the type of aromatic chain extender and its stoichiometry and the diol weight distribution. They found that the SS crystallization occurs at higher diol molecular weights. Dynamic mechanical properties were well correlated with the SS glass transition. The diol molecular weight influenced the PUs dynamic mechanical properties by affecting the degree of phase separation and hence the transition temperature. The diol molecular length distribution had no significant effect on the transition temperatures and dynamic mechanical properties.

Both the mechanical and thermal PUs properties are affected dramatically by phase mixing. Interaction between HS and SS may increase the T_G of the SS and decrease the T_G of the HS. The T_G of the *polyether* polymer was found to be lower than that of *polyester* based elastomer. This is due to more flexibility of the ether bond and higher phase segregation in comparison with the ester bond.

Numerous systems of PUs have been investigated using the dynamic mechanical techniques [211, 212]. An earlier review (year 1965) of the PUs mechanical data was made by McCrumb et al. [213]. In the earlier study made by Ferguson [211], a series of PUs were studied with systematically varying the *HS/SS ratio*. Their dynamic mechanical spectra were studied over a wide temperature range. The materials were prepared using polyether SS and the diisocyanate MDI which formed the bulk of the HS. The chain extender was 1,3-diaminopropane. The samples were studied at 2% static strain. They showed $\tan\delta$ maxima in three temperature regions, but each sample did not necessarily show all three transitions. In general, the transitions occurred at about −130 to −20°C and over a range from about 11° to over 225°C. These transitions were referred to as the α, β and γ, transitions in order of decreasing temperature. The transitions were interpreted in terms of molecular motion. The α transition, which only occurred when MDI was also present, was found in the 15–20° range for the materials with the higher SS concentrations as a result of the phenyl group rotation. As the HS concentration was increased, spherulites were detected and the α transition shifted to considerably higher temperatures, but a shoulder remained at about 20° until the HS concentration reached a high level. As shown [212] time (frequency) and temperature have opposite effects on the molecular motion.

The effect of the HS and SS degree on the morphology and dynamic mechanical behaviour of semicrystalline segmented PUs was investigated by Korley et al [214]. A series of high molecular weight PUs with SS macrodiols containing poly(ethyleneoxide) (PEO) of molar masses 1 000 and 4 600 g/mol and poly(ethyleneoxide)-poly(propyleneoxide)-poly(ethyleneoxide) (PEO–PPO–PEO) (molar mass 1 900 g/mol) were undertaken. The HS content was varied. The presence of the SS crystallinity (PEO 1 000 g/mol) resulted in an improvement of the E' modulus below the melting temperature of the soft block and in an enhanced toughness compared to the materials based of the SS macrodiols PEO–PPO–PEO. This was due to the crystalline soft regions acting as an additional load bearing component during deformation. In PUs achieved with the SS macrodiols PEO–PPO–PEO, the $\tan\delta$ peak broadened due to restrictions on the SS flexibility imposed by the hard domains, but no appreciable change in peak position was observed as the HS content increased. As shown by Korley [214], the crystalline regions of the soft domain served as reinforcing fillers and contributed to the mechanical integrity of the polymer. Upon melting of the PEO crystallites, the HS aggregates supported the micro-phase segregated morphology.

Eceiza et al investigated a series of thermoplastic PUs synthesized in bulk by two-step polymerization, obtained with the isocyanate-chain extender couple MDI–BG and *various SS macrodiols*. The chemical structure, the SS molecular weight and the *HS content were varied*. The effects on the thermal and mechanical properties were investigated [77]. The changes in the macrodiols in the case of PUs based on MDI–BG, resulted in a modulus curve that showed a plateau indicating the existence of physical crosslinks because of the increase in the size and inter-connectivity of HS domains. The plateau extended more in temperature with increasing the HS content, until sample softening occured, associated with the starting of physical crosslink disruption. For materials with HS contents less than 40 wt%, the E' modulus droped abruptly and no plateau was observed [215-217]. The HS were not able to form a sufficient amount of individual domains stable under stress as physical crosslink.

Christenson et al [208] reported research on a series of PUs synthesized with the couple MDI/BDO. The SS macrodiols were: PTMO (molar mass 1000 g/mol) or *a mixture of SS* based on ethylene oxide (EO) / propylene oxide (PO). The MDI/BDO polymers were compared with a MDI based material extended with mixed diamines [208, 218].

Both materials had a similar hard-to-soft segment ratio. The T_G was identified by a drop in the storage modulus E' and prominent peaks in $\tan\delta$ and in the loss modulus E''. Considerably higher T_G of the polymer with PTMO (−18°C) compared to poly(ether urethane urea) (−60°C) was attributed to greater constraint on PTMO soft segments, which resulted from their shorter length and the higher concentration of constraining HS. A second transition at −10°C appeared in the poly(ether urethane urea) spectra, as a small drop in the E' curve and as shoulders on the $\tan\delta$ and E'' curves. The transition temperature was consistent with the melting temperature of PTMO reported by Niesten et al [219]. The absence of this transition in the PTMO:MDI:BDO material spectrum indicated that constraint on the SS effectively prevented their crystallization.

Dynamic mechanical properties of PUs of *variable HS content* were also studied by Crawford et al [220], to determine the relationship between $\tan\delta$, loss modulus and storage modulus and the morphology of *strain aged and unaged model polymers*. A series of model PUs were obtained with the polyester macrodiol PBA or polyether PTMO, using the diisocyanate MDI and the chain extender BDO. The obtained data showed that the $\tan\delta$ peak magnitude and the T_G were sensitive to the material composition. As the HS content of the ester polymers was increased, their $\tan\delta$ magnitude decreased. This trend was consistent with polymers that have increasing amounts of crystallinity [221]. The increasing HS content resulted in larger hard microcrystalline domains that restricted the molecular motion of the SS thereby increasing the T_G. The DMA data indicated that the interaction between the hard and soft domains as a result of strain aging is quite complex. This was especially true with the polyester PUs which did not show a consistent trend in $\tan\delta$ or E'' modulus with increasing the strain. This was assumed to be due to the lower degree of phase separation present in the polyester PUs as compared to the polyether materials based on the same HS percentage; and to the greater potential of HS–SS interaction of the polyester based PUs due to potential hydrogen bonding between the urethane HS and the SS carbonyl. The data relating to $\tan\delta$ showed an increase in T_G up to a 200% strain elongation for the polyester PUs. Heat aging the PUs without strain elongation was shown to enhance domain ordering and phase separation and resulted in improved material abrasion resistance [220]. The polyether PUs exhibited a decreasing trend in tan peak magnitude with increasing levels of strain aging. Although the temperature associated with the $\tan\delta$ transition initially increased, beyond 100% elongation the transition temperature started to decrease without showing a clear trend. Although the $\tan\delta$ peak magnitude decreased for the polyether PUs with increasing the levels of strain aging, $\tan\delta$ of the same material showed the opposite trend. As a result of increasing strain levels, the magnitude of the E'' modulus increased and broadened. The polyester PUs did not exhibit notable changes in $\tan\delta$ or E'' peak magnitude as a result of strain aging. However, higher T_G values were observed after stretching of the materials to 100% and 200% levels of elongation [220].

The DMA data indicated that the interaction between the hard and soft domains as a result of PUs stretching is quite complex. This was especially true with the *polyester* PUs which did not show a consistent trend in the loss factor ($\tan\delta$) or loss modulus (E'') with increasing strain. Crawford et al suggested this could be due to the lower degree of phase separation present in the polyester materials as compared to the polyether polymers based on the same HS percent, and the greater potential of HS/SS interaction of the polyester based PUs due to potential hydrogen bonding between the urethane HS and the soft segment carbonyl. It concluded that the data relating to $\tan\delta$ showed an increase in the glass transition temperature (T_G) up to 200% strain elongation.

A DMA study was made by Bagdi et al [222], for a series of PUs where the *isocyanate/hydroxyl (NCO/OH) group ratio was varied.* The family of the materials studied consisted of polyether PTMO materials based on the diisocyanate-chain extender couple MDI/BDO, synthesized by melt polymerization. The –OH functional group ratio of the polyol/total diol was kept constant at 0.4, while the ratio of the isocyanate and hydroxyl group (NCO/OH) changed between 0.94 and 1.15 [222]. The DMA measurements indicated several transitions. The three T_G values observed were assigned to the relaxation of the $-CH_2-$ groups of the polyol, and to that of the SS and HS respectively. The T_G temperature of the soft and hard phase changed with the NCO/OH ratio, indicating modifications in the material structure and composition as also confirmed by the maximum of the number in the relaxing SS. Changes in the relatively small number of end groups resulted in considerable modifications of the PUs properties. T_G depended on the mobility of the relaxing species. The changes in the T_G with composition were also influenced by the interaction between the two phases, due to their mutual miscibility [222]. For the PUs studied by Bagdi, the T_G of the SS went through a slight maximum around the equimolecular ratio of the components. This was explained by the decreasing mobility of the SS with increasing the isocyanate content [222, 223]. As shown, with increasing the NCO content, the T_G of the hard phase increased continuously, but not linearly. The T_G non linear dependence on the phases and composition, suggested changes in the phase structure with changing the NCO/OH ratio.

A systematic DMA study was carried out by Marcos et al on a series of *aromatic and aliphatic* PUs chain extended with diamine functionalised poly(oxypropylene)-**tipped** poly(oxyethylene) of various molar masses [224]. Four diisocyanates, MDI, 2,4-toluene (TDI), hexamethylene (HMDI) and isophorone (IPDI) were employed. The *symmetry/asymmetry of the diisocyanate strongly influenced the dynamic mechanical properties.* The symmetric diisocyanates (MDI, HMDI) had more phase mixing (lower T_G) than the asymmetric diisocyanates. In the asymmetric diisocyanates (TDI, IPDI), the broad transition above T_G was ascribed to the domain boundary mixing [225, 226], in which the phase-mixed material in the interphase between domains relaxed. In the copolymer IPDI the large proportion of interphase produced the failure.

The HS relaxation temperature was affected by the symmetry of the reactants and by the E' value in the rubbery plateau for the urea group concentration and steric hindrance. As shown by Marcos [224], the copolymers were very sensitive to the linkage between HS and SS segments. The use of asymmetric diisocyanates lead to a large domain boundary mixing while symmetric diisocyanates determined a sharper boundary. Higher HS percent crystallinity has been observed in PUs with linear, aliphatic HS compared to those made of aromatic diisocyanates [177, 188].

Recently (2010), Hood et al. [227] reported research on a series of PUs, with high HS content. They were synthesized with the *aliphatic* diisocyanate HDI, and chain extended with BDO. The SS macrodiol poly(ethyleneglycol) (PEG) was chosen because it has low transition temperatures and desirable mechanical properties. A variety of molecular weights was chosen, in order to provide a wide temperature window to conduct confined crystallization studies.

The thermal transitions and material stiffness were investigated. Average E' ranged from ~1.3 to 4.1 GPa for −100°C and from 108 to 417 MPa for room temperature. The first decrease in modulus was broad, spanning from −50 to ~50°C. The decrease in storage moduli was approximately one order of magnitude, which was significantly less than for most PUs reported. This could be ascribed to the high HS contents which were still robust over the range of temperatures in the first transition. After melting of the SS, the materials exhibited elastomeric properties. A slight decrease in moduli, with increasing temperature, was attributed to PEG chain deformation with the higher molar masses, PEG showing the smallest decrease in slope in the elastomeric range.

By using greater than 50 wt.% HS, Hood et al were able to study the confining of the SS by a matrix of HS, in opposition to the traditional PU model in which HS are dispersed within the SS. Hood reported the morphological changes that take place when controlling the crystallization of either the HS or SS, by tailoring the PU composition.

The following general remarks can be made with regard to the PUs thermal behavious as revealed by the DMA experiments: (a) *Hard segment effects*: the HS type influences the E' values; higher E' values are characteristic for aromatic diisocyanates as compared to materials with aliphatic diisocyanates; the increase of *the HS content* results in an increase of $E'$and T_G of the polymer; (b) *Soft segment effects*: the increase in the SS length determines the increase of the degree of crystallinity and causes an increase in E'; the increase of molecular weight increases the PUs degree of phase separation by the increase of mobility; the increase of molecular weight leads to the decrease of the T_G, as less energy is needed for motion.

3.1.2 DMA behaviour of polyurethane elastomers based on mixtures of diisocyanaytes

We have also reported a series of studies on the DMA behaviour of PUs, by investigating materials derived from single or mixtures of the diisocyanates MDI (non-crystallizing) and DBDI, (crystallizing). As shown above, in the DMA experiments frequency can be varied at a constant temperature and vice versa.

In the studies made by others and by us it has been convenient to use a single frequency with varying the temperature. A low frequency such as 1 Hz is often chosen since secondary transitions and other structural features can be easily detected. Moreover, it is easier to relate the results obtained by using low frequency measurements, to the data obtained from other techniques.

For our materials derived from single diisocyanates, (i.e. the rigid MDI or the flexible DBDI), the three components (hard segment (HS), macrodiol (MD), and chain extender (CE)) were always mixed in the molar proportions HS:MD:CE = 4:1:3, giving hard segment mass fractions in the region of 40%, and isocyanic index I = 100. It should be noted that the stoichiometric proportions used in these polymers (I = 100) means that they are truly thermoplastic. They do not have the potential for further reaction with ambient humidity to produce chain lengthening and allophanate cross-linking, seen in similar polymers but with excess isocyanate groups (e.g. I = 110) [17, 127, 135, 228].

Shown in Fig. 3.1 and 3.2 are two example DMA curves obtained by us for two materials based on the single diisocyanates MDI or DBDI, with analogous structures consisting of the diol chain extender BDO and the *polyether PTHF*.

For the material with DBDI, a weak glass transition occurred at −70°C and the viscous flow related to the melting process started from 200°C. For the analogous structure but achieved with MDI a clear and strong glass transition was observed at −65°C and the viscous flow related to the melting process was observed starting from 180°C. The differences were explained in terms of the more mobile DBDI structure resulting in HS with a higher tendency to self associate evidenced by higher melting points. As seen the first two regions were well-defined for the MDI material and were smeared for the dibenzyl PU. The T_G value, defined as maximum of $\tan\delta$, was lower for the MDI material (64.6°C) than for the polymer with DBDI (69.8°C). There was evidence of more intensive molecular mobility in the flexible part of the PU structure, i.e. in the glycol component [228]. A more pronounced decrease of the storage modulus was observed for the MDI sample after transition from the glass to the elastic state, although at the starting temperature (−100°C) the values of storage modulus were approximately equal, (300 MPa) for both materials [228].

In the high-temperature region (50°C) the value of E' storage modulus was much higher for the material with DBDI than for that with MDI (50 MPa), revealing the presence of more rigid molecular chains or higher density of crosslinking in the DBDI material. The transition from the elastic to plastic state (third region) with sharp drop of the storage moduli occured at a significantly lower temperature for the

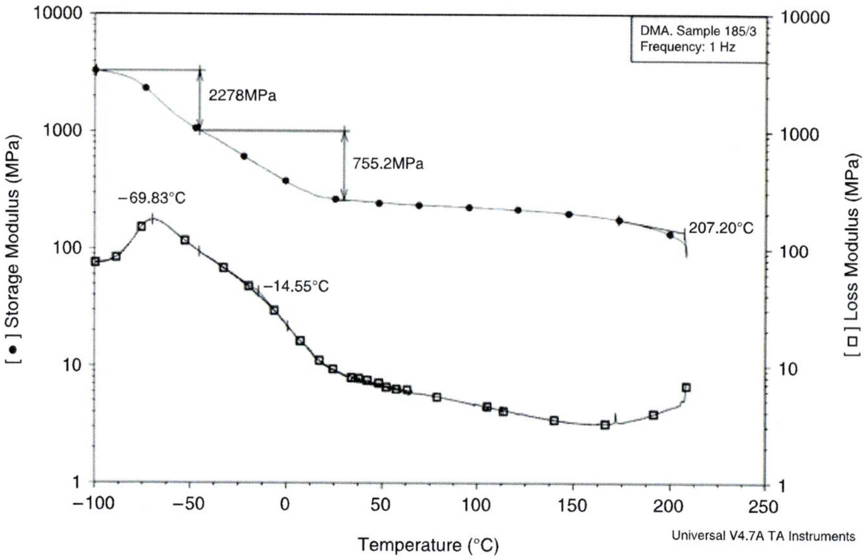

Fig. 3.1 Storage modulus and Loss Factor as a function of temperature for a segmented PU based on DBDI with the structure DBDI:PTHF:BG = 4:1:3. [228]

MDI based material (177°C) than for the sample with DBDI (>210°C). Since such a transition is a result of thermal-oxidation destruction, the difference between the storage moduli of the MDI and DBDI materials, revealed a higher thermal stability of the DBDI material.

The DMA investigations have been also employed by us for PUs *obtained with mixtures of diisocyanates*. The materials were obtained in the Romanian laboratory. *The two isocyanates were employed, MDI and DBDI, alone or as mixtures.* The materials were synthesized *with* ethylene glycol (EG) as chain extender. The SS macrodiol was poly(ethylene adipate) PEA or polytetrahydrofuran (PTHF) of molar masses 2 000 g/mol.

Changes induced by varying the type and the number of isocyanates and the order of their introduction were followed in a family of materials as described in Tables 3.1 and 3.2.

Note that the series of materials is the same as in section 2.3.2.3, where AFM measurements are presented for the same series of PUs.

In all cases the prepolymer obtained from the reaction between macrodiol and diisocyanate, was reacted with EG in a quantity calculated so that there should remain theoretically no excess of unreacted NCO groups. This corresponds to an isocyanic index I = 100 where: $I = (100 \times [NCO])/([OH]_M + [OH]_{CE})$. The description of the poliadditon procedures are detailed elsewehere [127, 207].

Shown in Fig. 3.3 are the schematic of the synthesis routes to obtain the materials with mixtures of diisocyanates listed in Tables 3.1 and 3.2:

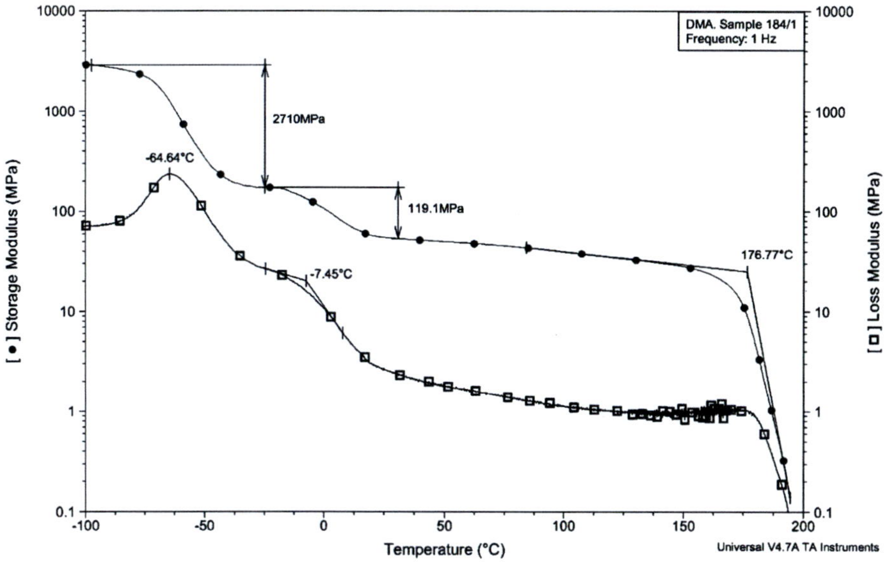

Fig. 3.2 Storage modulus and Loss Factor as a function of temperature for a segmented PU based on MDI with the structure MDI:PTHF:BG = 4:1:3. [228]

Table 3.1 Compositions of the family of polyurethane elastomers prepared and studied in this work

Recipe	PU structure	Isocyanic Index (I)
PU_{C1}	EG–PEA–(DBDI–MDI)	100
PU_{C2}	EG–(PEA–DBDI)–MDI	100
PU_{C3}	EG–(PEA–MDI)–DBDI	100
PU_D	EG–PEA–DBDI	100
PU_M	EG–PEA–MDI	100

In our experiments, the DMA cooling conditions were: room temperature to −60°C cooling rate of 10 K/min, −60°C to −120°C, 3.5 K/min and −120°C to

Table 3.2 Molar compositions of materials studied [127]

PU material	PU_1	PU_2	PU_C^*
Moles of polyol	1	1	1
Moles of MDI	0	4	2
Moles of DBDI	4	0	2
Moles of EG	3	3	3

* PU_C refers to PU_{C1}, PU_{C2} and PU_{C3} from Table 3.1

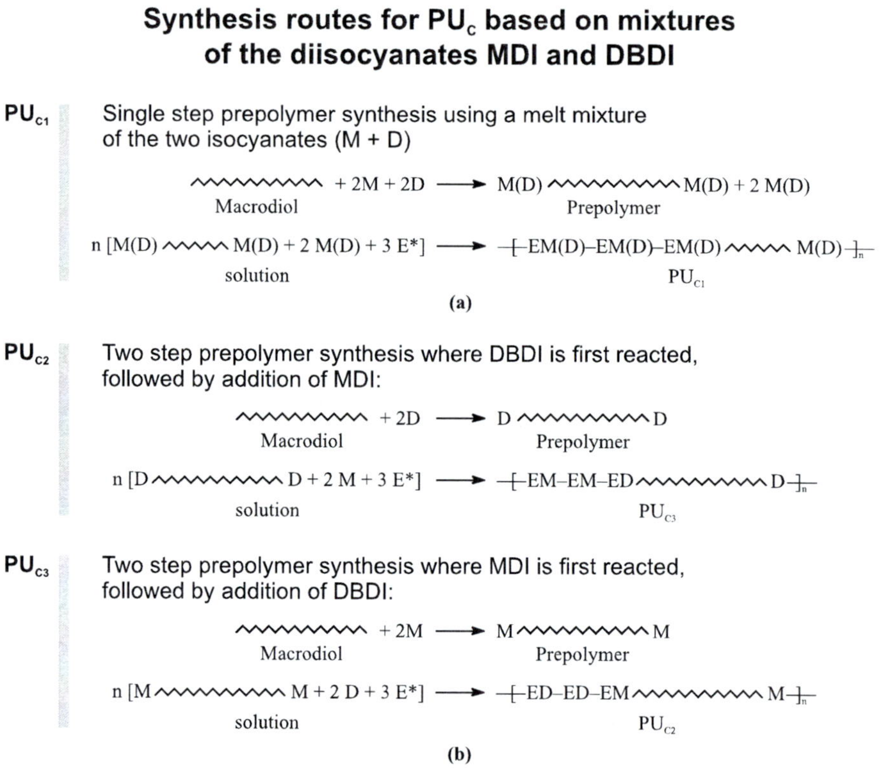

Fig. 3.3 Schematic of the synthesis routes to obtain materials with mixtures of diisocyanates [127]

$-140°C$ 2 K/min. The DMA test started at $-140°C$ and the complex modulus E^*, storage modulus E', loss modulus E'' and the loss factor $\tan\delta$ were measured as functions of temperature at a heating rate of 1 K/min. DSC measurements were also made for these materials by using a heating rate of 20 K/min, in nitrogen. The DMA instrument was operated with controlled sinusoidal force with a frequency of 1 Hz. Depending on the type of synthesis and PU structure, significant differences of E', and consequently differences in softening were found. The glass transition temperature depended on the mobility of the macromolecular chain. PU_D based on dibenzyl structures and derived from the single DBDI type of diisocyanate showed a higher tendency to crystallize.

Example DMA results are presented in Fig. 3.4 and 3.5 for the materials listed in Table 3.1, obtained with the polyesteric macrodiol PEA.

The E' storage modulus in the temperature range above the glass transition region was the highest for PU_D and was controlled primarily by the HS crystalline content. The DMA measurements have shown that modulus E' in the temperature range above the glass transition region was the highest for the DBDI based material (PU_D) of the type PEA–DBDI–EG where only a single type of isocyanate (DBDI)

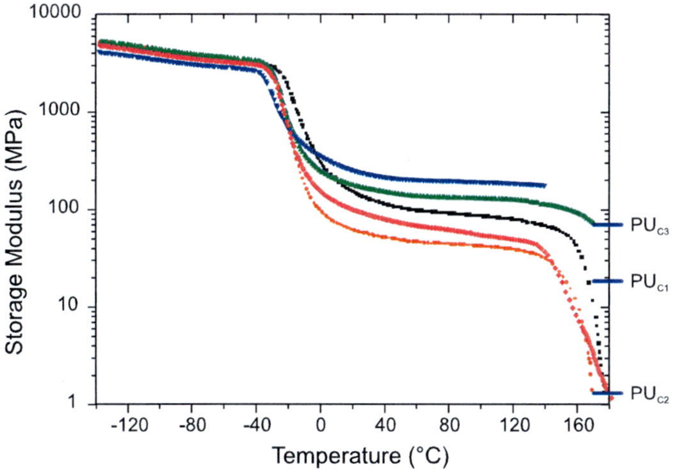

Fig. 3.4 Storage modulus as a function of temperature for materials listed in Table 3.1 [127]

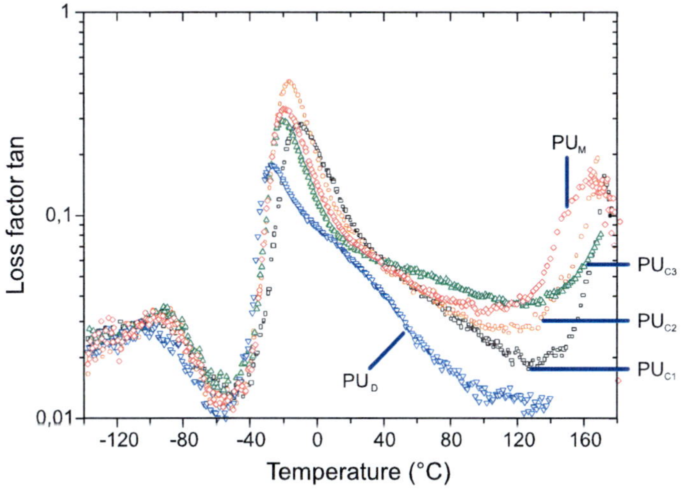

Fig. 3.5 Loss factor (tan δ) as a function of temperature for materials listed in Table 3.1 [127]

was used in the material synthesis. This is because of the higher content of DBDI based hard segments with a coplanar packing. For the series of elastomers the T_G temperatures as defined by the maximum of tan δ at f = 1 Hz was found to range from −11°C to −27°C, the lowest T_G value corresponding to PU_1 (EG–PEA–DBDI), which displays a variable geometry. The T_G values are shown in Table 3.3. The DSC curves are presented in Fig. 3.6.

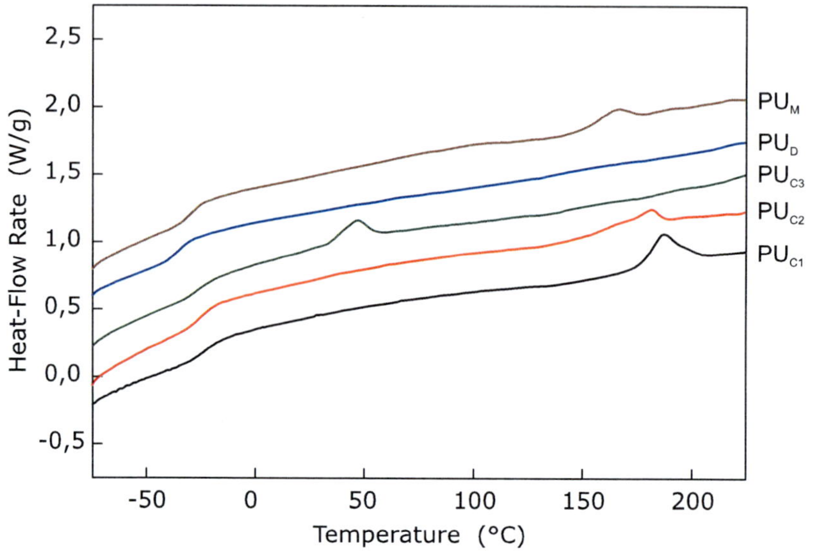

Fig. 3.6 DSC curves for polymers listed in Table 3.1, for a heating-rate of 20 K/min: low temperature region [127]

Table 3.3 DMA measurements: values of T_G (from $\tan\delta$) and peak $\tan\delta$ [127]

PU	PU_{C1}	PU_{C2}	PU_{C3}	PU_D	PU_M
T_G, °C*	–11	–16	–21	–28	–19
$\tan\delta$	0.275	0.448	0.287	0.168	0.333

* T_G defined here as the position of the $\tan\delta$ peak at 1 Hz

As shown elsewhere, a general improvement in PUs properties was obtained after polymer annealing at 160°C for 30 minutes [60] when two diisocyanates (MDI + DBDI) were included at once in the synthesis.

The above results are in agreement with the results from WAXS below (see Fig. 3.7), and from other previous studies [60, 61]. The polymer PU_D which displays a 50% degree of crystallinity has the lowest value of loss factor at T_G and the highest modulus above T_G. PU_{C3} and PU_D melt and degrade at a temperature above 250°C in DSC. As seen in Fig 3.6, PU_D based on DBDI alone, displays the highest thermal stability with two pronounced endotherm maxima occurring at 310°C–315°C. The DSC mesophase maxima were more pronounced in the two step prepolymer synthesis and were governed mainly by the type of the last introduced diisocyanate which was responsible for the hard segment EG–MDI or EG–DBDI formation. As we reported elsewhere [60] in the one step prepolymer procedure when the two MDI and DBDI isocyanates were introduced at once, the so-obtained copolyurethane

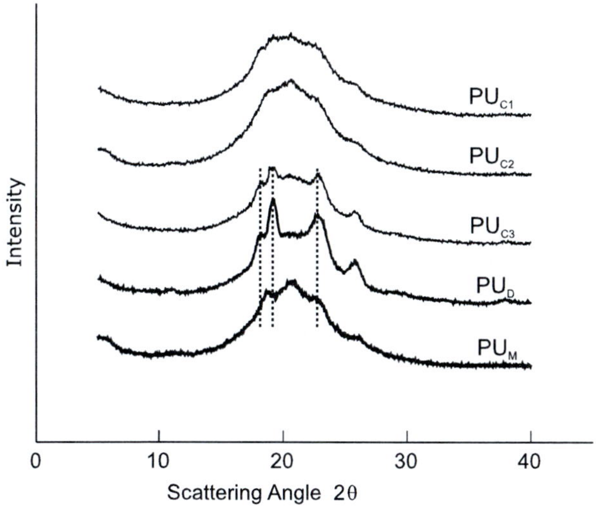

Fig. 3.7 Wide-angle X-ray diffractograms from materials listed in Table 3.1 [127]

displayed a more irregular structure and no more evident mesophase DSC maxima were revealed.

Fig 3.7 shows wide-angle X-ray scattering (WAXS) patterns of the PU series from Tables 3.1 and 3.2. As seen, the various materials displayed different degrees of crystallinity, as revealed by the presence of sharp diffraction peaks. PU_{C1} barely shows any evidence of crystalline reflections, and must be considered as almost but not completely amorphous. PU_{C2}, PU_{C3}, and PU_M exhibit slightly higher crystallinities than PU_{C1}, which is still less than 20%. The sample PU_D based on DBDI alone is the most crystalline sample and displays a crystallinity of 50%.

The crystallinity of DBDI-based PUs was strongly reduced when DBDI was mixed with the conventional rigid isocyanate MDI. The degree of crystallinity was found to be dictated by the diisocyanate introduced first during synthesis. Improvement in elastomeric properties, by reducing the crystallinity in PUs based on DBDI, was obtained especially when both diisocyanates were included and reacted together in a random fashion, instead of sequentially by using a prepolymer stage.

The results were in agreement with the observations which we made on analogous PUs but based on *polyetheric macrodiols[60]*.

3.2 DSC experiments

Methods and means cannot be separated from the ultimate aim (Emma Goldman)

3.2.1 DSC behaviour of polyurethane elastomers based on single diisocyanates

DSC maintains the sample and reference at the same temperature through varying the heat flow. When the differential heat flow supplied is plotted as a function of temperature, it is possible, after normalizing the sample mass, to obtain enthalpic information, such as the heat of fusion.

Although the DSC technique is less sensitive to molecular transitions than DMA, it generally supports results regarding the thermodynamical changes which the material undergoes upon heating or cooling. As in the case of DMA, DSC allows the estimation of the PUs characteristics of microphase separation or phase mixing.

The DSC experiments to investigate the PUs phase separation are based on the fact that the T_G of the homogeneous mixtures in general obey the following copolymer equation:

$$\frac{1}{T_G} = \frac{w_1}{T_{G_1}} + \frac{w_2}{T_{G_2}} \tag{3.2}$$

where w_1 and T_{G_1} are the weight fraction and the glass transition temperature (T_G) of the fist component and correspondinly, w_2 and T_{G_2} are the weight fraction and the glass transition temperature of the second component. An interphase layer has a T_G which is a function of the relative amounts of the compounds[4].

Since the phase mixing is partial, three separate phase transitions usually occur. In practice the amount of mixed material is small compared to the mass of pure phases, so that the third T_G is not observable. Usually some increases of the soft segment T_G and decrease of the hard segment T_G is observed [4].

Modifications in the PUs microphase separation characteristics through changes in the glass transitions of the components can be observed in the phase transformation from the DSC endothermic and exothermic maxima of the polymer depending on the *nature of the hard domains* (diisocyanate and chain extender couple), HS content or ordering of the elements constituting the HS [60,127].

Another approach to estimate the PUs amount of *phase mixing* is based on *measuring the specific heat*, C_p, at the glass transition temperature and comparing it with the C_{p0} for the pure component $\frac{\Delta C_p}{\Delta C_{p0}}$ assuming that the chains of the soft phase mixed with the hard phase will be immobilized and will not participate in the cooperative motions at the glass transition of the pure soft phase. The C_p value jump at the glass transition is proportional to the mass of the segments taking part in the transition [4].

The PUs thermal transitions are directly related to *the degree of mixing* between HS and SS. The shorter HS can dissolve within the soft microphase if their length is below the critical length for microphase separation. This dissolution process increases the T_G of the soft microphase and degrades the low-temperature response of the material. The SS may also dissolve into the hard microdomains or become trapped therein during structure formation [180]. Mixing of this nature decreases the strength of the hard microdomain to plastic flow, leading to a decrease in the heat distortion temperature.

As shown [180], the *softening behaviour* provides another basis for the determination of the compositional transition in the PUs domain morphology. PUs with 40% or lower HS contents, soften at considerably lower temperatures than materials with higher HS contents. For these compositions, softening occurs at teperatures below the DSC high temperature endotherm, indicating that the HS ordering does not provide significant structural reinforcement for the PUs with a low HS content at high temperatures. Yet, for PUs achieved with a higher HS content (50% and above), softening coincides with the onset of high temperature DSC endotherms [180].

Wilkes and Wildnauer [229] proposed a model based on classical rubber elasticity which predicted that as the temperature increases, the entropic stress caused by the slightly extended SS is increased and the compatibility is decreased, causing the break up of some of the smaller domains. Blackwell and Lee [230] observed polymorphism in the MDI based PUs. This also has to be considered in identifying the DSC endotherms.

The existence of multiple endotherms was described in numerous studies on the DSC thermal behaviour of PUs [231–239].

In general, *three distinct endotherms* were observed in the DSC experiments. The lowest temperature endotherm was found at temperatures ca. 20°C higher than the annealing temperature and was attributed to a local restructuring of the HS within the hard microdomains. An intermediate temperature endotherm, found generally in the range 140–200°C, has been associated with the destruction of long-range order of an unspecified nature. A higher temperature endotherm observed above 200°C was generally ascribed to the melting of microcrystalline regions within hard microdomains.

With regard to the *first low temperature endotherms*, Seymour and Cooper [232] indicated that its behaviour can be improved by annealing. Later Cooper et al [154] also found that the first temperature endotherm was absent in the DSC thermogram, when taken immediately after quenching to room temperature from the annealing temperature at 120°, 150° and 170°C, but it reappeared some time after quenching, and at a long time, shifted to some higher temperature and grew larger in size. They indicated that this later phenomenon was an achievement of better short range order. Cooper et al further indicated that the first low temperature endotherm also appeared in pure HS materials (polymers models made up only by HS). They suggested that the induced ordering would occur within the hard microdomain rather than in the interphase between domains. Alternatively, Koberstein et al [151], in their study on conventional polyether PUs showed that the temperature dependence for the first temperature endotherm may be explained by solubility effects embodied in the microdomain model developed by Koberstein and Stein [136]; according to which there is a critical HS sequence length below which the HS are dissolved within the soft microphase. At low annealing temperatures, short and long HS separate from the soft microphase. Because of the limited mobility of long HS at low annealing temperatures, only short HS are able to align to form ordered structures. As the annealing temperature increases, the shorter HS become soluble in the soft microphases and the ordering process is due to progressively longer HS.

Materials with *MDI/BDO hard domains* have been extensively investigated. A series of polyether PUs based on the diisocyanate MDI and macrodiol PTMO and extended with BDO was investigated by Abouzahr and Wilkes who reported that the T_G increased with the increase of the HS content [176].

At lower HS percentages (26–33%), a broad hard domain transition was observed. Increasing the HS content from 33% to 43% produced a multiple melting endotherm. A further increase in the HS composition (47%) was proved to lead to HS melting transitions

For PUs obtained with MDI/BDO based HS, Saiani also showed that PUs usually display three endotherms [186, 187]. A first "annealing endotherm" was observed at 20–30°C, higher than the applied annealing temperature. The two other melting endotherms were observed at higher temperatures (up to 250°C). This is consistent with the results obtained by us on the MDI based PUs achieved with various diol chain extenders [60].

Chen et al [236] investigated the origin of the so-called "annealing endotherm". The authors suggested that this endotherm was related to some relaxation effects of the polymer chain in the HS phase due to the physical aging of the sample.

The PUs morphology shows a complex dependence on the sample history, preparation method and chemical composition. The *effect of annealing* on the DSC endotherms were investigated by Wilkes et al [174, 229], Van Bogart et al [237] and Hesketh et al [234]. When the HS were annealed below the crystalline melting point, a DSC endotherm was observed at 20–50° above the annealing temperature, which was attributed to long range HS and SS ordering. Van Bogart et al. attributed the effect to interdomain rather than interphase phenomena, because it was also observed in the corresponding homopolymers. Wilkes et al and Hesketh et al studied samples that were annealed for varying times and at different temperatures, and then quenched. They observed an increase in the phase mixing in the quenched samples. The increased temperature of the annealing endotherm resulted from the HS association with more stable domains.

Saiani et al [186] also investigated the origin of the high temperature melting endotherms for a series of *polyether PUs by varying the HS content*. They used annealing temperatures higher than the T_G of the HS, so to avoid physical aging. The DSC phase behaviour of a set of high hard block content (50% to 100%) was investigated. The MDI based HS were chain extended with 2-methyl-1,3-propanediol. The observed high temperature endothermic transitions were attributed to the disruption of an ordered structure appearing in the hard phase [186] and to the microphase mixing of the SS and HS. The results suggested a two-step melting process: (1)

melting of the ordered structure present in the hard phase; (2) microphase mixing of the SS and HS. Investigation of the melt-quenched samples showed that for a HS concentration lower than 65% a homogeneous mixed phase was obtained. A HS concentration higher than 65% resulted in a two-phase system, one pure HS phase coexisting with a mixed phase with a HS/SS weight ratio of ~1.8 corresponding to a 65% HS concentration.

As shown by Clough et al, in the MDI based PUs a wide range of HS structural organizations may occur depending on the composition and on the *thermal treatment* [82]. Two endothermic transitions characteristic of noncrystalline domain structure occuring at 80° and 130° to 150° were identified by DSC. By means of annealing procedures the 80° transition could be moved to progressively higher temperatures until it merged with the endotherm at 150°C. In the case of the materials with sufficiently long HS length, the transition could be increased to above 200°C, which would correspond to the melting temperature for crystalline HS domains.

As shown by Bagdi and by ourselves [222, 60], two glass transitions correspond to the relaxation of the HS formed from a non-crystallizing isocyanate and the BDO chain extender and to that of polyether SS respectively. Both transitions shift toward higher temperatures *when increasing the NCO/OH ratio* [222].

In general, for the PUs based on conventional non-crystallizing diisocyanates, the endotherm peak observed at around 210°C was assigned to the beginning of the melting. As observed by ourselves, the melting process in the case of PUs obtained with the crystallizing diisocyanate DBDI, may occur at higher temperatures (about 230°C) [60]. The melting peak changes dramatically with thermal history.

The DSC technique was used by Li et al [42] to investigate a series of segmented PUs with different HS flexibilities, based on the *aliphatic* diisocyanate HDI or aromatic MDI, as the HS showed a folded-chain conformation. The chain extenders were the diol BDO or 4,4'-diaminodiphenyl ether (DDE); the SS macrodiols were PTMO or poly(ethylene/propylene adipate) (PES) with a molar mass 2 000 g/mol. The materials were prepared by a solution polymerization.

The phase structure was found to be insensitive to the increase of the HS content and thermal treatment. Phase separation was very fast in these materials as the HS mobility was relatively high and the system viscosity was low. The DSC results showed a SS glass transition temperature which was only about 5°C above that of the pure SS, indicating that the separation between SS and HS was nearly complete. The results obtained by Li were discussed based on the viscosity-mobility-interaction argument suggesting the importance of kinetic effects in formulating a better understanding of the PUs structure-property relationships The PES/HDI/BDO systems showed low system viscosity, high HS mobility, and strong interaction between the HS chains. The phase separation was rapid and the kinetic barriers were not important. The MDI/DDE systems had high system viscosity, low HS mobility, and strong interaction between the HS chains. While the HS based on HDI/BD adopted a folded-chain configuration, the HS with MDI/DDE appeared to exist in coiled configurations.

Brunette et al investigated *the changes in the frequency, bandwidth, and intensity of the bonded N–H absorption band following heat treatment in model PUs and correlated them to structural changes as evidenced by DSC* [147].

The materials were based on 2,4- and 2,6 toluene diisocyanate (TDI) and on MDI respectively and the chain extender was BDO. The temperature dependence of the hydrogen bonding in these materials was also investigated. Annealing of the 2,6-TDI-BD and MDI-BD based materials, resulted in a change in distribution toward stronger hydrogen bonds increase in the melting temperatures. The half-width and intensity of the N–H stretching band also showed a strong dependence on the annealing time as did the shape and size of the melting endotherms: with longer annealing times, they narrowed and became more intense, which indicated uniformity in the hydrogen bonds and in the structure. The insensitivity to thermal treatment observed for the 2,4-TDI-BD copolymers was a result of a poorer HS organization and inability to crystallize.

Numerous model PUs were characterized *before and after varying degrees of strain to study the effect of strain on the polymer morphology by means of DSC.* Crawford et al investigated a series of model polyesteric and polyetheric PUs based on the conventional diisocyanate MDI and macrodiol BDO by varying the HS content [220]. They studied the effect of strain on the PUs by means of the DSC and DMA experiments along with the mechanical properties before, and after stretching materials to variable degrees of strain. Strain was induced by stretching the polymers at 100% intervals between 100% and 400% elongations. The samples were held at constant length throughout the experiments. DSC scans showed that all of the model polymers exhibited three prominent thermal transitions including a glass transition below 0°C and two broad endothermic transitions at higher temperatures. The lower temperature endothermic transitions had been previously identified with disruption of SS/HS bonds or disruption of short-range order within the HS microdomains. The higher temperature endothermic transitions were related to the breakup of inter-urethane hydrogen bonds.

The data obtained by Crawford et al [220] showed that stretching of the materials resulted in disruptions of the domain structure which involved phase mixing with, or without hydrogen bonding taking place between the hard and soft domains, a breaking up of the hard domain microstructure and dispersion of the HS within the soft domains, or plastic deformation of the hard domains. As shown the lower endotherm transition was identified with disruption of the SS-HS bonds [198, 226, 232] or disruption of short range order within the HS microdomains [178].

The higher temperature endothermic transitions is related to the break-up of the inter-urethane hydrogen bonds [198]. This series of transitions is representative of the two phases that are present in the polymer and reflect the relative amounts of the HS and SS present. As the percentage of the HS is increased, the peak endothermic temperatures increase. The T_G of the SS increases as the HS content increases.

Example DSC diagrams are shown below for two example materials based on the single diisocyanates MDI and DBDI but obtained by using BDO as a macrodiol.

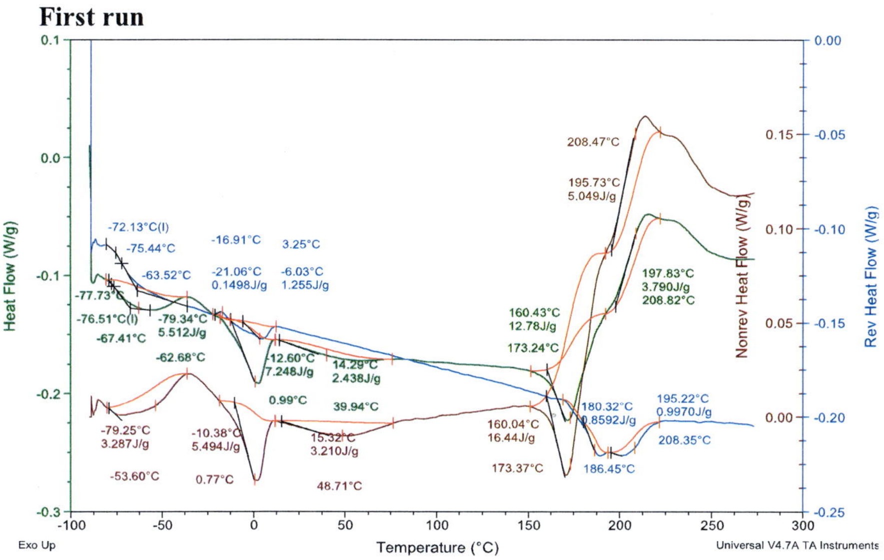

Fig. 3.8 DSC curves of a PU based MDI:BDO:PTHF [228]

Phase separation, crystallization phenomena revealing the PUs thermal behaviour of the materials based *on single* diisocyanates have been also extensively studied *by us, by using DSC measurements.*

Example first run DSC diagrams are shown in Fig. 3.8 and 3.9 for two PUs with similar structures. They differ only in the type of diisocyanate, (rigid MDI or flexible DBDI). The macrodiol PTHF of molar mass 2 000 g/mol was used. The three components (diisocyanate, macrodiol chain extender) were mixed in the molar proportions hard segment:macrodiol:chain extender =4:1:3, giving hard segment mass fractions in the region of 40%, and isocyanic index I = 100.

Such a variety of temperature regions can be explained from the perspective of the PUs microsegregated structure that is given by the HS flexible segments of the glycol and isocyanate components. Evidently, the volume part of the SS is bigger than part of rigid segments, so the flexible matrix has a defined distribution of rigid blocks in it. For such a structure the microsegregated separation of components is thermodynamically steady. In the first region, T_G is approximately the same for both of the materials (74–75°C), suggesting an identical intensity of the molecular mobility in the glycol component. In the second region, the low-temperature endothermic peaks are well-defined for both materials; the peak position is identical for both of the samples, (0.5–1.0°C). This peak can be related to the melting process of the rigid blocks. The values of the melting heat for both materials are almost equal (7.24 J/g for the MDI sample and 7.33 J/g for the DBDI material). In the third region, the high-temperature endothermic peaks are present only on the thermogram of the MDI polymer with maximum at 172°C. The exothermic peak can be related to thermal-oxidation destruction of PUs which starts at 180°C. *The positions*

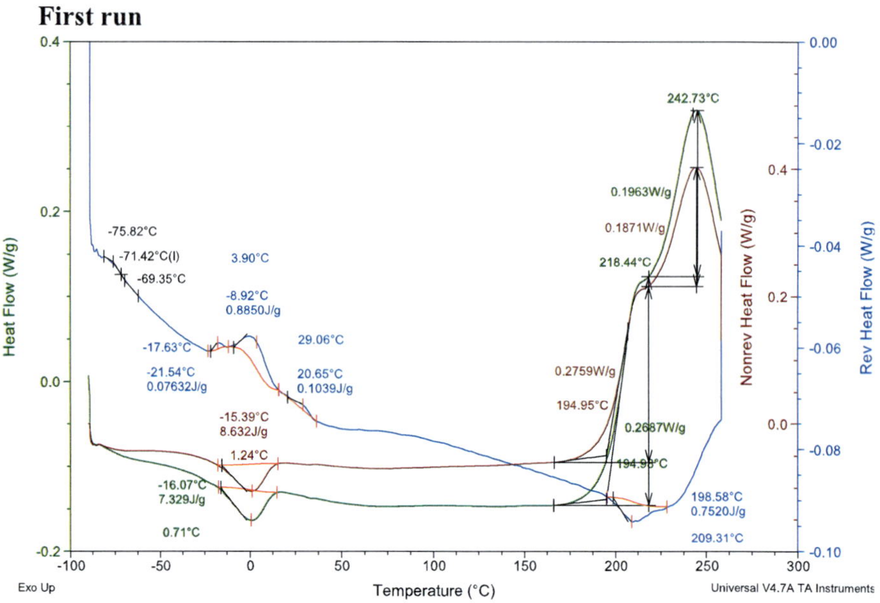

Fig. 3.9 DSC curves of a PU based DBDI:BDO:PTHF [228]

of maxima are 198°C for the material with MDI and 243°C for the sample with DBDI, which makes evidence of higher thermal-oxidation stability of the chemical structure for the material achieved with DBDI.

3.2.2 DSC behaviour of polyurethane elastomers based on mixtures of diisocyanates

By using DSC measurements, we also have studied the thermal behaviour of a series of PUs based on *mixtures of diisocyanates. The compositions of the family of PUs prepared and studied, is given in section 2.3 (Tables 2.2 and 2.3) where we also report SEM results obtained by us for this family of materials achieved with mixtures of diisocyanates.* The molar proportions were hard segment:macrodiol:chain extender = 4:1:2.64, giving HS mass fractions in the region of 40%, and isocyanic index I = 110.

The experimental DSC proved to be a useful tool to show *evidence of the difference in structure brought about by changing the copolymeric ordering* (Fig. 3.10 and 3.11). Remarkable changes could be observed in the phase transformation from the DSC endothermic and exothermic maxima of the polymer depending on the ordering of the elements constituting hard segments.

In order to obtain a better attribution of the DSC maxima, simple macromolecule models were used which reproduce parts of the copolymer chain as hard sequence $(EG–MDI)_n$ and $(EG–DBDI)_n$ and simple elastomers with only one diisocyanate as PU_M and PU_D abbreviated $PU_{M\ PEA}$ and respectively $PU_{D\ PEA}$ (in the case of PEA soft segments) and $PU_{M\ PTHF}$ and $PU_{D\ PTHF}$ (in the case of PTHF segments).

As observed by us, all the PUs obtained by using mixtures of diisocyanates ($PU_{C1\ PEA}$ and $PU_{C1\ PTHF}$) did not display significant DSC maxima in the region between 20°–150°C characteristic mainly to the ordered soft segment structure as observed in the case of stretched $PU_{D\ PTHF}$ model when the small maximum at 50°C reveals the PTHF moiety crystallization tendency which maintains also after removing the stress, Fig. 3.10 (e), curve 2. Such a tendency was observed also in many other PUs.

The region 150°–230°C characteristic of a mesophase transformation was very sensitive to the changes in the copolymeric ordering. So, the mesophase in which there are involved MDI and PEA soft segment, displayed two maxima at 195° and at 217°C (Fig. 3.11 d), whereas when using DBDI and PEA there did not appear clear significant similar interactions (Fig. 3.11 e). A comparable situation appeared in the case when PTHF was used as a SS. The two above-mentioned maxima displayed some shifts toward lower temperatures and appear at 185° and at 206°C respectively (Fig. 3.10 d), indicating that the interaction of the (MDI–EG) hard segments with PTHF is somewhat slower. This situation was due to the weaker NH···O hydrogen bonding in PTHF relative to NH···O=C in PEA, where the carbonyl ester group is involved.

As shown in Fig. 3.10 and 3.11 (solid and dotted lines), *all of our copolyurethanes were sensitive to supplementary thermal treatments. After annealing at 160° for 30 minutes all the mesophase interactions tend to diminish and, as will be shown in a subsequent section, the mechanical properties are influenced significantly.*

The result of DSC measurements made by us, also confirmed the fact that in *the two step prepolymer synthesis* the latest introduced diisocyanate determines the preponderant nature of the HS. Thus the maximum characteristic for the $(EG–MDI)_n$ homopolyurethane at 246°C (Fig. 3.11 f) can be found again in the range of 240°–248°C. This is the case for $PU_{M\ PEA}$ (248°C) (Fig. 3.11 d), $PU_{C2\ PEA}$ at 244°C (Fig. 3.11 b) and $PU_{C2\ PTHF}$ at 240°C (Fig. 3.10 b). This maximum also visible yet in a smaller measure, in the case when MDI and DBDI mixture was introduced at once in the reaction as for $PU_{C1\ PTHF}$ at 238°C, (Fig. 3.10 c).

A special situation appeared in the case of DBDI when the $-[EG–DBDI]_n$ *segments displayed an exceptional tendency to crystallize and to self associate.* The respective homopolymer EG–DBDI presented hard self interactions characterized by a high endotherm temperature maximum at 287°C (Fig. 3.11 g). This maximum can be found again also in the $PU_{D\ PEA}$, at 284°C (Fig. 3.11 e).

The peculiar behaviour of the EG–DBDI sequence was revealed also when studying the DSC curves of the simpler elastomer type $PU_{D\ PTHF}$, (Fig. 3.10 e) which included PTHF soft segments on the macromolecular chain. In this case there was a lower tendency of the ether group to participate in the mix hydrogen bond association with the urethane EG–DBDI segments than that observed in the case of the respective PU containing PEA soft segments ($PU_{D\ PEA}$). As a result, during the PU synthesis in melt, in the last step of the process related to the EG–DBDI segment appearance, the melt viscosity increased quite quickly and part of the formed EG–DBDI segments did not have time to segregate in self associations, remaining in a dissolved state in the vitrified polymer matrix (the maximum characteristic for EG–DBDI at 287°C did not appear anymore).

During the DSC measurements, once the temperature was raised sufficiently the polymer matrix became more mobile and allowed the EG–DBDI segments to move and crystallize in self-associations. The crystallization was evidenced by the exothermic maxima appearing at 225°C, (Fig. 3.10 e, curve1 — see also 230°C in Fig. 3.10 c — solid line, characteristic of $PU_{C3\ PTHF}$ which also has segments of the ED type). *On repeating 3 times the DSC scanning of $PU_{D\ PTHF}$ till 250°C this exotherm maximum disappeared completely and there was a significant increase in the EG–DBDI characteristic crystalline maximum at 277°C* (Fig. 3.10 e, curve 4).

Note that in Fig. 3.10 and 3.11, the solid lines are for the initial PUs samples, denoted by (I)), whereas the dotted lines are for the same materials after annealing, denoted by (A).

Interpretation of the DSC curves above 250°C in the case of the polymers containing EG-DBDI segments should be made with prudence, due to the fact that as we found on employing the thermogravimetrical analyses (see following section), all of the studied PU started to decompose at about 250°C and at about 275°C their loss in weight represented approx. 5%. So in this region some variation in the DSC curves is more or less brought about by the incipient degradation thermal effects.

Fortunately, a similar partial crystallization effect as mentioned before could be obtained on the same polymer at lower temperatures (180°C – 1 hr) (Fig. 3.10 e, curve 3), and also at normal temperatures as a result of the polymer stretching (Fig. 3.10 e, curve 2). Evidently, in these situations the thermal degradation process could not be involved.

With regard to our purpose related to the effect of ordering in copolyurethane based on two isocyanates, the DSC curves exhibited significant differences depending on the copolyurethane ordering. The DSC mesophase maxima were more pronounced in the step by step prepolymer synthesis and are governed mainly by the type of the last introduced diisocyanate which is responsible for the hard segment EG–MDI or EG–DBDI formation. In the one step prepolymer procedure when the two MDI and DBDI isocyanates were introduced at once, the so-obtained copolyurethane displayed a more irregular structure and no more evident mesophase DSC maxima were revealed.

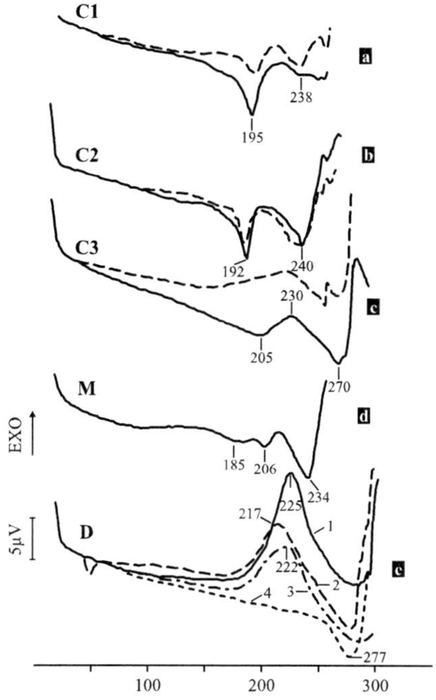

Fig. 3.10 DSC curves of copolyurethanes based on PTHF: (a) $PU_{C3\ PTHF}$ –12.6 mg (I), 13.5 mg (A); (b) $PU_{C2\ PTHF}$ – 13.0 mg (I), 11 mg(A); (c) $PU_{C1\ PTHF}$ – 11.7 mg (I), 11.9 mg (A); (d) $PU_{M\ PTHF}$ – 10.7 mg (A); (e) $PU_{D\ PTHF}$: (1) – 9.4 mg initially; (2) – 9.4 mg (after stretching till rupture); (3) – 9.6 mg after annealing at 180°C for 1 hour; (4) – 8 mg after annealing on repeating 3 times the DSC scanning of the sample till 250°C [60].

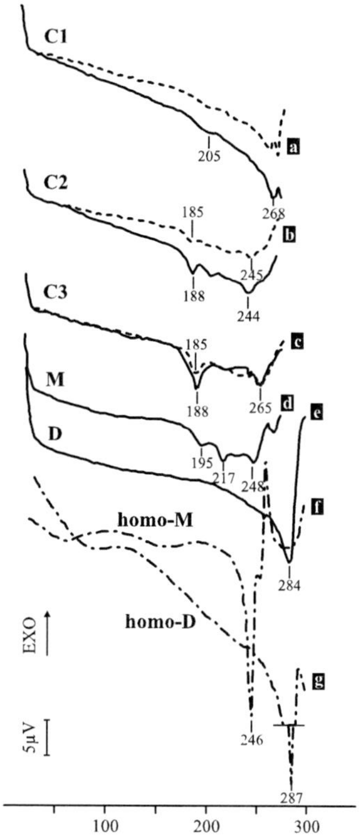

Fig. 3.11 DSC curves of copolyurethanes based on PEA: (a) $PU_{C1\ PEA}$ – 9.9 mg (I), 8.0 mg (A); (b) $PU_{C2\ PEA}$ – 8.1 mg (I), 8.0 mg (A); (c) $PU_{C3\ PEA}$ – 9.9 mg (I), 7.8 mg (A); (d) $PU_{M\ PEA}$ – 9.9 mg (I); (e) $PU_{D\ PEA}$ – 2.2 mg (I) -(A); (f) homopolyurethane – $[EG–MDI]_n$ 3.6 mg (A); (g) homopolyurethane – $[EG–DBDI]_n$ 4.2 mg (A); solid lines – initial samples (I); dotted lines – after annealing (A) at 160°C for 30 min [60].

3.3 Thermogravimetry

The only way to lose weight is to check it as airline baggage. (Peggy Ryan)

Well, it seems Mrs Ryan forgot about thermogravimetry ☺ In the thermogravimetric analysis (TGA), the weight loss of a sample is continuously monitored as the temperature is raised. The transitions observed by using DMA or DSC do not result in weight changes [240–242].

Thermal methods as it is the DMA technique are not well suited to investigate the thermal degradation of PUs, as long as thermal degradation involves chemical reaction processes as it is the scission or crosslinking of the PUs chains. On the other hand, thermal degradation which often involves the release of volatile degradation products, complicates the DSC analysis. Therefore, PUs thermogravimetry (TGA) is a better method to quantify the PUs thermal degradation as long as it takes advantage of the tendency to create volatile degradation products.

In the previous two sections we reported research on the block structures of PUs and their consequences for the phase transitions (recorded by the DMA and DSC methods) within T_G of the SS and HS, and for the melting temperatures of crystalline structures formed within the phases. The effects of those parameters are also essential for the PUs thermal stability. What is especially important in this respect is the initial degradation temperature which is dependent on the thermal history of the weakest points within the PU macromolecule - urethane bonds and ester bonds.

Numerous studies have been reported on the PUs thermal stability. As shown by Barikani [243], *the polyester based PUs are more stable than the materials obtained with polyether macrodiols.*

For materials based on MDI and polyester SS, the initial decomposition point is 227°C [244]. Much higher decomposition temperature, i.e. 319°C was recorded for polyester PUs achieved with TDI and polyesters derived from adipic acid, ethyl glycol and neopentyl glycol, and trimethylolpropane [245]. Decomposition of the urethane bonds in polyester PUs based on the diisocyanate MDI and chain extended with EG begins at 240°C, and it reaches its maximum rate at 337–356°C [246]. Decomposition of PUs obtained form TDI was observed to start as early as at 200°C and to run through three steps with polybutadiene derived PUs being more stable [247].

PUs thermal decomposition is a multi-stage process which depends on the presence of the HS. The first stage of degradation is fast but its rate decreases with increasing the SS content. As shown, example values of the rate constants for the first order decomposition in the case of polyester PUs synthesized with MDI, are 60 to 429 10^6 s^{-1} at 300°C, depending on SS content.

The degradation is mainly initiated within the HS. Following the first stage of decomposition when the weakest bonds in the polymer have been used up, the second and third stages are much more slow [248]. The main process which occurs during degradation is the PU decomposition to yield the diisocyanate and polyols. The presence of the aromatic rings in the HS reduces the volume of volatile degradation products released [248].

Hood et al [227] reported research on a series of PUs, with high HS content. They were synthesized with the *aliphatic* diisocyanate HDI, and SS macrodiol poly(ethyleneglycol) PEG, by using the chain extender BDO. TGA scans of the PU series were collected under nitrogen at 10°C/min to detail the thermal degradation. The materials showed a 5 wt% loss at ~285–314°C. Differential weight loss curves showed two main peaks with T_{1max} at ~340–370°C and T_{2max} of ~423–351°C. The former was attributed to the greatest rate of mass loss of the HS while the later accounted for the greatest rate loss of SS mass [73].

The decomposition of PUs also depends on the adopted synthesis route. The decomposition of the materials obtained by using the prepolymer polyaddition procedure takes place much easier than in the case of analogous materials but obtained via the 'one-shot' technique. As reported [249], the activation energy for the first stage of decomposition of PU obtained from the single stage method is 77–101 kJ/mol, depending on the composition of the material, whereas the second stage needs an energy of 111–316 kJ/mol.

The PUs thermal stability has been also investigated by us for a series of MDI and DBDI based materials. The effect of the HS *(crystallizing or not), has been followed.* As observed, the thermal stability is strongly affected by the type of chain extender and diisocyanate used in the PUs synthesis [250, 251]. As seen in Fig. 3.12 and in Table 3.4, for the *polyester* materials based on PEA, the presence of the different DTG maxima showed that the degradation process is made up by subsequent processes which are different from a PU to another. The TGA analyses have shown that the dibenzyl-based PUs start to decompose at higher temperatures (up to 290°C) as compared to analogous materials but derived from classical isocyantes with rigid structures. As seen in Fig. 3.12, the limits of the thermal stability for the three types of PEA based PUs are closed as follows: PU chain extended with 4,4'-methylene-bis-(ortho-chloroaniline) (MOCA) (290°C) > PU chain extended with the diol DEG (280°C) > commercial Adiprene L 100 (100–270°C).

The TGA analyses were also made by us for a series of intermediates which were synthesized with two identical functional groups situated in different rings: 2,2'-DBDI, 2,4'-DBDI and 4,4'-DBDI [59] — *see section 1.2.7.* For these structures the melting point of the polymer was 312°C and the initial decomposition temperature of the polymer was 220°C (Fig. 3.13).

Table 3.4 Thermogravimetric data of PU-Adiprene L 100, PU with CE DEG and PUE with CE MOCA [250]

PU type	$T_0{}^a$(°C)	$T_5{}^a$ (°C)	$T_{10}{}^a$ (°C)	MaxDTG$_1{}^b$	MaxDTG$_1{}^b$
PU – Adiprene L 100	270	305	325	405	-
PU extended with DEG	280	315	320	334	430
PU extended with MOCA	290	322	337	366	415

^{a} T_0, T_5, and T_{10} represent the start of degradation, the temperatures of 5% and 10% weight loss; ^{b} MaxDTG represents the peak(s) of DTG curves corresponding on the inflexion points of T_G curves as determined by TGA.

Shown in Fig. 3.14 and 3.15 are example TGA/DTGA curves for two *polyetheric* PTHF materials with analogous structures. Only the type of diisocyanate differs (MDI or DBDI). The molar proportions is hard segment:macrodiol:chain extender=4:1:3, giving HS mass fractions in the region of 40%, and isocyanic index I = 100.

As seen in Fig. 3.14 and Table 3.5, for the MDI based materials three main degradation processes can be seen: at approximately 340°C it occurs the decomposition of urethane groups, at 420°C the destruction of ether groups takes place and at 560°C the destruction of carbon chains and rings begins. For the DBDI based material (Fig. 3.15), three main degradation processes are also observed: at approximately

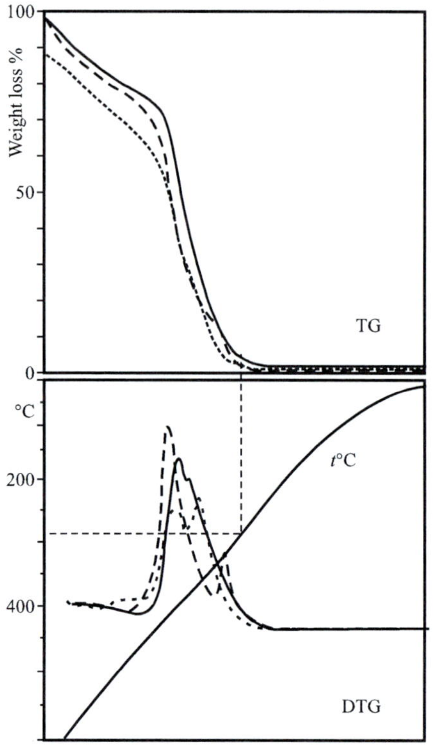

Fig. 3.12 TGA and DTG curves for 3 PUs: (——) commercial PU-Adiprene L 100; (- - - -) casted dibenzyl-based PU derived from polyester macrodiol (DBDI:PEA$_{2000}$:DEG, I = 110); (· · ·) casted dibenzyl-based PU with polyether macrodiol methylene bis o-chloroanyline (MOCA): (DBDI:PEA:MOCA) [250].

350°C it occurs the decomposition of urethane groups; at 420°C the destruction of ether groups occurs and at 570°C the destruction of carbon chains and rings begins. Increasing the HS block content leads to the increase of the decomposition temperature [60, 185, 251].

Table 3.5 TGA/DTGA paramaters for the two PUs depicted in Fig. 3.14 and 3.15

Parameters of spectra	MDI:BG:PTHF			DBDI:BG:PTHF		
	I	II	III	I	II	III
Temperature intervals of the decomposition regions, °C	260-380	380-480	480-700	260-380	380-510	510-800
Maximum of DTGA, °C	336	420	565	347	425	572
Loss of weight for every region of decomposition, %	46	37	17	45	41	14

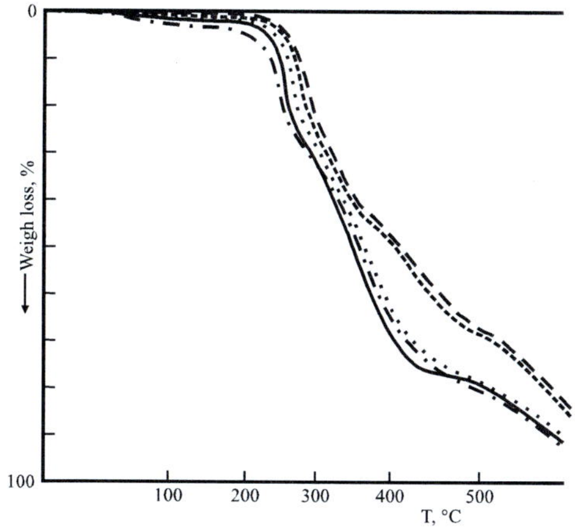

Fig. 3.13 TGA curves for PUs based on EG, and 2,2'-DBDI (——), 2,4-DBDI (– · –) and 4,4'-DBDI (- - - -); TGA curves for PUs based on BDO and 2,4-DBDI (· · ·) and 4,4'-DBDI (– – – –); [see Fig. 3.6]

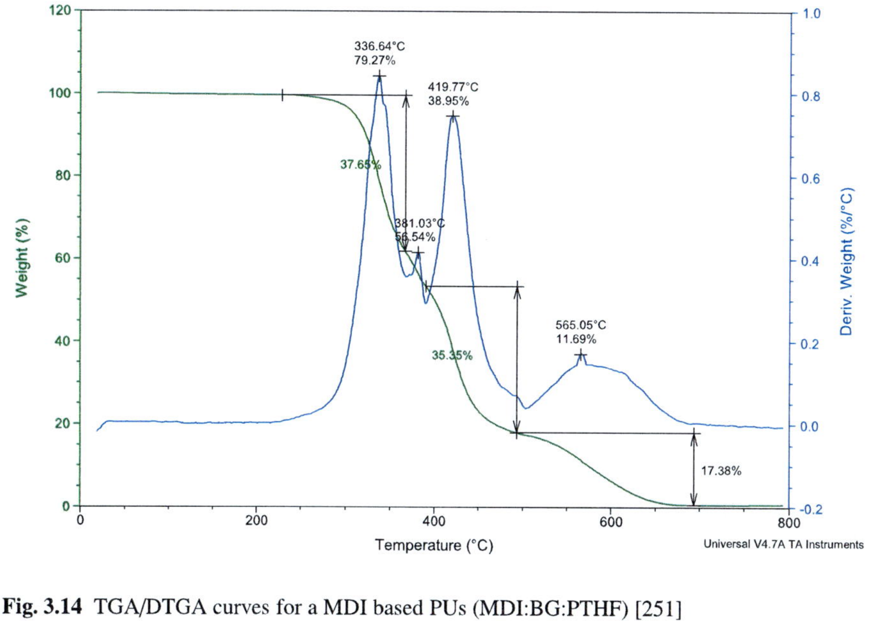

Fig. 3.14 TGA/DTGA curves for a MDI based PUs (MDI:BG:PTHF) [251]

The chemical structure of the samples allows us to relate the observed regions of decomposition to following processes: region I is related to decomposition of urethane groups (links); region II can be connected to destruction of ether groups, region III is related to destruction of carbon chains and rings.

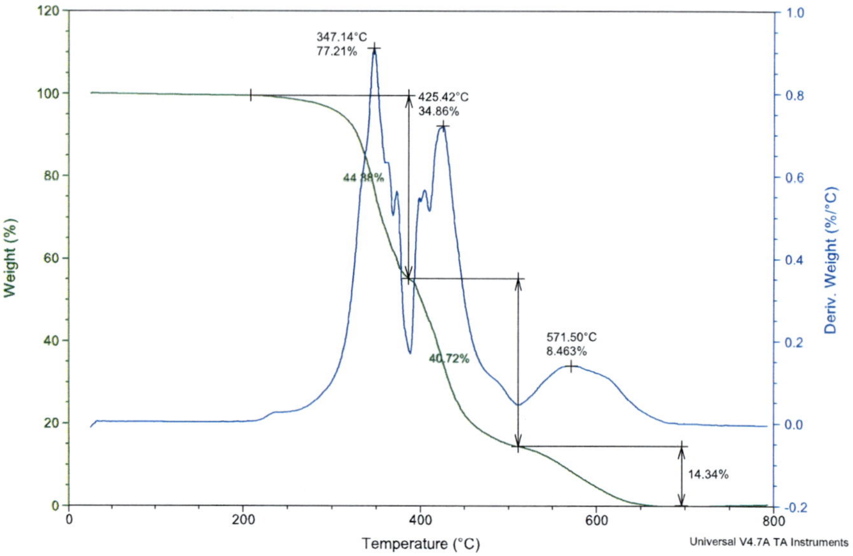

Fig. 3.15 TGA/DTGA curves for a DBDI based PUs (DBDI:BG:PTHF) [251]

The analysis of the data shown in Table 3.5 indicates that for both materials the temperature intervals of region I are identical whereas the temperature intervals of regions II and III are shifted to higher temperatures for the material with DBDI. The temperatures of the DTGA maxima are 5–11°C higher in every region for the DBDI polymer. The loss of weight (indicating the fraction of structure which is destroyed and evaporated as a result of thermal-oxidation decomposition), is close for both materials in the region I. The loss of weight is lower for the material with MDI in the region II and correspondingly higher in the region III than for the material with DBDI.

Although in region I the temperature intervals and loss of weight are close for both materials, the DTGA peak (maximal rate of decomposition) is higher for the material with DBDI (347°C vs 336°C) which is evidence of a higher thermal stability of DBDI structure. The same conclusion can be drawn for the DBDI polymer, from the shift of the temperature intervals and DTGA peaks to higher temperatures in the regions II and III. The lower value of loss weight in the region II for the MDI polymer may be caused by a lower number of ether groups in PU structure originating from the MDI product.

In general, the DBDI material has a higher thermal-oxidation stability than the similar polymer achieved with MDI.

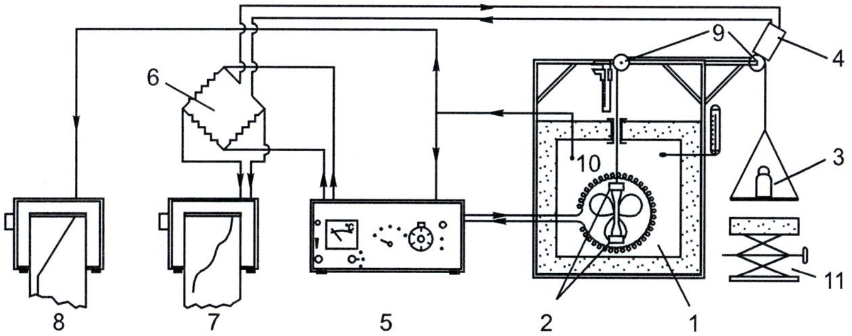

Fig. 3.16 Schematic of the in-house thermal creep testing machine with temperature increasing gradient: (1) – heating chamber; (2) – jaws; (3) – weights; (4) – multi-turn potentiometer; (5) – linear temperature programmer; (6) – Wheatstone bridge; (7) – creep deformation recorder; (8) – temperature recorder; (9) – fixed pulley; (10) – thermal resistance [252]

3.4 Thermomechanical behaviour of polyurethane elastomers obtained with mixtures of diisocyanates

Everything should be made as simply as possible, but not simpler. (Albert Einstein)

The effect of ordering in the copolyurethane elastomers was also assessed by us on employing thermomechanical analyses with the aid of a home made thermal creep instrument of own conception (Fig. 3.16). When we built this instrument, we realised Einstein was right ☺. We followed the course of elongation under constant load of a pre-tensioned sample so to start from a constant initial 50% strain level. As seen in Fig. 3.16, as the specimen deforms the wire cable moves and rotate the potentiometer axle which constitutes one of the Wheatstone bridge's arms. It results a disequilibrium in the bridge, which is converted into electric current; the deformation curve is recorded on the creep deformation recorder and simultaneously the temperature curve is recorded on the temperature recorder. The temperature programmer has 11 positions for different rates of temperature increase in the heating chamber: 1; 1.5; 2; 3; 4; 6; 8; 12; 24; 33°C/min. The gradient of temperature is controlled by a thermal resistance which is located in the heating chamber [252].

Initially the sample was stretched to a 50% strain amplitude (as revealed by our IR dichroic measurements). Then the strain was kept constant and the variation of sample elongation was followed by increasing temperature with constant increments of 1°C / min.

Note that the compositions of the family of materials prepared and studied in this work, is similar to that of the PUs achieved with mixtures of diisocyanates which were characterized by WAXS and SEM in section 2.3 (Tables 2.2 and 2.3) and also by DSC in section 3.2.2. So, for a better understanding of the PUs types investigated in this work, the readers should see Tables 2.2 and 2.3 in section 2.3.

The results which are portrayed in Fig. 3.17 and 3.18 show the dependence of thermomechanical properties on the sequential ordering of the block copolyurethane blocks. As seen in Fig. 3.17 in the case of PUs based on PTHF soft segments, the course of the thermal creep curves achieved on original PU as obtained after synthesis did not depend essentially on the sequential block ordering.

All of these polymers (based on mixtures of the diisocyanates MDI+DBDI), $PU_{C1\ PTHF}$, $PU_{C2\ PTHF}$ and $PU_{C3\ PTHF}$ – (*listed in section 2.3, Tables 2.2 and 2.3*) kept their elasticity modulus very well on heating. Differences only appeared with regard to the temperature at which rupture in the polymer occured. Thus, the material with mixtures of the diisocyanates MDI+DBDI ($PU_{C2\ PTHF}$) characterized by the presence of accumulated EG–MDI blocks displayed a higher resistance at temperatures situated around 130°C, whereas in the case of polymer $PU_{C3\ PTHF}$ where the EG–DBDI blocks were preponderant the temperature of break decreased to 112°C. After annealing of these polymers at 160°C for 30 minutes, the thermomechanical properties changed and the elongation of the polymer specimen increased sensibly along with the temperature increase. The higher deformation in dependence of temperature was remarked in the case of polymer $PU_{C2\ PTHF}$ whereas the most stable annealed material was $PU_{C1\ PTHF}$, which possess a statistical distribution of the EG–MDI and EG–DBDI blocks [60].

Note that polymers PU_{C1}, PU_{C2} and PU_{C3} are achieved with mixtures of diisocyanates (MDI+DBDI) as shown in Tables 2.2 and 2.3 section 2.3. Polymers PU_{C1} were achieved from the two diisocyanates randomly distributed (see Table 2.3).

The synthesis of the PU_{C2} polymers involved a two step prepolymer synthesis where the first introduced diisocyanate was DBDI, followed by the distribution of MDI in the prepolymer. The synthesis of PU_{C3} polymers was achieved similarly, except that the first introduced diisocyanate was MDI, followed by the distribution of DBDI in the prepolymer.

A similar situation was observed by us for similar materials i.e. also achieved with mixtures of MDI+DBDI, but where the SS macrodiol was the polyester PEA, (Fig. 3.18). The enhanced accumulation of EG–MDI blocks ($PU_{C2\ PEA}$) induced a higher resistance to heating with T_{break} = 140°C whereas in the case of $PU_{C3\ PEA}$ with preponderant EG–DBDI block accumulations, T_{break} decreased to 117°C. In this case it should be mentioned also the more pronounced enhance of elongation against temperature [60].

As shown in Fig. 3.17 and 3.18, a significant difference in the thermomechanical curves appeared in the case of the annealed PUs at 160°C for 30 minutes when all the break temperatures increased. The elongation increment increased significantly too once with the temperature increase. A higher degree of elongation versus temperature was ascertained in the case of $PU_{C3\ PEA}$ with accumulated EG–DBDI hard segments. It should be also mentioned that the annealing of $PU_{C3\ PEA}$ improved its properties from all points of view. So, the temperature at break increased from 117°C to 138°C, in parallel with a lower dependence of deformation against temperature.

Again, the materials denoted PU_{C1}, PU_{C2} and PU_{C3} are polymers with mixtures of the diisocyanates (MDI+DBDI) as detailed in Tables 2.2 and 2.3 section 2.3.

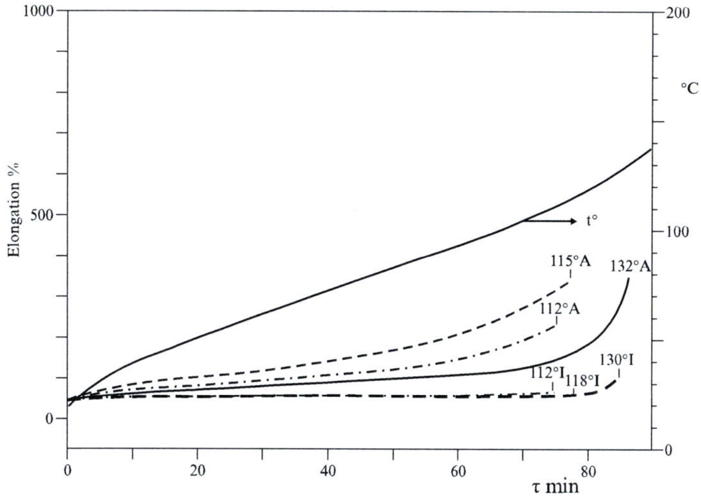

Fig. 3.17 Thermomechanical curves of copolyurethanes with PTHF: (——) – $PU_{C1\ PTHF}$; (– – –) – $PU_{C2\ PTHF}$; (– · –) – $PU_{C3\ PTHF}$; (I) - initially; (A) - after annealing.

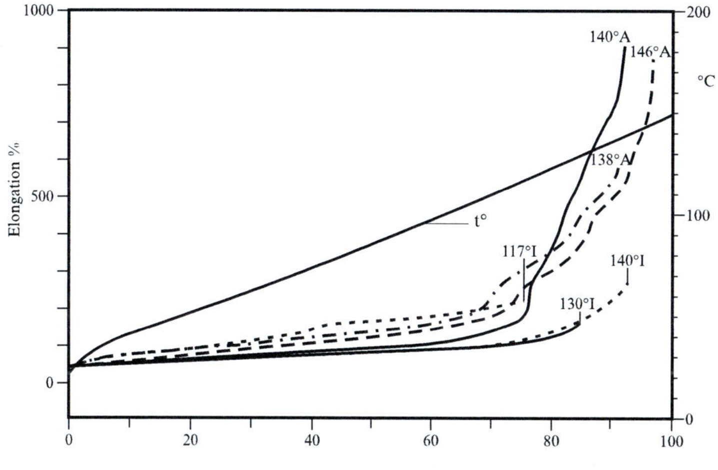

Fig. 3.18 Thermomechanical curves of copolyurethanes with PEA: (——) – $PU_{C1\ PEA}$; (– – –) – $PU_{C2\ PEA}$; (– · –) – $PU_{C3\ PEA}$; (I) - initially; (A) - after annealing.

Other example thermomechanical curves *but for materials based on the single diisocyanate DBDI and SS macrodiols of variable molecular weights* are shown in Table 3.6. The isocyanic index I ranged between 100–110 [17, 250].

As seen in Table 3.6, the following criteria were used by us: (a) σ_{50} – the stress to produce a 50% strain level; (b) T_R – the temperature at which the polymer rupture occurred; (c) E – elongation at break; (d) R – residual elongation; (e) t_{100} – the temperature at which the sample reached a 100% level of elongation.

Table 3.6 Thermomechanical properties of some DBDI-based polyurethanes [253]

	Structure					$\sigma_{50\%}$	Thermomechanical curves [°C]	T_R	E	R	t_{100}
	Chain Extender	% I	Macrodiol	Diisocyanate	Index	MPa	20° 40° 60° 80° 100° 120° 140°	[°C]	[%]	[%]	[°C]
1.	DEG	40	$PTHF_{2000}$	DBDI	100	4.14	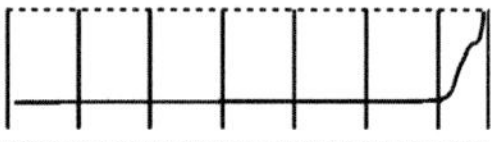	153	826	145	132
1.	DEG+TMP	40	$PTHF_{2000}$	DBDI	100[a]	4.48	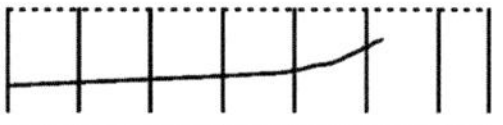	116	205	35.5	98
3.	DEG	40	$PTHF_{1000}$	DBDI	110	3.88	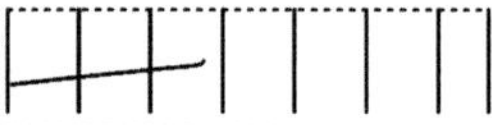	70	152	0	40
4.	EG	40	$PTHF_{650}$	DBDI	105	6.83		60	574	6	30
5.	BDO	40	$PTHF_{1000}$	DBDI	110	6.64	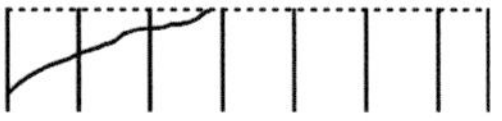	76	510	80	26
6.	DEG	40	$PTHF_{1546}$	DBDI	110	5.34	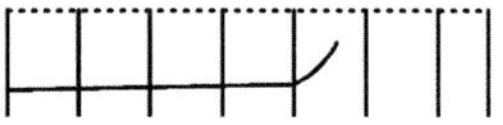	114	210	6	111
7.	DEG	40	$PTHF_{1000}$[b]	DBDI	110	3.45	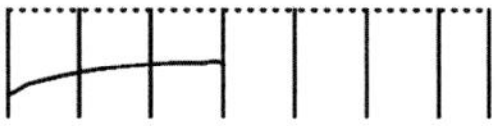	79	233	10	42
8.	DEG	40	PBA_{1000}	DBDI	110	6.18	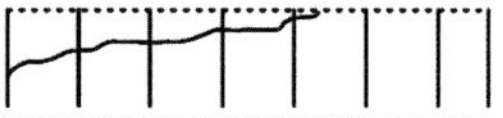	112	999	319	26
9.	DEG	40	$PTHF_{2000}$	DBDI	110	3.78	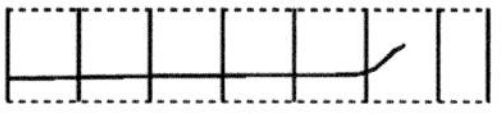	129	267	10	122

[a] Crosslink; [b] (2000 + 650)

The best thermal behaviour was found for the thermoplastic DBDI based material with $PTHF_{2000}$ and DEG (curve 1), followed by a similar material but achieved as crosslinked (I = 110) ($PTHF_{2000}$, curve 9). Changing the chain extender type lead to significant changes as it observed the case of the material with BDO (curve 5), which displayed a poor thermal behaviour with T_R = 76°C. A similar situation was found in the case of the crosslinked PU with I = 105 (curve 4); this material was obtained with the SS macrodiol of low molecular weight ($PTHF_{650}$), and was chain extended with EG.

The molecular weight influenced the PUs thermomechanical behaviour. The best results were obtained with Mw = 2000. The type of SS macrodiol (polyether PTHF or polyester PBA) also influenced the PU thermomechanical behaviour. Numerous phase transformations were observed in the material obtained with PBA_{1000} (curve 8) as compared to the material with $PTHF_{1000}$ (curve 3). Both materials were based on the same chain extender, DEG and diisocyanate DBDI.

The differences observed in the thermal behaviour of the materials were also attributed to the chain extender-diisocyanate couple EG–DBDI or BDO–DBDI, where the HS were observed to crystallize [60, 61, 127, 135, 253, 254]. As shown in Chapter 4, in such materials, the stiffness, hysteresis, residual elongation and stress relaxation were all increased relative to the corresponding MDI based materials.

3.5 Thermal behaviour as revealed by compression tests

Things get worse under pressure (Murphy's law of thermodynamics)

Fortunately, this is not always a rule, although we had this experience while doing PUs compression experiments achieved at high temperatures ☺. In our previous compression experiments carried out on PUs we have shown that (as expected), for the same material, it obtained different load-deflection curves if changing the shape and/or size of PUs specimens [250]. A Shape Factor (θ) function was calculated by us, from the ratio between the charged area and the sum of free areas of the PU sample when compressed. The stress-strain diagrams of compression were determined by using a large scale of θ numbers ranging between 0.1 to 4. The variation of the Young Modulus as a function of the θ factor was determined for MDI and DBDI materials, as shown elsewhere [250].

In the study reported here, the compression set at variable temperatures was followed for a series of PUs with dibenzyl structures as compared to analogous materials but obtained with the diisocyanates MDI and 1,5-naphthalene diisocyanate (NDI).Temperatures were ranged from 20°C to 80°C. The compression set data were determined as percentages of the original compression (20%) that was not recovered. The tests were conducted on cylindrical specimens of diameter Ø= 20 mm and height H = 20 mm according to the ISO 815-1972 standard. At the end of the test, the samples were removed and allowed to cool at room temperature for 30 minutes before measuring the differences between the final and original dimensions of the sample. The experiments were made on employing an in-house made hydraulic press previously described [250].

3.5.1 *Effect of increasing the hard segment content on the compression properties*

A series of materials was studied, based on the flexible diisocyanate DBDI and different chain extenders. The effect of increasing the HS percentage on the residual elongation and on the modulus at a 20% compression (E_{20}) was studied [250]. As shown in Table 3.7 in the case of PEA/BDO based PUs, increasing the HS content (at temperature $t = 20$°C), from 32% to 40% had no sensible effect on the compression set values.

The compression set values decreased with increasing temperature to $t = 80$°C as follows: for a 32 wt% from 4.3% at 20°C to 1.8% at 80°C; for a 37 wt% from 4.8% to 2.8% at 80°C; and for a 40 wt% from 4.47% to 2.7% at 80°C. At temperature $t = 20$°C the highest E_{20} values (60 MPa) were found for PUs of 40 wt%.

As seen in Tables 3.8 and 3.9, replacing the chain extender BDO with DEG resulted in variations of the E_{20} modulus. The modifications of the E_{20} modulus data were followed as a function of temperature. A dramatic decrease of E_{20} at temperatures of 80°C was observed. Increasing the HS percentage determined the

Table 3.7 Compression set and modulus at a 20% compression (E_{20}) for PUs based on DBDI and chain extender BDO of variable hard segment weight percentage

Type of CE	hard segment percentage, (wt%)	T [°C]	Compression set [%]	E_{20} [MPa]
BDO	32	20	4.3	44.0
BDO	32	50	4.3	42,0
BDO	32	80	1.8	34,8
BDO	37	20	4.8	37,2
BDO	37	50	4.2	35,8
BDO	37	80	2.8	33,7
BDO	40	20	4.47	60.0
BDO	40	50	3.6	40.0
BDO	40	80	2.7	36.2

increase of polymer toughness by the enhance of the number of hydrogen bonds [250].

Table 3.8 Influence of the hard segment concentration on E_{20} modulus at variable temperatures, for a series of DBDI-based PUEs obtained with DEG and macrodiol PEA; (CM) – compression under medium conditions; (CD) – destructive compression.

CE type DEG hard segment percentage (wt%)		T [°C]	E_{20} [MPa]
		20	31.2
DEG 32	CM	50	21.2
	CD	80	-
		20	22.8
DEG 40	CM	50	20.0
	CD	80	13.7
		20	47.6
DEG 47	CM	50	29.6
	CD	80	24.3
		20	38.1
DEG 50	CM	50	31.5
	CD	80	20.0
		20	29.1
DEG 60	CM	50	27.7
	CD	80	18.3

CD – destructive compression at 313 MPa and t° = 80°C;
CM – medium compression at t° = 20÷80°C.

As seen in Table 3.9 below, *for a 40% HS percentage*, significantly higher E_{20} values were found in the case of PUs based on the chain extenders BDO and EG

(60.0 MPa and 55.9 MPA respectively), in contrast with similar polymers but derived from DEG (22.8 MPa) which as shown, inhibits crystallization [61].

The higher compression set data are due to the couples BDO–DBDI or EG–DBDI, where the variable geometry of DBDI favors the polymer crystallization. In polymers based on the couples BDO–DBDI or EG–DBDI (especially polyether PUs) there perform high phase separations and differences in hard phase plastic flow stress, resulting from crystallinity in the DBDI phase.

Table 3.9 Compression set and E_{25} Young Modulus at t= 20°C for DBDI based PUs and various chain extenders [250]

Type of CE hard segment percentage (wt%)	T (°C)	Compression set after 30 mins [%]	E20 [MPa]
BDO 40	20	3.7	60.0
DEG 40	20	2.29	22.8
EG 40	20	3.75	55.9

3.5.2 *Effect of type of diisocyanate on the compression properties*

The compression set data obtained for the DBDI based materials were compared to those of analogous PUs but obtained with the conventional diisocyanates MDI and 1.5-NDI. The materials depicted in Table 3.10 were synthesized with a 40% HS percentage, PEA of molar mass 2 000 g/mol, and chain extender BDO.

Table 3.10 Compression set at t= 20°C for DBDI based PUs as compared to other conventional PUs [250]

	T (°C)	NDI	MDI	DBDI
Compression set [%]	20	21.3	13.2	4.47

As seen, the compression set of the polymer with BDO/DBDI is significantly smaller than the data for the analogous materials but derived from MDI and NDI. The lower compression set for the DBDI materials were attributed to the BDO-DBDI couple which favors the polymer crystallization.

3.6 Polyurethane blends and their thermal behaviour

Let your mind be applied... to the study of the mixtures (Theophilus)

3.6.1 Steric conformations

The introduction of a rotary ehtylenic bridge between the phenyl rings and of a macromolecular chain entails a series of specific phenomena. As shown in our previous investigations we found that in contrast to other common diisocyanates as it is MDI, the diisocyanate DBDI reacts with glycols, leading to insoluble products which exhibit a higher melting point temperature and high crystallinity [59].

The PUs based of the couple DBDI–EG precipitates from solution after dissolving in DMF. Due to the high association of the crystalline lattice [59], the obtained powder cannot be redissolved in DMF even by heating.

As reported earlier, the X-ray diffraction spectrum of this product shows a periodicity of 19.7±0.2 Å and it is similar to that of the triclinic form of nylon 66 [255], nylon 6, and nylon 8 [256]. This suggests a change in the polymer conformation in solution from the disordered form into the extended form, which is due to the lower entropy level of the crystalline structure and to maximum hydrogen bridge formation. As seen in section 3.3, the melting point of this polymer was found to be 312°C and the initial decomposition temperature of the polymer was 220°C.

In the case of PUs based on MDI and EG, the X-ray measurements evidenced a periodicity of 15.7±0.1 Å with a quantitatively similar arrangement [59, 256], but belonging to the hexagonal lattice crystalline system.

The orientation effects produced by the introduction of the HS derivatives from DBDI and EG into polyester based PUs were followed. We investigated the copolyaddition of 3 mol 4,4'-DBDI and a mixture of 1 mol PEA or PBA of Mw = 2 000 with OH terminal groups, and 2 mol EG in DMF at 60°C. In this cases, we observed an unexpected phenomenon: although initially, as a result of the initial reaction between DBDI and EG, the mixture was homogeneous, even in the incipient first phase of the reaction an oligomer resulted which, due to association and crystallization, separated from solution as a very fine suspension. The reaction progressed in two phases: solution and suspension [59].

The mixture was disproportionate, which resulted in a white, viscous, and stable solution-suspension. By intense centrifugation, a precipitate could be separated [59]. By precipitation in water of the remaining clear solution, a second polymer with a different structure could be separated [59].

The suspension became clear when diluted with 4 times the volume of DMF and heated at 60-70°C. The crystalline structure was destroyed, and it did not precipitate by cooling or standing at room temperature [59].

These facts, and the apparent constant composition of the polymers obtained when starting from different molar ratios of the reactants, suggested that in copolyaddition, the growth of the macromolecules in suspension is governed by the formation of the crystalline structure. Among others, this behaviour was earlier used for the processing of artificial leather.

In the case of DBDI, the main characteristic of the polymers is their capacity to self-orientate in crystalline structures.

In the light of these observations, we proposed to study what are the limit situations that can be reached when approaching the ‘one shot’ synthesis route in two distinct cases, e.g. when employing MDI as one of the most utilized commercial diisocyanates and when using the isocyanate of variable geometry DBDI obtained by us. The following aspects were envisaged: (a) – to elucidate why an apparently small change in the diisocyanate structure (one or two $-CH_2-$ spacers between two aromatic nucleus) can change dramatically the course of polyaddition, by splitting the process course into two phases (in solution and in parallel in suspension); (b) – to study the difference in the molecular suprastructure of the two copolymers mixture which may appear simultaneously under different reaction conditions. In principle this system of a bicomponent macromolecular structure can be regarded as a composite system at different degrees of dispersion ranging from a monodispersion to a molecular nanodispersion state.

It is already well known that the PUs are obtained by reacting a diisocyanate (I) with a macrodiol (MD) of a polyesteric or polyetheric type, and with a glycol as a chain extender (CE). The simplest approach seems to be the simultaneous mixing of the three components on approaching the ‘one shot’ synthesis route:

$$(n+1)I + 1MD + nCE \quad \text{in solution or in melt.} \tag{3.3}$$

If the I reactivity against the MD and CE is not equal, there is the possibility that the reaction *firstly* begins with the most reactive partner, which is the CE, by the formation of blocks of the type –[I–CE]–I– that are finally performed in the middle of the final macromolecule.

We have evidenced the difference between the reactivity of the MD and CE by measuring the reaction rate of I with CE and MD separately. We have shown the existence of a supplementary synergetic effect due to which, in the case of the reaction between I with the mixture MD–CE, the reactivity of the glycol increases in comparison to that of MD [44–46, 89]. With this aim, high pressure liquid chromatography (HPLC) measurements where employed to monitor the evolution of every molecular species which can appear in the reaction, on the model reaction which employs a monoisocyanate instead of the diisocyanate [89].

Within the *second* part of the macromolecule building, the contribution of the MD increases as well by the formation of intermediaries mixed segments. Finally in the *latest stage* of the macromolecule increase process, there are inserted only –[I–MD]– blocks without the CE participation.

$$\begin{array}{c} \text{MD I}-[\text{I}-\text{MD}]_y-\text{MD I}---[\text{ICE}]_x-\text{I}---\text{I}-\text{MD}-[\text{I}-\text{MD}]_y-\text{I}-\text{MD} \\ \text{hard middle block} \qquad \text{soft end block} \end{array} \tag{3.4}$$

To avoid the appearance of terminal blocks of the type –[I–MD]–, the prepolymer technique is usually used in most of the cases. In this type of synthesis, the MD is first reacted with an excess of I so that the whole amount of MD is transformed in a macrodiisocyanate prepolymer:

$$(n+1)I + 1MD \rightarrow \underset{\text{macrodiisocyanate}}{1I\text{–}MD\text{–}I} + (n-1)I \quad n = 1 \div 9 \tag{3.5}$$

In this case the further reaction of the newly appeared intermediate of the type I–MD–I performs only by the CE contribution so that it assures permanently the presence of a HS as an element of connection between the SS which leads to the appearance of a more homogeneous molecular structure.

$$\underset{\text{hard middle block}}{MDI\text{–}[I\text{–}MD]y\text{–}MDI\text{–––}[ICE]_x\text{–}I}\text{–––}\underset{\text{soft end block}}{I\text{–}MD\text{–}[I\text{–}MD]_y\text{–}I\text{–}MD} \tag{3.6}$$

3.6.2 DBDI based PUs blends

Two PUs systems were obtained with MDI or DBDI, with the following structures:

$$\begin{matrix} (MD) & (CE) & (DI) \\ PBA & — EG — & D \\ 1 & n & n+1 \end{matrix} \quad (n = 1 \div 10) \tag{3.7}$$

where the MD was the macrodiol PBA of molar mass 2 000 g/mol and the CE was EG. The materials were obtained by using the one shot technique in melt and in solution. The isocyanate index was I = 100. The ratio between the reactants is summarized in Table 3.11. Details regarding the PU synthesis were reported elsewhere [25].

When MDI was used in the polymer synthesis, irrespective of the synthesis route it obtained homogeneous soluble polymers inseparable in fractions, by selective dissolution in DMF. Certainly, the sequence of ordering on the macromolecular chain differs from a PU synthesis route to another. As mentioned, the reaction rate between phenyl isocyanate (FI) and chain extender, ethylene glycol (EG) was found by us to be 4 times higher than the reaction between FI and PEA when measured separately, and 11.6 higher respectively if they are in a mixture (EG + PEA) (synergetic effect) [45, 89].

The situation changed dramatically when using DBDI in the PU synthesis. As already mentioned, in both of the cases when using MDI or DBDI, due to kinetic reasons in the first stage it obtained oligomers with preponderant $-[I\text{–}EG]_n-$ structures. Due to the high interaction between *oligomer–DMF*, the oligomers corresponding to the –[MDI–EG]– type remained in solution of DMF. The further steps of reaction performed also in solution till all the reactants were consumed. In the synthesis with DBDI the situation changed: due to the possible intramolecular

Table 3.11 PU type and adopted synthesis routes [257]

Type of PU	Synthesis Route	CE	MD	DI	Observations
PU 1	One shot DMF T = 60°C	EG 3	PBA 1	DBDI 4	insoluble in DMF (27.9%) soluble in DMF (72.1%)
PU 2	One shot melt T = 60°C	EG 3	PBA 1	DBDI 4	13.12% insoluble 86.88% soluble
PU 3	One shot DMF T = 60°C	EG 3	PBA 1	MDI 4	100% soluble
PU 4	One shot DMF T = 110°C	EG 3	PBA 1	DBDI 4	insoluble in DMF (28.86%) soluble in DMF (71.14%)
PU 5	One shot NMP T = 115°C	EG 3	PBA 1	DBDI 4	insoluble in DMF (2.85%) soluble in DMF (97.15%)
PU 6	Prepolymer melt	EG 3	PBA 1	DBDI 4	soluble in DMF
PU 7	Prepolymer melt	EG 3	PBA 1	MDI 4	soluble in DMF

rotation of the DBDI skeleton, the incipient growing oligomer –[DBDI–EG]– adopted a plan parallel structure which allowed a more compact intermolecular packing of the growing oligomers. Thus, the interaction between *oligomer–oligomer* became preponderant towards the interaction between *oligomer–DMF* and in solution appeared life species in suspension which continued to grow by interphasic reactions.

In parallel, the kinetic-statistically determined incipient reaction of DBDI with the MD, although rare, leaded to soluble oligomers with a good solubility, with sufficiently long SS so to maintain the so-obtained polymer in solution till the end of the process.

The crystallinity of the copolymer resulted in suspension (separated by centrifugation), was evidenced with the aid of X-ray measurements. The X-ray diagram of the so-obtained insoluble copolymer was similar to the pure homopolymer –ED–DBDI [59]. The similitude was as closer as the 'n' number in the schemes from above was higher.

Replacing the DMF solvent with 1-Methyl-2-pyrrolidinone (NMP), the situation returned to the "normal" behaviour similar to the MDI based PUs reaction in solution. In this case the interaction between oligomer-solvent was always higher than the interaction between oligomer-oligomer (polymer-polymer), and it obtained only a single type of soluble copolymer [257].

Thus, when employing the 'one shot' synthesis in solution (PU 4, Table 3.11), it obtained a suspension containing 2 copolymers simultaneously one of which was soluble in DMF (71.2%), and the other was insoluble in DMF (28.85% — Table

3.11). Replacing DMF by NMP, it obtained a homogeneous viscous solution which contained one single copolymer. A similar situation appeared when the syntheses were performed in melt [257].

As a result of the polyaddition of the DBDI based materials by the 'one shot' synthesis in melt, it obtained opaque solid elastomer blends consisting of two copolymers, one of which was soluble in DMF (87%), while the other was insoluble in DMF (13%), (PU_{100} — Table 3.11).

3.6.3 DMA measurements

The DMA characterization was made of the series of polymers denoted PU 1 to PU 5 as listed in Table 3.11. The samples were tested in the tensile mode with a starting distance between the clamps of 45 mm. The cooling conditions were: room temperature till −60°C cooling rate of 10 K/minute, −60°C to −120°C 3.5 K/min and −120°C to −140°C 2 K/min. The test started at −140°C and the complex modulus E^*, storage modulus E', loss modulus E'' and the loss factor were measured as function of temperature at a heating rate of 1 K/min. The instrument was operated with controlled sinusoidal force with a frequency of 1 s^{-1}.

Depending on the material structure and on the conformation of diisocyanate, there were observed large differences between the E' modulus and $\tan\delta$ loss factors respectively. The material 3, obtained with MDI via the one shot technique in melt (Table 3.11), showed an amorphous structure. The E' modulus above T_G was the lowest of all the materials, and correspondingly the loss factor $\tan\delta$ in the glass transition region was the highest. For the samples PU 1, 2, 4 and 5 (DBDI, one shot), the T_G values ranged between −28°C (PU 2) and −14°C (PU 3), while $\tan\delta$ varied between 0.127 (PU 2) and 0.394 (PU 3).

Example selected DMA curves of selected materials listed in Table 1.1 are shown in Fig. 3.19–3.22.

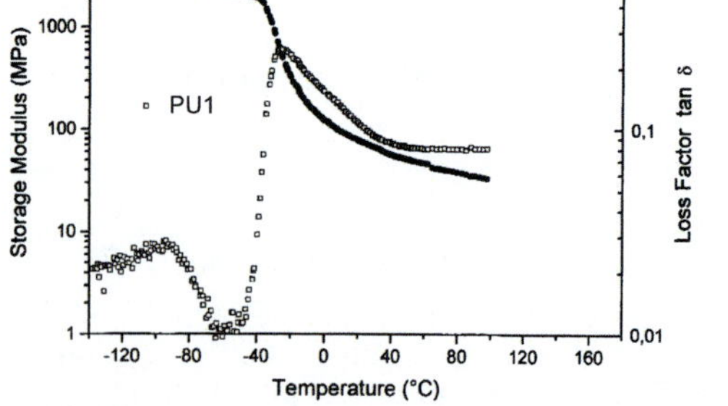

Fig. 3.19 Storage Modulus and Loss Factor as a function of temperature for PU 1

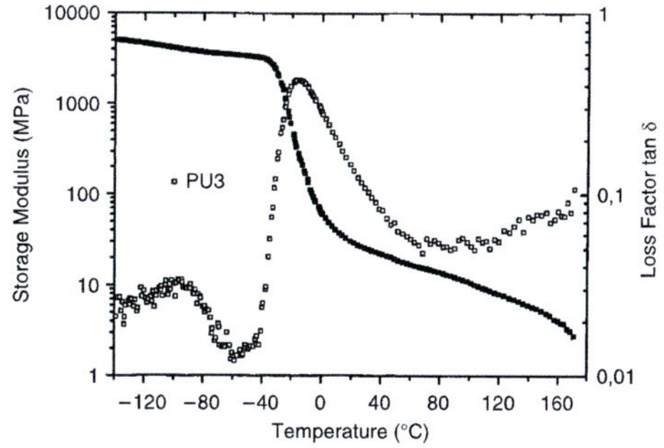

Fig. 3.20 Same as in Fig. 3.19 but for material PU 3 [257]

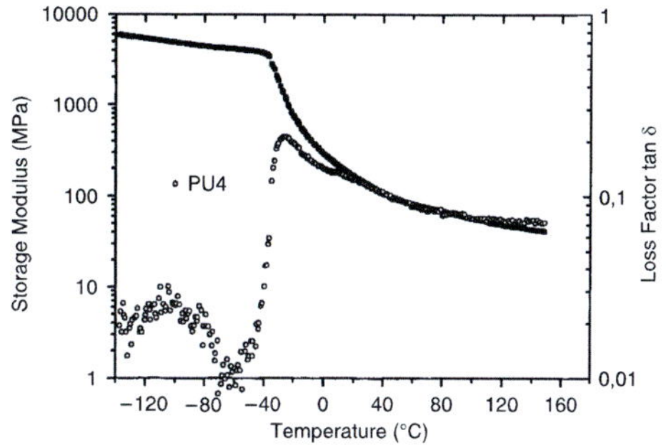

Fig. 3.21 Same as in Fig. 3.19 but for material PU 4 [257]

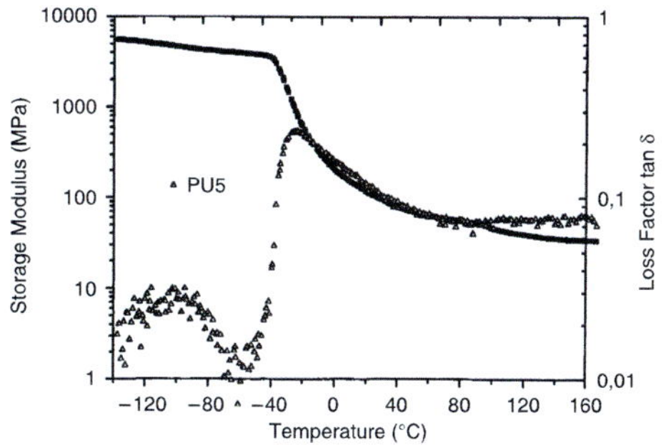

Fig. 3.22 Same as in Fig. 3.19 but for material PU 5 [257]

Chapter 4
Mechanical aspects of polyurethane elastomers

4.1 Tensile properties

Nothing ruins the truth like stretching it (unkown source)

But "truth" is not a material... suggest we see if it applies to polymer stretching also ☺?

Tensile tests are used to control product quality and for determining the effect of chemical or thermal exposure on an elastomer.

As shown [127], the *tensile stress or tensile modulus* is the force per unit of original cross-sectional area required to stretch the specimen to a stated elongation. 50%, 100% and 300% extensions are commonly used. Usually, the test is performed on a dumbbell-shaped specimen [258].

Elongation is the percentage increase in original length of an elastomer as a result of tensile force being applied to a PU specimen. Usually we talk about percent elongation, which is the length the polymer sample after it is stretched (L), divided by the original length of the sample (L0), and then multiplied by 100.

Ultimate elongation or elongation at break is the elongation at the point that the sample breaks [127, 258].

Certainly *residual elongations* in materials have nothing to do with the the human reminiscences that *"make one feel so deliciously aged and sad". (George Bernard Shaw).* Or yet, maybe do they? ☺ *Residual elongations in materials or permanent set* is determined as follows: tensile force is applied to break a benchmarked dumbbell specimen. A period of relaxation follows after which the two parts are fitted together and the distance between the marks is re-measured. The difference between the new measurement and the original is expressed as a percentage of the original distance [258].

Tensile strength is the maximum tensile stress which a material is capable of developing. It is the force per unit of the original cross-sectional area which is applied at the time of rupture of a specimen. It is known variously as "breaking load", "breaking stress" and "ultimate tensile strength". Tensile stress, and tensile strength

C. Prisacariu, *Polyurethane Elastomers*, DOI 10.1007/978-3-7091-0514-6_4,

are both measured in units of force divided by units of area, usually N/cm^2, or megapascals (MPa) [127, 258].

In PUs, the stress is not linear with strain. Therefore *the modulus* is neither a ratio nor a constant slope, but rather denotes a point on the stress-strain curve.

4.1.1 Effect of polyol type and molecular weight on the tensile properties

The mobility of the polyol influences the PUs properties as it results in low temperature properties and the variation in the tensile strength properties. Thus it is obvious that features related to the mobility of polyols, such as the glass transition temperature or the melting temperature and the ability to crystallize under deformation, certainly impacts the mechanical properties. The polymer mobility is strongly influenced by the type and the molecular weight of the polyol. To prepare products with typical rubber elasticity, an average molecular weight between 1 000 and 4 000 (corresponding to a chain length between 180 and 300 Å is desirable [28]. Higher molecular weights result in materials with better tensile properties but with an increased tendency to cold-harden, a phenomenon caused by gradual crystallization of the flexible blocks during storage. This process can be avoided by incorporating copolyesters to provide structure irregularity [259]. The primary consequences of increasing the SS molecular weight for a given overall molar ratio of the polyol block to hard block (diisocyanate plus chain extender), are a decrease in modulus and an increase in the elongation at break, as well as a decrease of the glass transition temperature of the soft block [28].

Example values showing the influence of molecular weight on the stress-strain data of a polyester dibenzyl based PU (DBDI:PEA:EG) are shown in Table 4.1.

Table 4.1 Example values showing the influence of molecular weight (Mw) on the stress-strain data of a polyester dibenzyl based PU (DBDI:PEA:EG).

Mw of PEA	Strength Stress [MPa]	Elongation at break [%]	Hardness [Sh°A]	300% Modulus [MPa]
4 000	45	750	74	9
3 000	39	670	87	13
2 000	34	525	98	19
1 000	33	450	99	19.5

It is widely known that a polyester segment with a molecular weight of about 2 000 confer optimum properties to a PUs. Below this level the elastomers may become too weak, while above this level they may tend to undergo crystallization while being stored. As shown by others [4] and by us in section 1.2.4., because the ester groups in particular form strong attractions to urethane groups by means of hydrogen bonding, they also contribute to the strength of the material. In polyester PUs, the cohesive energy between ester groups in the chain is greater with PEA or PBA than with PCD because of greater distance between ester groups in the latter [135]. High concentrations of flexible ether groups results in very weak elastomers. Polyethers usually have a lower glass transition temperature and a weaker interchain force than polyesters, therefore rendering PUs with reduced mechanical properties. As shown in section 1.2.4, this is attributed to stronger hydrogen bonding between the NH and the ester carbonyl group, rather than urethane NH-ether oxygen bond. But, as shown by us [60], annealing may significantly improve the polyether based PUs elastomeric properties.

4.1.2 Effect of type of diol chain extender on the tensile properties

The study of the variation in polymer properties obtained by using different glycols with any given polymeric diol emphasized that increasing the distance between urethane groups (when using longer chain glycols) decreases the density of cohesive force between chains. This effect is coupled with the "*odd-even*" effect of chain fit [260]. The "zigzag" pattern in tensile strengths and elongations is particularly evident in polyester elastomers.

As also reported by others and by ourselves [261], irrespective of the type of the HS, *crystallizing or not*, the PUs mechanical response vary with the length / type of the chain extender. The best mechanical behaviour was observed in PUs derived from diol chain extenders with *even* numbers of $-CH_2-$ groups. The stress-strain and abrasion data decreased dramatically in the case of the materials obtained with chain extenders based on *odd* numbers of $-CH_2-$ groups (Fig. 4.1) [261].

A PU chain extended with "even" diols adopts the lowest energy fully extended conformation that allows hydrogen bonding and, therefore, a higher crystalline order, which explains why the materials based on BDO and higher diols have better elastomeric properties. The HS based on these chain extenders crystallize easier, therefore promoting phase separation [28].

Similar observations were made by Born et al [262], for PUs extended with *aliphatic glycols or diamines*. The model compounds studied by them were prepared by the reaction between diphenylmethane-4-mono-isocyanate and glycols of the $HO-(CH_2)-OH$ structure with $n = 2$ to 6. The series of materials showed a characteristic fluctuation in their thermal and hydrothermal properties which was dependent on the number ("*even*" or "*odd*"), of CH_2 groups in the chain extender.

The causes of this behaviour were examined in by using X-ray crystallographic analyses of single crystals of model urethanes. It observed that whilst strain free hydrogen bonds could form between neighboring molecules in urethanes with "*even*" chain extenders, significant strains occurred in urethanes containing "*odd*" chain extenders which resulted in reduced stability of the physical crosslinking system.

In earlier studies Bonart et al [156] observed a great fluctuation in the thermal and hydrothermal properties of elastomer polyurethane fibers as a function of the

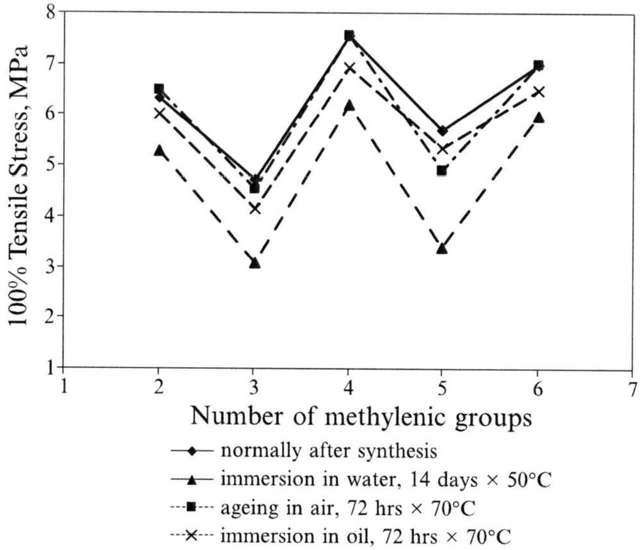

Fig. 4.1 Stress to produce a 300% strain as a function of number of CH_2 groups for a DBDI based PU

structure of the amine chain extender. A characteristic fluctuation in the properties, which was dependent on the number of CH_2 groups, was particularly prominent when chain extenders of the $H_2N{-}(CH_2)_n{-}NH_2$ structure were used. A similar fluctuation was observed by Bonart et al [263] using thermomechanical measurements.

Large supermolecular structures, in the form of spherulites, were often found to be associated with *crystallization of the polyether segment.* Spherulites were observed in PCl:MDI:BDO [264, 265], PTMO:MDI:BD [266, 267], PEO:HMDI:para-phenyldiamine systems [268] and other segmented PUs [269–271].

Comparison of the effect of diol chain extenders on tensile strength shows that BDO allows superior properties. The PUs strength was found to decrease in the following order for these chain extenders and MDI polyester polyol materials: BDO > EG > 1,5-pentanediol > HD > 1,3-propanediol [4]. If different polyols are used, then the following order has been suggested [272] of chain extenders according to their effect on strength: BDO > HD > 1,5-pentanediol > 1,3-propanediol > EG. However the differences are not very large.

The effect of *excess of glycol* on the PUs mechanical properties was largely studied. As shown by Rausch [24], the use of excess glycol reduces the PUs molecular weight The shorter chains in the polymer, being less entangled and having less total force acting on them, slip more easily. Excess glycol results in lower tensile properties and higher residual elongations. As the excess glycol approaches zero, the change in molecular weight is minimal. Hence, as shown by Rausch, the PUs properties do not significantly differ between a 0-2%, excess.

4.1.2.1 Non-chain extended polyurethane elastomers and their tensile properties

Until recently there were no reports in the literature on the preparation of segmented PUs without chain extenders. Fairly attractive thermal and mechanical properties have been recently reported, displayed by non-chain extended poly(ether-urea) segmented copolymers prepared from symmetrical diisocyanates, such as para-phenylenediisocyanate (PPDI), 1,6-diisocyanatohexane (HDI), and trans-cyclohexanediisocyanate CHDI [273]. Due to its small size and symmetry, the PPDI molecule provides a compact HS and excellent phase separation in PU formulations. Starting from this, a series of PUs with no chain extenders was obtained by Yilgor et al [273], with analogous poly(ether-urethane)systems, through stoichiometric reactions of PTMO of molar mass 1 000g/mol and the above mentioned diisocyanates. In addition to symmetrical diisocyanates, high molecular weight PUs obtained with MDI, TDI, and hexamethylene diisocyanate (HMDI) were also prepared. The PUs based on PTMO and symmetrical diisocyanates PPDI, HDI, and CHDI formed mechanically strong films at ambient temperature. However, the PUs based on the asymmetrical diisocyanates HMDI, TDI and MDI yielded very sticky films that showed no mechanical integrity at room temperature. The AFM studies made on these materials demonstrated that while the PUs based on PTMO of molar mass 1000g/mol and PPDI, HDI, and CHDI showed microphase morphology, the materials based on HMDI, MDI, and TDI did not show any significant microphase separation at room temperature. Reducing the SS molar mass, for example, in the case of PTMO – from 2 000 to 1 000 g/mol, resulted in materials with enhanced phase mixing.

Such polyether materials were expected to have the same hydrogen bonding capacity and if hydrogen bonding were the main driving force for microphase separation, they all would have displayed microphase morphology and good material properties. However, this was not the case. Only the materials based on symmetrical diisocyanates were film forming with good mechanical properties at room temperature [273]. These observations showed that the diisocyanate symmetry (and resulting HS crystallinity) is a much more important factor that influences the morphology development in non-chain extended polyether PUs as compared to the hydrogen bonding.

4.1.3 Effect of isocyanate-polyester ratio on the tensile properties

Increasing the diisocyanate-polyester ratio leads to an enhance of the concentration in the high-melting PUs hard domains which affects the physical properties of the elastomer. This results in greater tensile strength, lower elongation, higher melt temperature, and increased hardness [24]. The greater tensile strength and lower elongation can be discussed in terms of the glass transition temperature, T_G. Studies have shown that T_G increases as the urethane concentration increases, most probably as a result of increased intermolecular hydrogen bonding. An increase in the *isocyanate-polyester* ratio determines a higher urethane concentration and hence a higher T_G.

Due to the numerous factors which influence the PUs mechanical behaviour, it is also of interest to refer to the structure-property relations in segmented hydroxyl terminated *polybutadiene based PUs* in which the polar and hydrogen-bonding interactions between the HS and SS units are eliminated. Huang et al investigated a series of hydroxyl terminated polybutadiene based PUs that were synthesized by solution polymerization with different types of diisocyanates and HS contents, by using variable NCO/OH ratios and changing the polymerization method. The MDI based materials with higher HS content and NCO/OH ratios, achieved via the one-stage polymerization method exhibited higher stress at break and lower residual elongations. There was no greater extent of hydrogen bonding between the HS and SS for these PUs. A higher value of hydrogen bonding index [274] was observed for higher NCO/OH values when the materials were obtained by the one-stage method.

4.2 Stress-strain curves and tensile strength properties

My curves are not crazy (Henry Matisse)

Nor are those describing the PUs tensile behaviour ☺. The amount of stress is gradually increased while the elongation which the sample undergoes is measured at each stress level until the sample breaks. It obtains a plot of stress versus elongation. In general, the harder the elastomer, the smaller the elongation. Yet, the ability of PUs to achieve higher hardness while maintaining good elongation permits them to withstand higher tensile loads.

Uniaxial tensile stress-strain curves of PUs generally exhibit sigmoidal shapes characteristic of rubber-like elastomers. A relatively steep increase in stress (σ) at low strain (ε) is followed by a gradual decrease in slope (i.e., strain-softening), a less steep section of curve at intermediate strain and a second region of increased slope (i.e., strain-hardening) at higher strain [275–278]. The Young's modulus (E) generally increases with the HS content, which has been attributed to HS microdomains acting as reinforcing fillers [127, 259, 275–279]. Strain hardening has been attributed to strain-induced crystallization [135].

4.2.1 Effect of type, number and order of diisocyanates on the stress-strain behaviour

Example stress-strain curves and their tensile properties for six materials obtained by us are compared in Fig. 4.2 and 4.3. Two families of PUs are included. The two families differ in the soft phase (PEA, Fig. 4.2 or PTHF, Fig. 4.3). The SS molar mass was 2 000 g/mol. The chain extender was EG. Two HS were compared, based *on the diisocyanates MDI, DBDI, and mixtures of them.* Rotation around the central $-CH_2-CH_2-$ bond in DBDI allows alignment of successive aromatic rings and hence crystallisation within the DBDI hard phase [60, 61, 127, 135]. The two families of PUs consisted of three materials each denoted as: A, B, C, where hard segment molar compositions were (%MDI / %DBDI): A 100/0; B 0/100; C 50/50. In the case of material C, equal molar quantities of MDI and DBDI were pre-mixed before the combination with macrodiol, followed by chain extension. Further details of the synthesis are given elsewhere [280, 281]. The molar proportions used in the syntheses were: diisocyanate:macrodiol:chain extender (EG) = 4:1:3. WAXS measurements showed mass fraction degrees of crystallinity in the three families: A 0%; B 14%; C 8%. DSC measurements showed evidence for some phase separation [280, 281].

To obtain the true stress at break, (in Fig. 4.2 and 4.3), tensile strength values were multiplied by the extension ratio $\lambda = l/l_0$, where l and l_0 are the extended and original PU sample lengths and the true strain true λ elongation at break, ε ($\varepsilon = \lambda + 1$).

As seen in Fig. 4.2 and 4.3, in particular, polymer B with DBDI hard segments, showed higher stiffness and strength than the conventional MDI-based polymer A, but with lower strain recovery and strain energy recovery on cycling—a primary consideration for elastomers. Both features of the response were attributed to differences in hard phase plastic flow stress, resulting from crystallinity in the DBDI phase [60, 61, 127, 135], absent in MDI.

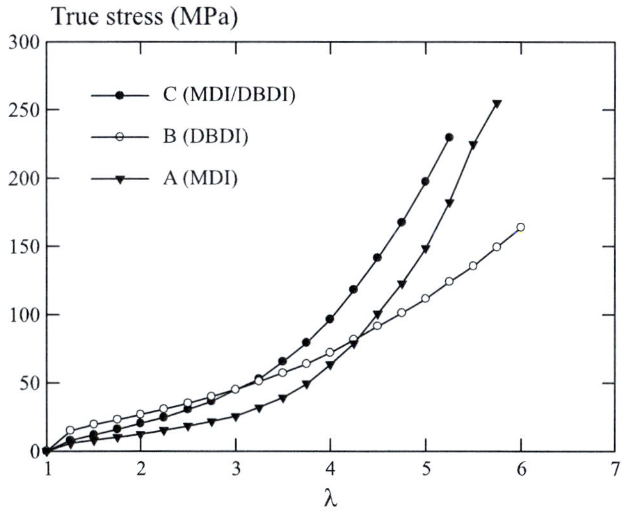

Fig. 4.2 Stress-strain curves during loading at nominal strain-rate 0.0042 s^{-1} for PUs with PEA as a soft phase. Material C was achieved by using a single step prepolymer synthesis using a melt mixture of the two isocyanates (MDI+DBDI) [280]

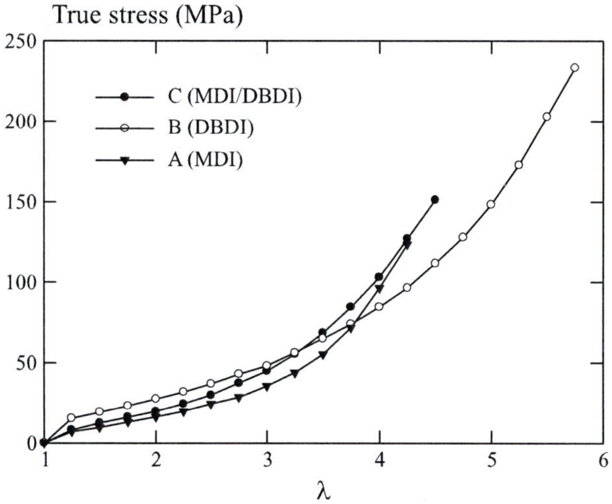

Fig. 4.3 As in Fig. 4.2 but for materials with PTHF as soft phase [69]. Material C was achieved by using a single step prepolymer synthesis using a melt mixture of the two isocyanates (MDI+DBDI) [69, 280].

Polymer C (with randomly mixed MDI/DBDI hard phase), however, showed advantages of increased strength and stiffness, plus *increased* strain recovery compared to polymer A, thus offering improvements in all these properties [61, 254].

Of great interest for use of the polymers as elastomers is the degree of strain-recovery on unload. Strain-recovery data for the three polymers are compared in Fig. 4.4. It may be seen that polymer C, obtained with mixtures of diisocyanates MDI+DBDI shows he highest degree of strain-recovery.

Shown below in Table 4.3 *are the stress-strain data of a larger series of PUs obtained by us, with single or mixtures of the diisocyanates MDI and DBDI.* They are

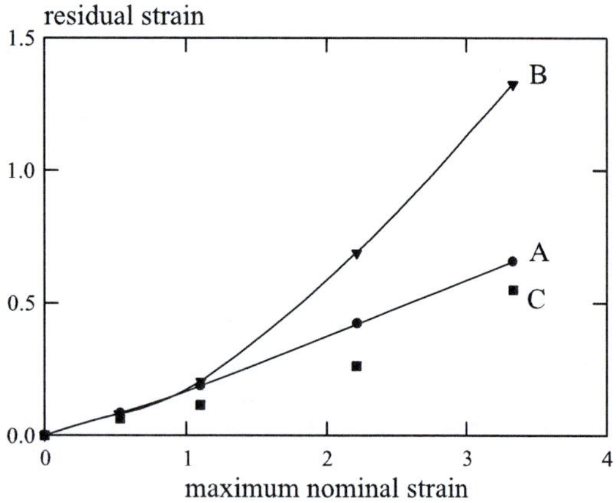

Fig. 4.4 Residual elongation from one load-unload cycle at nominal strain-rate 0.0042 s^{-1}, from the data in Fig. 4.2–4.3 comparison of polymers of types A, B and C. Material C is polymer $PU_{C1\ PTHF}$ achieved by using a single step prepolymer synthesis using a melt mixture of the two isocyanates (MDI+DBDI) [61].

all three-component systems combined in stoichiometric proportions, and consisting of: (1) a diisocyanate—either MDI or DBDI; (2) a macrodiol—PEA or PTHF; and (3) a small molecule diol as chain extender—anhydrous EG. The macrodiols were all of molar mass Mw = 2 000±50 g/mol. The three components were mixed in the molar proportions HS:MD:CE=4:1:2.64, giving HS mass fractions in the region of 40%, and isocyanic index I = 110 which means that they have the potential for further reaction with ambient humidity to produce chain lengthening and allophanate crosslinking.

In Table 4.3, materials with single diisocyanates, (MDI or DBDI) are compared with a family consisting of 6 polymers obtained *with mixtures of MDI+DBDI* as follows: (1) $PU_{C1\ PEA\ or\ PTHF}$ obtained by using a single step prepolymer synthesis using a melt mixture of the two diisocyanates MDI+DBDI; (2) $PU_{C2\ PEA\ or\ PTHF}$ based on mixtures of MDI+DBDI where DBDI is first reacted followed by addition of MDI; (3) $PU_{C2\ PEA\ or\ PTHF}$ also based on mixtures of MDI+DBDI but where DBDI is first reacted followed by addition of MDI. The composition of the family of elastomers prepared and studied is given in Table 4.2. *Also note that previously, in section 2.3.2, SEM results are presented for the same series of materials.*

Table 4.2 Compositions of the family of polyurethane elastomers prepared and studied in this work

Recipe	PU structure
$PU_{C1PEA\ or\ PTHF}$	CE[a]–MD[b]–(DBDI–MDI)
$PU_{C2PEA\ or\ PTHF}$	CE–(MD–DBDI)–MDI
$PU_{C3PEA\ or\ PTHF}$	CE–(MD–MDI)–DBDI
$PU_{D\ PEA\ or\ PTHF}$	CE–MD–DBDI
$PU_{M\ PEA\ or\ PTHF}$	CE–MD–MDI

[a]CE – chain extender; [b]MD – macrodiol

The stress-strain testing performed on the initial polymer (I—Table 4.3) or on the same polymer after annealing for 30 min at 150°C, 160°C (A—Table 4.3), 170°C and 180°C. Fig. 4.5 reveals noticeable differences in the mechanical performance of different ordered copolyurethane elastomers.

Strong variations started to occur above 150°C, and as noted from DSC results there was a good case for attributing this to variations in copolymeric order and a reduction in mesophase interactions [60]. As seen in Table 4.3, *as a rule the higher the disorder the better the elastic properties of the elastomer. This fact is characterized by a lower variation of the Young's Modulus during the increase of deformation, and also by a lower residual elongation.*

DBDI has a variable geometry which allows crystallinity to develop, and leads to an increase in mechanical properties (tensile stress, ultimate strength stress and also hardness) whereas the residual elongation is also dramatically increased.

Table 4.3 PU stress-strain data of the materials depicted in Table 2.3 of section 2.3.2. It includes two analogous families of PUs with single or mixtures of diisocyanates. The two PUs families vary in the type of macrodiol (PEA or PTHF)[60]

Recipe	Type of PU structure* and moles		H [Sh°A]	100% Tensile Stress [MPa]	300% Tensile Stress [MPa]	Strength Stress [MPa]	Elongation at Break [%]	Residual Elongation [%] (a) (b)
$PU_{C1\ PEA}$	EG–PEA–(DBDI–MDI)	I	98	11.04	25.3	48.7	450	24 20
		A	96	6.29	13.0	45.6	575	25 —
$PU_{C2\ PEA}$	EG–(PEA–DBDI)–MDI	I	94	9.3	24.7	41.6	375	60 50
$PU_{C3\ PEA}$	EG–(PEA–MDI)–DBDI	I	99	11.3	24.0	30.3	350	50 40
$PU_{C1\ PTHF}$	EG–PTHF–(DBDI–MDI)	I	98	11.0	24.3	24.3	300	5 3
		A	96	8.80	14.9	33.6	450	— 30
$PU_{C2\ PTHF}$	EG–(PTHF–DBDI)–MDI	I	99	10.3	21.3	21.3	300	20 15
		A	95	7.70	13.8	34.2	450	— 80
$PU_{C3\ PTHF}$	EG–(PTHF–MDI)–DBDI	I	100	11.5	—	18.9	250	10 7.5
		A	98	11.3	15.7	33.9	475	— 70
$PU_{D\ PEA}$	EG–PEA–DBDI	I	98	15.8	19.5	33.7	525	155 155
			96	12.8	17.3	27.7	600	240 —
$PU_{M\ PEA}$	EG–PEA–MDI	I	92	8.1	22.0	47.0	425	45 35
$PU_{D\ PTHF}$	EG–PTHF–DBDI	I	96	13.6	14.8	27.6	525	160 160
$PU_{M\ PTHF}$	EG–PTHF–MDI	I	94	8.7	21.4	21.4	300	10 10

H = Hardness [Sh° A]; I = initially; A = after annealing at 160°C for 30 minutes;

(a) = immediate residual elongation

(b) = residual elongation after 1 week

* The structure is shortly represented by the nature of the monomers used in synthesis. A more exact representation of the structure including and the molar ratios is given in the schematic of PU synthesis depicted in Table 2.3, section 2.3.2.

As seen in Table 4.3 from above, *the general improvement in properties with DBDI was retained in polyester PEA type polymers when both diisocyanates are included, especially when reacted together in a random fashion ($PU_{C1\ PEA}$), rather than sequentially, in the prepolymer stage (PU_{C2} and PU_{C3} series).* Under the conditions, the residual elongation returns to that of a more conventional polyurethane (Figure 4.4), eliminating the major inconvenience derived from the high permanent deformability and high crystallinity which is undesirable for an elastomer.

For the PUs family achieved with PTHF (Table 4.3), the best mechanical properties were achieved only after an additional annealing performed by heating the samples at 160°C for 30 minutes. Annealing favored stronger hydrogen bonds and increased the melting points as it determined a greater uniformity to the PUs network.

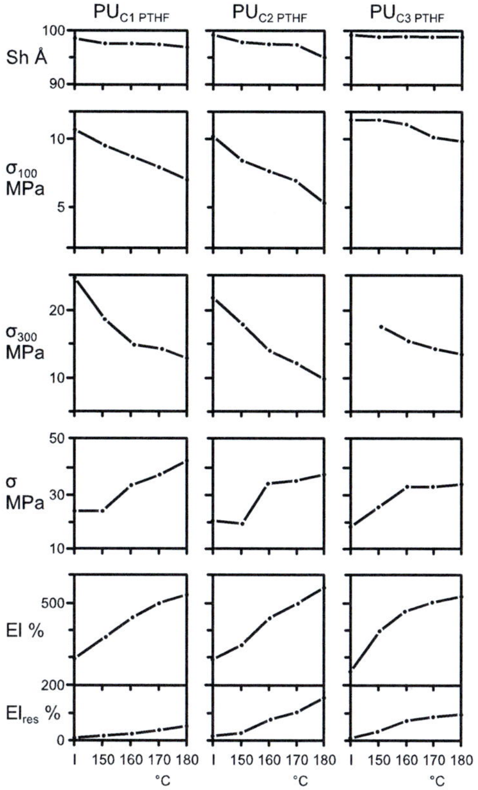

Fig. 4.5 Influence of the annealing on the mechanical properties of PTHF based PUs: initially (I) and after annealing for 39 min at 155, 160, 170 and 180 [60]

4.3 Stress relaxation

The only reason for time is so that everything doesn't happen at once (Albert Einstein)

This is true as well for PUs subjected to stress relaxation ☺. And it applies here too. As is widely known, PUs are not perfectly elastic. Thus, when deformation is held constant, the induced *stress relaxes gradually*.

Figure 4.6 shows example data, for stress relaxation over 100 hours, for a DBDI material based on PEA and BDO (denoted as PU3 in Table 4.4) stretched to 300% elongation. It also reveals that over the range shown there is a linear dependence of stress on the logarithm of time, as expressed by equation (4.1).

$$\sigma = \sigma_0 - a \log t \tag{4.1}$$

where σ_0 and a are constants for a given material. This relation was found to hold for all the materials tested by us [61].

Values of percentage stress relaxation, 10 minutes after straining to 300% elongation at 3.1×10^{-2} s^{-1}, are collected in Table 4.4 for several of the PUs studied,

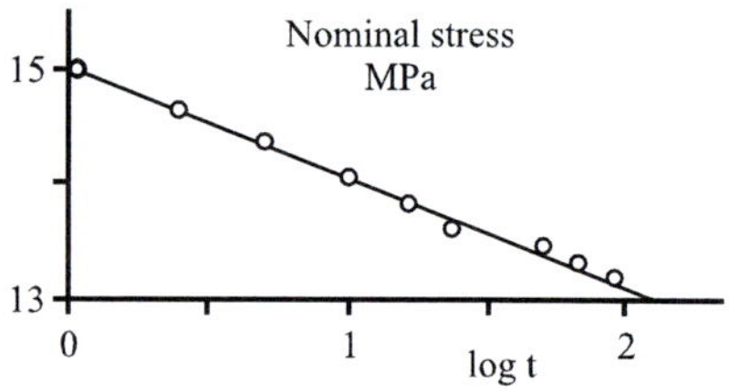

Fig. 4.6 Stress relaxation of PU3 (DBDI-PEA-BDO, I = 110) following 300% elongation at nominal strain-rate 3.1×10^{-2} s^{-1}. Time t is expressed in hours [61]

together with the initial nominal stress on reaching 300% elongation $\sigma_{300\%}$. Also shown, for comparison, are the values of first and second cycle hysteresis E^*_{1H} and E^*_{2H} and the residual elongation after unloading from 300% elongation.

Table 4.4 Percentage stress relaxation in various PUs, 10 minutes after elongation to 300% at a nominal strain rate 3.1×10^{-2} s^{-1} [61]

Polymer	PU structure	$\sigma_{300\%}$ MPa	Stress relaxation %	E^*_{1H} %	E^*_{2H} %	Residual elongation %
PU1	DBDI–PEA–DEG	11.3	21.0	84.4	46.5	20
PU2	DBDI–PTHF–DEG	10.5	19.2	81.1	47.2	75
PU3	DBDI–PEA–BDO	17.2	18.6	83.3	67.7	190
PU4	DBDI–PTHF–BDO	14.0	16.8	86.9	63.0	240
PU5	MDI–PEA–BDO	15.3	16.3	74.7	39.1	30

The three components were mixed in the molar proportions hard segment:macrodiol:chain extender = 4:1:2.64, giving isocyanic index I = 110.

Also shown is the initial stress $\sigma_{300\%}$. For comparison, the table also shows the residual elongation and relative hysteresis in first and second cycles of elongation at the same rate E^*_{1H} and E^*_{2H} [61].

The measured values of percent relaxation over 10 minutes at 300% elongation were found to lie in the range 17-21% in the case of DBDI based PUs, which may be compared with the range 16-24% mentioned by Abouzahr and Wilkes for similar MDI based polyurethane elastomers [282]. As may be seen, the range of stress relaxation for the DBDI and MDI based PUs do not differ greatly. Nevertheless, some interesting differences are visible in Table 4.4. Comparing the results for PU3 and PU5, it may be seen that there is an increase of 14% in stress relaxation when the MDI hard segment is replaced by DBDI, for the same macrodiol/chain extender combination and urethane group concentration. This is consistent with the increase in residual elongation and hysteresis also shown in Table 4.4. This table also shows some influence of changing the macrodiol (10% increase in stress relaxation from PTHF to PEA) or the chain extender (13% increase from BDO to DEG), when DBDI was the HS. But it is interesting to note that these increases do *not* correlate with increases in residual elongation or hysteresis. Other factors are at work here,

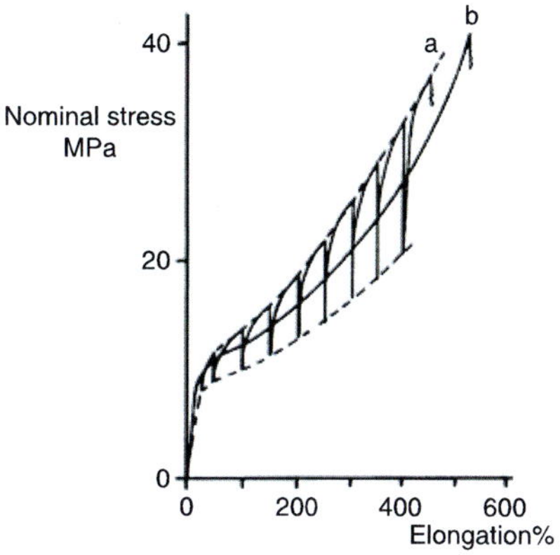

Fig. 4.7 Stress-strain curves for polymer PU3 (DBDI-PEA-BDO) at nominal strain-rate 3.1×10^{-2} s^{-1}: (a) interrupted test, with 10 min. stress relaxation at intervals of 50% elongation; (b) continuous test [61]

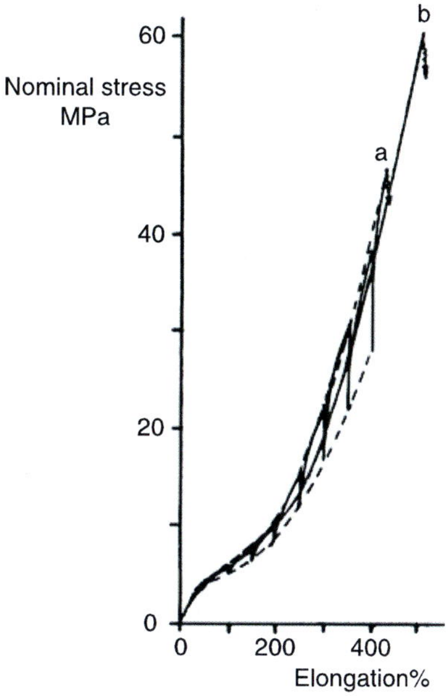

Fig. 4.8 As Fig. 4.7 but for polymer PU5 (MDI-PEA-BDO) [61]

including the fact that use of DEG as chain-extender inhibits crystallization of the DBDI [60, 61].

Further tensile stress relaxation tests were conducted by interrupting constant strain-rate tests (again 3.1×10^{-2} s^{-1}) at intervals of 25% elongation, to observe the relaxation of stress at constant strain over 10 minutes, before resuming extension. Sample results are shown in Fig. 4.7 and 4.8, where materials based on DBDI and MDI may be compared (PU3 and PU5). In both cases, the stress-strain curve for continuous straining is included for comparison. A common pattern of behaviour can be seen. When straining is resumed after stress relaxation for 10 minutes, the stress rises rapidly before continuing a steady increase with strain, resuming approximately the original curve. The fact that the stress-strain curves for continuous straining lie slightly below the corresponding envelope curves for the interrupted tests, is believed to reflect a small degree of adiabatic heating in the continuous tests. The nominal strain-rate employed (3.1×10^{-2} s^{-1}) is at the upper end of the range where isothermal conditions are likely to apply in the testing of an elastomer [283]. Any adiabatic heating is likely to be absent in interrupted tests, where sufficient time was available for heat to dissipate. As may be seen once again, Fig. 4.7 and 4.8 show that, with DBDI as the hard segment, the extent of relaxation is greater than with MDI.

4.4 Cyclic tensile responses. Hysteresis

Magic lives in curves not in angles (Mason Cooley)

People prefer curves. ☺ Why curves? Well, sharp contours might convey a sense of threat (*Bar, M. & Neta, M. – Psychological Science, 2006*). They are associated with negative emotions, whereas curved contours are linked with positive reactions, as symbols of continuity...

Hysteresis is a concept that has become involved in a lot of intellectual fields, from economy to philosophy or from history to science: from the economic hysteresis describing the physics of unemployment, to hysteresis in hand-eye coordination, or from hysteresis of brain-mind duality to engineering hysteresis used to determine the different routes taken by systems when they change in two directions.

All of these (so various) fields have *a common feature: to determine different modifications taken by systems when they change in two directions.*

Hysteresis describes changes that elude our attempts to make predictions of future behaviour. Thus, why not, metaphorically speaking, hysteresis might even be considered as a genuine tool capable to describe people's patterns of thinking, their feelings, the changes and chances in life, in terms of ups and downs given by load/unload cycles. In this scenario, hysteresis may show deterioration or success evidenced by variations in the human being energy loss, acting as a measure of the extent to which people are affected *after* facing various situations. From such a metaphoric perspective, an "elastic" and tolerant perspective of life would certainly lead to a minimum hysteresis. ☺(Not being able to do so, gradual change may suddenly become catastrophic (?)

Hysteresis occurs in many fields of science. Hysteresis is observed in various substances and physical processes. Magnetic hysteresis occurs when a permeable material like soft iron is magnetized by being subjected to an external magnetic field. Thermal hysteresis occurs when the value of a given property of a body depends not only on the body's temperature but also on whether the temperature is rising or falling.

The material damping, or *mechanical hysteresis*, is related with the energy dissipation in the volume of a macro-continuous medium. This phenomenon takes place when a more or less homogeneous volume of material is subjected to cyclic stresses. As a result of inelasticity, a phase shift occurs between strain and stress and in the stress-strain coordinates, the geometrical locus of operational point becomes a loop, known as *hysteresis loop*, which evolves with the number of stress cycles and offer useful information upon the material state.

4.4.1 General considerations on hysteresis and Mullins effect

Hysteresis, which *is characteristic of most polyurethanes*, has been attributed to a number of factors including nonaffine deformation, plastic deformation of the hard domains, irreversible disruption of the microstructure (particularly in PUs with interlocked HS domains), irreversible orientation of the HS domains, and irreversible strain induced SS crystallization. Therefore, often it is very difficult to determine the dominant mechanism responsible for the PUs hysteresis.

PUs display high hysteresis possibly due to irreversible structuring of the morphology and high energy losses caused by the HS orientation. PUs hysteresis and the associated mechanical loss processes have been attributed to the breakdown and reformation of the mesophase morphology during mechanical cycling. HS domains act as crosslinking volumes with high functionality and simultaneously act as fillers, thus raising the elastic modulus. However, in contrast with fillers in conventional rubbers they do not reduce the breaking strain, since being in a material under macroscopic load, they will be partly deformed. Since the domains undergo plastic deformation, the retractive force will decrease. This causes a stress-strain hysteresis and a so-called permanent set, which is in fact a frozen-in deformation of the material [4].

Numerous studies have been reported on the investigation of the PUs and other elastomeric materials hysteresis behaviour [284–293]. As shown, the hysteresis and the associated mechanical loss processes can be partly attributed to the breakdown and reformation of the mesophase morphology during mechanical cycling [284].

Harwood et al [285] noticed that a pure natural rubber and a carbon-black filled natural rubber exhibit a similar softening when both materials are stretched up to the same stress level. According to this result, the Mullins softening was related to a stress level than to a strain level, which is why often the Mullins effect has been referred to as a stress-softening effect [286]. Applying load/unload cycles up to various stretch levels, Diani et al [286] noted that there is no general stretching lower limit to observe some Mullins softening, but this depends on the materials.

Muller-Riederer et al [287], and Bonart et al [288] developed the concept of molecular orientation and inhomogeneous transfer of force throughout the matrix of segmented PUs, which assumed the existence of stretched and unloaded chains due to imperfect morphology. Upon the first stretching some chains bear high local stresses. Since the chains pass through several hard domains, the pre-stretched chains cause pulling of HS out of some domains until the homogenization of the stress transfer, i.e. plastic deformation of the hard domains results. High stresses are necessary to extend prestretched chains, whereas in the repeat run only entropic forces or uncoiling of the soft segments are involved.

Abouzahr and Wilkes investigated the hysteresis behaviour of segmented polyetheric PUs based on MDI/BDO and PTMO of molar mass 2 000 g/mol [176]. They revealed that different mechanisms account for stress softening and showed that particularly, at high HS concentrations, domain disruption dominates starting at low extensions. As shown by Abouzahr et al [176], at high HS concentrations, the domain disruption dominates starting at low elongations. PUs also display high hysteresis possibly due to irreversible structuring of the morphology and high energy losses caused by the HS orientation. As observed by Abouzahr, the materials with distinct domain texture showed the lowest hysteresis while those with no domain texture showed an increasing level of hysteresis with increasing the strain. This was explained by Abouzahr in terms of a morphological model according to which the domains act as effective crosslinks and filler particles that physically crosslink and reinforce the SS segment matrix [176].

PUs exhibit an appreciable change in their mechanical properties resulting from the first extension. This property, observed by Bouasse and Carrière [289] was firstly reported in filled and non-filled rubber-like materials. In order to provide a better understanding of the stress-softening resulting from the Mullins effect, several physical interpretations were proposed, from chain breakage at the interface between the rubber and the fillers, slipping of molecules, rupture of the clusters of fillers, chain disentanglements, to more complex composite structure formation [286]. In the carbon black filled rubbers the explanation of the Mullins phenomenon involved breaking or detaching "prestrained" chains from the surface of filler particles in the first cycle. In the following cycles the detached chains did not support the stress, giving rise to the observed softening bonds during stretching [290].

PUs load/unload cycles were found to be sensitive to the temperature dependent morphologies. Such observations were reported, for example, by Gorce et al [291] for segmented PUs based on BDO, $PTMO_{2000}$, and MDI. The mechanical hysteresis was found to be sensitive to the temperature dependent morphologies of these materials and to the morphological changes induced by the strain level and deformation energy during a deformation cycle.

The mechanical hysteresis *ofpolyether* PUs investigated by Gorge et al was studied as a function of *temperature, percent strain, and deformation energy*. Hysteresis values remained small at low temperatures when the extent of the sample deformation did not disrupt the glassy matrix. This was readily evident at temperatures below the glass transition temperature (T_G) of the polymer where the material did not formally yield. At temperatures above the T_G of the polymer, hysteresis remained small even at substantial strains levels and demonstrated the capabilities of the HS domains to act as physical crosslinks. At high emperatures, the percent hysteresis increased as the hydrogen-bonded HS domains weakened.

Nallicheri et al [292], studied the *influence of crosslinking on the hysteresis* behaviour of poly(urethane-diacetylene) segmented copolymers. They developed a new class of segmented PUs that contain reactive diacetylene groups within their HS. In such materials, the diacetylene groups remain inactive during material formation but can be subsequently reacted in the solid state to produce HS domains covalently link together with fully extended poly(diacety1ene) chains. The conversion of the diacetylene groups to poly(diacety1ene) chains is a topochemical, lattice-controlled process and therefore occurs without significant disruption of the hard domain organization or the morphology of the segmented copolymer. This has made possible to introduce well-defined covalent crosslinks within the hard domains without modification of the molecular or supermolecular organization of the polymer [292].

Nallicheri examined the influence of hard domain cohesion on the hysteresis behaviour. The hard domains of these materials were selectively crosslinked in the solid state by polymerizing the diacetylene groups contained within their HS. For materials with weakly associated HS, it oberved that low levels of crosslinking significantly reduced the amount of hysteresis which they exhibited at low strains. Premature break-up of the hard domains due to weak hard domain cohesion was partially responsible for the high hysteresis values observed at low strains. The energy loss was eliminated by diacetylene crosslinking [292].

The thermodynamics of stress-softening and hysteresis in polybutadiene polyurethanes (PBUs) with different compositions were studied by Godovski et al [85, 294], by using *deformational calorimetry*. The intramolecular energy contribution of the soft blocks was investigated. The PBU based PUs are typical poly block copolymers composed of alternating soft and hard blocks which differ from conventional (polyester) PUs by the chemical structure of the PBU block: it cannot crystallize and form hydrogen bonds. This leads to a high degree of segregation between the hard and soft blocks [279]. The absence of interphase interaction makes PBU a suitable model for studying the deformation behaviour of a polyblock system.

In such materials, deformation causes significant intermolecular changes in PBUs which raise the internal energy. As shown by Godovski [294], irrespective of composition (for 24-79% hard block content), the initial strain (up to 10%), was accompanied by heat absorption and was characterized by bulk elasticity. This indicated that the hard phase in samples was continuous. Further extension, leading to yielding points and necking, was exothermic, the amount of evolved heat being dependent on the hard block content. These results refered to elongations of up to 60%.

As revelead by Godovski, irrespective of temperature, the extension of PBU based materials curves for the first cycles differ from those of subsequent deformations. Similar conclusions were drawn by us for other polyester of polyether PUs derived from MDI, DBDI or mixturews of them [135].

The maximum mechanical hysteresis and residual strain, which were strongly dependent on the hard block content, occurred during the first cycle and the residual elongations and hysteresis decreased with temperature.

The thermodynamic analysis of stress-softening and hysteresis made by Godovski has shown that the softening of PBU was accompanied by considerable energy change [294]. The energy expeditures in urethane hard blocks could be caused by rearrangements of hydrogen bonds, and by the disruption and rearrangement of the hard domains structures under strain. They are responsible for the stress-softening and hysteresis losses. Recently (2010), similar conclusions were drawn by us for other *polyester and polyether* PUs derived from MDI and DBDI [135].

Overall, the study made by Godovski et al showed that the PUs softening and hysteresis were accompanied by energy absorption reflecting intermolecular change in the hard domains. An increased hard block content resulted in increased energy changes, whereas increasing the temperature resulted in a reverse effect.

A series of studies was also made by us, of the PUs cyclic stress-strain response. The range of structures achieved by us was widened by inclusion of DBDI, as a diisocyanate with a very strong tendency to packing due to its constitutional mobility. A systematic investigation (as shown in Table 4.5), was made of the effects of varying HS and SS chemistry, crosslinking and preparation procedures, on the hysteresis behaviour and Mullins effect of melt-cast polyurethane elastomers, The materials were prepared with no excess of NCO groups, giving isocyanate index I = 100 [175].

Table 4.5 Compositions of the family of polyurethane elastomers PU1-PU14 prepared and studied in this work [175]

		PEA	PTHF
MDI	EG	PU1	PU7
MDI	DEG	PU2	PU8
MDI	BG	PU3	PU12
DBDI	EG	PU4	PU9
DBDI	DEG	PU5	PU10
DBDI	BG	PU6	PU11

The major structural feature which we have explored has been the relationship between the nature of the HS (crystallizing or not) and that of the SS. *The role of the HS structure was investigated by comparing mainly hard segments with MDI and DBDI.*

The range of polymers was designed to reveal the roles of choice of macrodiol and of chain extender in determining the performance as an elastomer. The SS were the polyether PTHF and polyester PEA or PBA, of molar mass 2 000 g/mol. The chain extenders used in the synthesis were EG, BG or DEG. Mechanical tests included load-unload cycles at constant rate of extension, with measurement of hysteresis and strain recovery, and stress relaxation tests.

Example curves of nominal stress versus nominal strain for three load/unload cycles between zero stress and a nominal strain of 3 are given in Fig. 4.9 and 4.10 below for two polymers based on DBDI and MDI. As seen, notable features were the pronounced hysteresis, and unrecovered strain.

The experiments were carried out at constant rate of extension. Irrespective of the HS crystallizing or not, PUs exhibited an appreciable change in their mechanical properties resulting from the first extension The presence of DBDI hard segments instead of MDI led systematically to increases in: the input strain energy to a given elongation, hysteresis and residual strain under cyclic loading, and stress relaxation.

Polymers based on DBDI, displayed higher stiffness and strength than the conventional MDI based PU. Both features of the response were attributed to differences in hard phase plastic flow stress, resulting from crystallinity in the DBDI phase. Lower strain recovery and strain energy recovery on cycling (hysteresis) were observed in the case of PU derived from DBDI. The material based on DBDI, displayed higher Young Modulus values and higher residual elongation than did the conventional MDI based PU. In such materials the stiffness, hysteresis, residual elongation and stress relaxation were all increased relative to the corresponding MDI based PU,

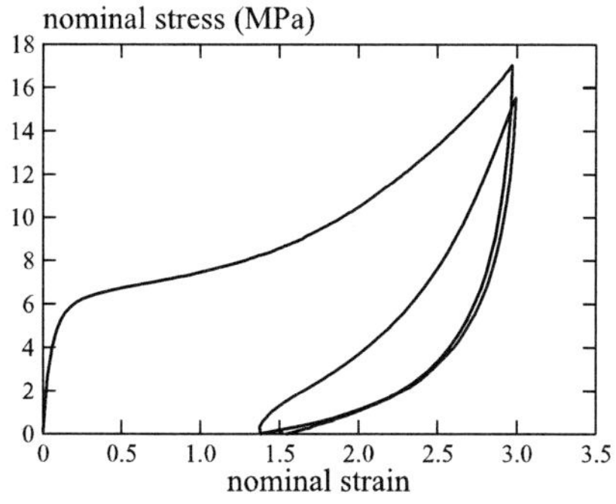

Fig. 4.9 Tensile load/unload cycles for a material based on MDI:PU1(EG-PEA-MDI) at strain rate 0.03 s^{-1} [69]

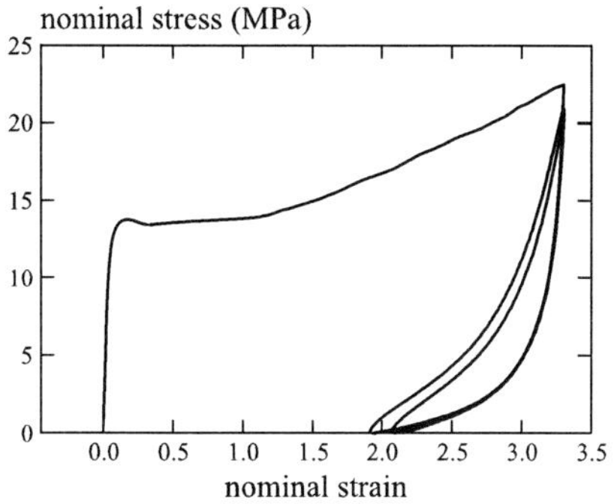

Fig. 4.10 Tensile load/unload cycles for a material based on DBDI:PU4(EG-PEA-DBDI) at strain rate 0.03 s^{-1}[69]

Comparing successive load-unload cycles we found that irreversible changes to the stress-strain response were confined essentially to the first loading cycle. In subsequent cycles the load-unload stress-strain curves remained almost unchanged. For example, the recovered energies in first and second cycles were equal in most cases, to a good approximation. Similarly, increasing the time interval between the end of one unloading cycle and the start of the next, from a few seconds to several hours (up to 24 hours), had no sensible effect on the measured recoverable strain energy. This was in contrast to the input energy, which was clearly dependent on the nature of the HS (crystallizing or not).

First loading to a given deformation was associated with higher stiffness and hysteresis than is typical for a homogeneous crosslinked elastomer. But the first loading caused damage to the microstructure, such that *subsequent* loading inside the envelope of previous deformations, was associated with lower stiffness and hysteresis. Consequently, re-loading followed a stress-strain path closer to the previous unloading path than to the first loading path: i.e. these materials typically exhibited the well-known Mullins effect [283, 295–297].

4.4.1.1 Cycling to fixed extension. Mullins effect

In the first phase of the study, the materials were cycled between an extension of 3, and zero load, for 3 cycles. An example curve of nominal stress versus nominal strain is shown in Fig. 4.11.

Notable features were the pronounced hysteresis, unrecovered strain and *Mullins effect (whereby re-loading follows a stress-strain path closer to the unloading path than the original loading path)*. From curves such as these we calculated several quantifiers of the inelasticity. Consider the first cycle for material PU1, shown above. It defines three zones A, B and C in the stress-strain diagram.

In this study we denoted by E_1 the energy density expended on first loading to a nominal strain of 3, thus from Fig. 4.11 [69, 175]:

$$E_1 = E_A + E_B + E_C \tag{4.2}$$

We indicated all relative quantities by asterisks. Thus we defined the relative first cycle residual strain, and relative first cycle hysteresis energy dissipation by

$$\varepsilon_r^* = \frac{\varepsilon_r}{\varepsilon_{max}}, \quad E_{1H}^* = \frac{E_A + E_B}{E_1} \tag{4.3}$$

An additional point of interest in cyclic straining was the Mullins effect. It was helpful to have a quantitative measure of how close the materials come to exhibiting an "ideal" Mullins response, as defined for example by Ogden et al (1999) [175, 296]. To this end we defined a "Mullins factor" M, such that an ideal Mullins response was characterized by $M = 1$ [175]:

$$M = 1 - \frac{E_B}{E_1} \tag{4.4}$$

As is usual for PU elastomers, all the materials showed substantial deviations from hyperelastic response, indicating the presence of relaxation processes. The

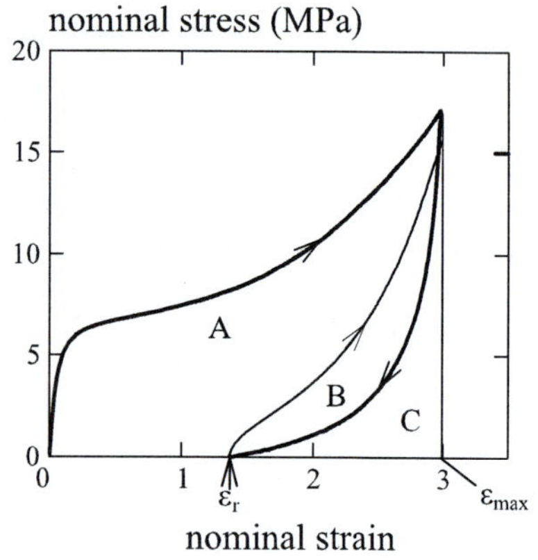

Fig. 4.11 Loading-unloading-reloading cycle for material PU1, with definitions of zones A, B, C of the diagram, maximum nominal strain ε_{max} and residual nominal strain ε_r [69, 175]

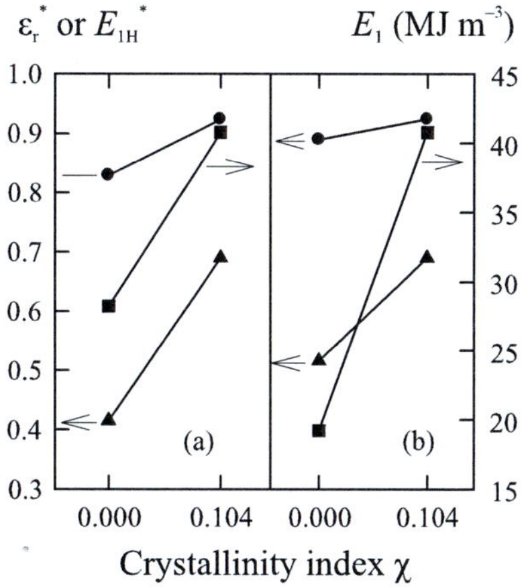

Fig. 4.12 Graph showing the effects of varying hard phase crystallinity on two measures of inelasticity — ε_r^* (▲) and E_{1H}^* (•) — and the first cycle energy input E_1 ( ). The data are for materials PU3 and PU6 (a), and PU5 and PU6 (b). The lines have no significance except to guide the eye [175].

hard domains clearly did not act simply as rigid crosslinks for the soft domains. Inelasticity measures ε_r^* and E_{1H}^* laid in the ranges 0.23–0.75, and 0.74–0.94, respectively. Similarly, all the materials showed a response on re-loading after the first cycle that had significant Mullins-type character, with M in the range 0.79–0.94.

With respect to varying PU composition, the dominant result from these tests was the differences between polymers based on the two hard segments. *For all combinations of chain extenders and macrodiols, residual strain ε_r^* and hysteresis energy dissipation E_{1H}^*, were highest for hard segments of DBDI than of MDI.*

Results showing the effect of varying crystallinity are given in Fig. 4.12. It compares results for the amorphous ($\chi = 0$) materials PU3 (a) and PU5 (b), with PU6 which has partially crystalline hard domains ($\chi = 0.104$).

Thus 4.12(a) shows the effect of introducing HS crystallinity by change of HS, while 4.12(b) shows the effect of introducing crystallinity by changing the chain extender. In both cases the effect of crystallinity is the same.

Now consider the role of phase separation. In this study this was quantified in terms of the SAXS peak area. A striking feature of the results was that all materials with PTHF as SS showed much higher intensity in SAXS than the corresponding polymers with the other soft segments. Moreover, in mechanical tests, polymers with PTHF soft segment showed reduced inelasticity, as expressed via ε_r^* and E_{1H}^*, and increased first cycle energy E_1. These effects are shown below in Fig. 4.13, which compares results for PU2 (MDI+DEG+PEA) with PU8 (MDI+DEG+PTHF) (a), and PU5 (DBDI+DEG+PEA) with PU10 (DBDI+DEG+PTHF) (b). Thus 4.13(a) shows the effect of greater phase separation in a MDI-based system and 4.13(b) shows the corresponding effect in a DBDI-based (but non-crystalline) system.

Also included in Fig. 4.13 is the Mullins factor M. It is interesting to note that in 4.13(a) it increases slightly, while in 4.13(b) it decreases slightly when the SAXS intensity increases. This is evidence for the fact that the Mullins effect, although

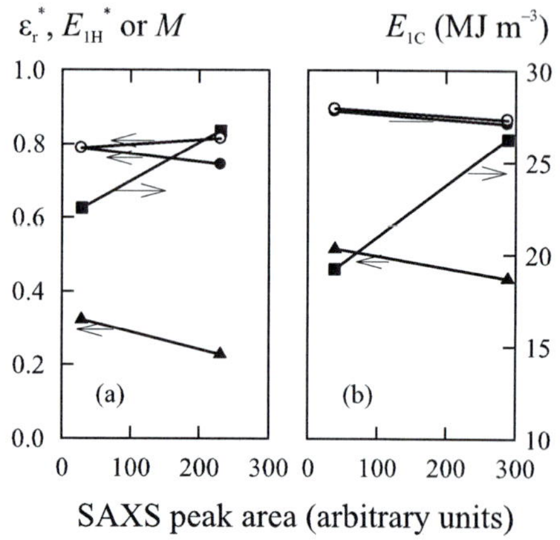

Fig. 4.13 Variation of inelasticity measures ε_r^* (▲) and E_{1H}^* (•), and the first cycle energy input E_1 ( ), with degree of phase separation as expressed by the integrated SAXS peak intensity. Results shown are for PU2 and PU8 (a), and PU5 and PU10 (b). Also shown is the Mullins factor in each case (○). The lines have no significance except to guide the eye [175].

usually associated with the presence of hard filler particles, is not sensitive only to the degree of phase separation in the present materials.

Since the Mullins effect appeared to be special to elastomers, it was of interest to us, to observe how its appearance relates to more usual manifestations of inelasticity such as E_{1H}^*. Data for M and E_{1H}^* are crossplotted in Fig. 4.14, therefore, for all the materials tested cyclically to an extension of 3 [175].

Fig. 4.14 revealed a remarkable, and unexpected, correlation between these two parameters, for the range of materials studied. All except two of the materials appear to lie on the straight line shown, which is simply

$$M = E_{1H}^* \tag{4.5}$$

The reader will note the two outlying points clearly did not lie on the same line, having exaggerated values of M for their values of E_{1H}^*. Interestingly, they are for

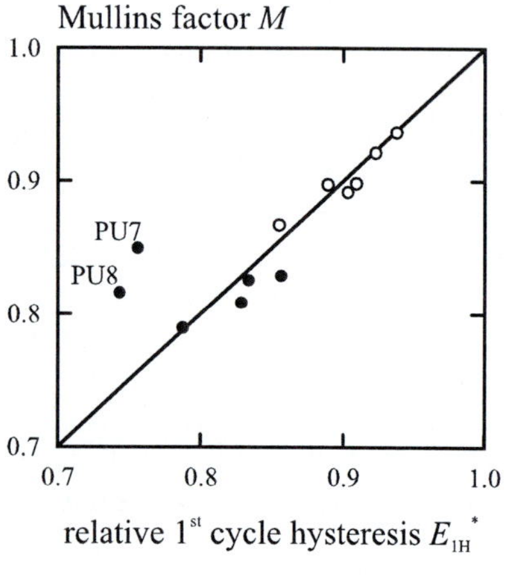

Fig. 4.14 Results from cyclic extension to an extension of 3, plotted as M versus E_{1H}^*. Filled symbols—data for MDI-based polymers; open symbols—data for DBDI-based polymers [175]

materials PU7 and PU8: the two MDI-based polymers showing SAXS peak areas much higher than the others, indicating a higher degree of phase separation. The evidence of Fig. 4.14 is that, at least in the materials studied here, the Mullins effect is linked to the mechanism of hysteresis, but is accentuated by a high degree of phase separation.

4.4.1.2 Cycling to increasing extension

Similar observations were made by us when PUs were cycled to increasing extension [175]. A further series of mechanical tests was designed to compare inelasticity of the various materials under cycling to increasing extension. Thus each specimen was subject to a continuous pseudo-cyclic straining sequence, such that on the nth cycle it was cycled between a maximum nominal strain $\varepsilon_{max} = n$ and zero load. The magnitude of extension rate was the same as in the previous experiments.

In view of the apparent partial correlation between M and E^*_{1H} revealed by Fig. 4.14, it has been of interest to represent the data from the second series of experiments in the same manner. This is done in Fig. 4.15.

Again the Mullins factor appeared to be related to the hysteresis in a remarkably similar manner for all materials. With increasing maximum extension the points moved closer to the line of equation (4.5). We noted that the two materials PU7 and PU8 that are outliers in Fig. 4.15 (where $\varepsilon_{max} = 3$) are here moving in the same sense, but by $n = 3$ they have not yet reached the line.

These experiments revealed clearly that inelastic features in the constitutive response of PUs are sensitive to microstructural detail on length scales of order nm. *One prominent result was the strong influence that hard-domain crystallinity exerts on inelasticity of the elastomers. Thus, there was a clear indication from data such as those of Fig. 4.15 that the physical origin of the flow stress must be relative displacement of the hydrogen-bonded HS.* Note that in PU4 and PU5 hard and soft

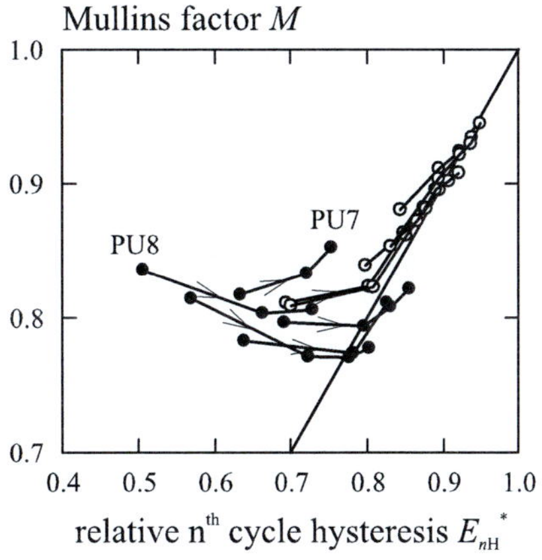

Fig. 4.15 Results from cycling to extension n on the nth cycle: M versus E^*_{1H}. Filled symbols—data for MDI-based polymers; open symbols—data for DBDI-based polymers. Arrows show the direction of increasing n [175]

segments are the same (DBDI and PEA respectively), the only difference being the chain extender that induces differing levels of crystallinity. This result is consistent with previous observations on similar materials with I = 110 [60, 61].

A second clear result from the present work is the influence of phase separation on the constitutive response. Evidence for some phase separation was seen in SAXS for all the materials studied. But in the PTHF-based materials it was particularly pronounced (as evidenced by SAXS peak intensity). Inelasticity as measured by hysteresis and unrecovered strain was reduced by increased phase separation.

By doing this study, a quantitative correlation was found by us, between the magnitude of the Mullins effect and the fractional energy dissipation by hysteresis under cyclic straining, giving a common relation that was approached by all the materials studied. Stress relaxation in interrupted tests also increased with hysteretic energy loss [175].

4.4.1.3 Influence of strain amplitude on Mullins effect and hysteresis energy loss

A more pronounced increase of the hysteresis number [69, 175] with increasing the strain amplitude was found in the case of the materials derived from DBDI as compared to materials based on MDI. The hysteresis number ranged between 0.56–0.75 (for the MDI type polymer), and between 0.93–0.99 (for the DBDI based materials) as seen in Fig. 4.16. Similar observations were made by us, with regard to the Mullins number, which increased with increasing the strain amplitude [69, 175]. Examples are shown in Fig. 4.17-4.19. As seen the Mulllins number increased with increasing the strain amplitude, irrespective of the type of chain extender (DEG or BG), or type of diisocyanate (MDI or DBDI).

Note that for the same diisocyanate, (DBDI or MDI), the Mullins response was not significantly influenced by the adopted polyaddition procedure, i.e. the prepolymer or the one-shot route.

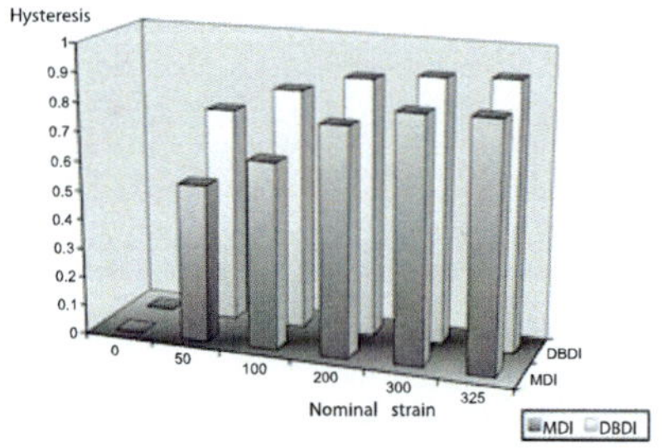

Fig. 4.16 Variation of hysteresis number with increasing the strain amplitude from 0% to 300%, for two PTHF-EG materials selected from Table 4.5: PU7 (MDI) and PU9 (DBDI)

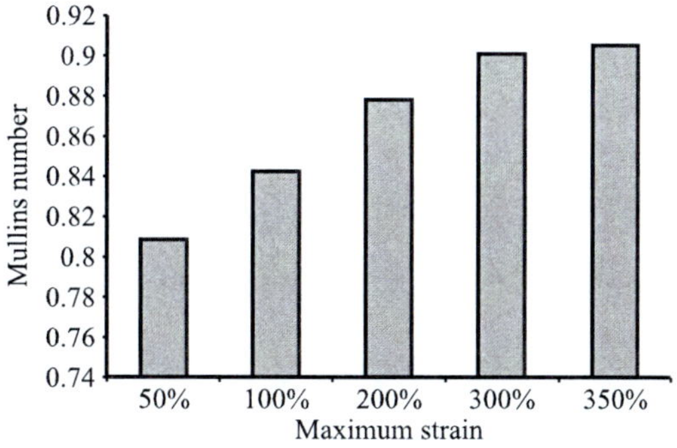

Fig. 4.17 Mullins number as a function of strain amplitude (strain rate = 0.03 s^{-1}) for PU10 (Table 4.5), with the structure DBDI:DEG:PTHF

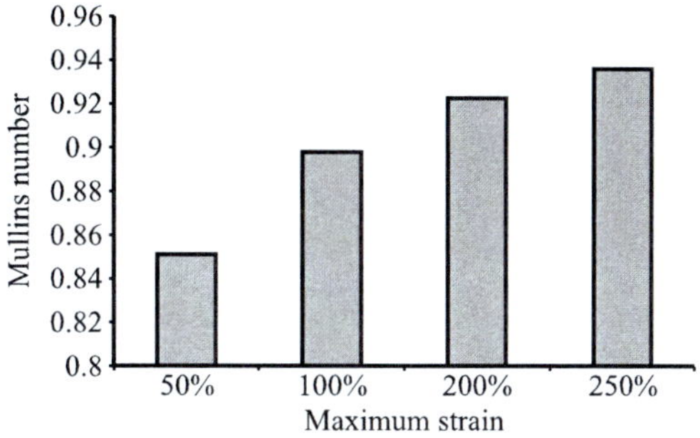

Fig. 4.18 Same but for PU11 with the structure DBDI:BG:PTHF

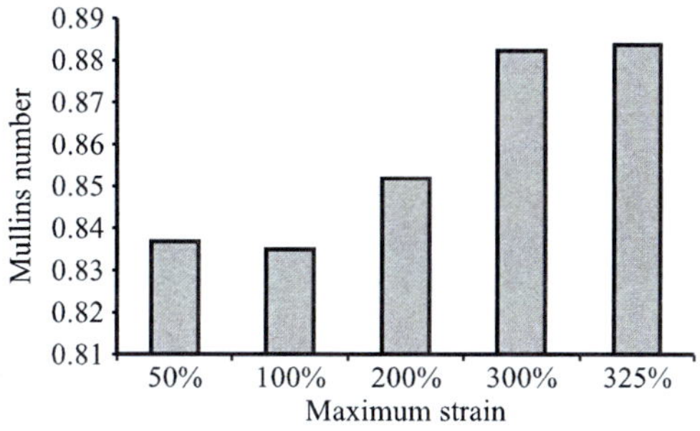

Fig. 4.19 Same but for a PU12 with the structure MDI:BG:PTHF

4.4.1.4 Influence of strain rate on Mullins effect and hysteresis energy loss

Neither M nor E_{1H} were influenced significantly by variation of strain rate, irrespective of the choice of combinations between diisocyanates, chain extenders and macrodiols. Shown in Tables 4.6 and 4.7 are examples data of M and E^*_{1H} with increasing the strain rate for material PU11 of Table 4.5.

Neither M nor E_{1H} were significantly influenced by variation of the type of SS macrodiol, PEA or PTHF. For the DBDI based polymers, the highest E_{1H} values were found in the case of polymers based on the couples DBDI-EG and DBDI-BG, while the smallest hysteresis values were observed for materials with DBDI and chain extended with DEG that inhibits crystallization [61].

Table 4.6 Mullins number (M) data at 300% maximum strain tests conducted at variable strain rate for PU11 (DBDI:BG:PTHF)

Strain rate	Maximum nominal strain, (300%)	Mullins number (M)
0.05	300	0.9212
0.03	300	0.9202
0.01	300	0.9199
0.001	300	0.9287
0.0003	300	0.9352

Table 4.7 Hysteresis energy dissipation (E_{1H}) data at 300% maximum strain tests conducted at variable strain rate for PU11 (DBDI:BG:PTHF)

Strain rate	Maximum nominal strain, (300%)	Hysteresis energy dissipation, (E_{1H})
0.05	300	0.9000
0.03	300	0.8897
0.01	300	0.8959
0.001	300	0.8993

4.4.2 Cycling tensile response with varying the hard segment content

As shown by Abouzahr and Wilkes [176] in the case of MDI based PUs extended with BDO, morphological changes take place as the HS fraction is increased. The texture changes from that in which little such domain content exists at low HS levels (15%), to that in which the polymer has an interlocking domain morphology at high HS content (35 and 45%). Preferable elastomeric properties (low hysteresis, high extension) can be obtained when isolated HS domains exist (25% HS).

Hysteresis behaviour revealed that different mechanisms account for the stress softening in these PUs. At high HS fractions, the domain disruption dominates starting at low elongations. The materials also display high hysteresis which is due to irreversible structuring of the morphology and high energy losses caused by the orientation of the HS.

Materials that possess rather distinct domain textures show the lowest hysteresis characteristics while those with no domain texture show an increasing level of hysteresis with strain. This was explained by Abouzahr [176] in terms of a morphological model according to which the domains act as effective crosslinks and filler particles that physically crosslink and reinforce the SS segment matrix.

As shown by Abouzahr, the kinetics of domain formation was found to be dependent on the driving force for segment segregation and the viscoelastic properties of the material.

Godovski et al [294, 298, 299] studied the stress softening and mechanical hysteresis of a series segmented polybutadiene polyurethanes (PBU). They mainly differed from conventional polyester polyurethanes by the chemical structure of the soft polybutadiene block which cannot crystallize and form hydrogen bonds. This lead to a high degree of segregation between the hard and soft blocks. Therefore the absence of interphase interaction makes PBU a suitable model for studying the deformation behaviour of such polyblock systems. Isocyanate MDI with a triazine-containing diol was used as the hard blocks [298]. *The hard block content was varied from 24% to 63%.* For all samples, irrespective of temperature, the load curves for the first cycles differed from those of subsequent deformations. The maximum mechanical hysteresis and residual strain, which were strongly dependent on the hard block content, occurred during the first cycle. The residual elongations and hysteresis decreased with temperature. The hysteresis losses during the second and subsequent cycles were significantly lower than that during the first one. After five load/unload cycles the losses were small and were less than 30%. Similar conclusions were reported by us for materials based on crystallizing hard segments [60, 61].

Russo and Thomas studied a series of PUs *with increasing percent of hard segment by mechanical testing at low strain rate* [300]. They found that an increase in the HS content results in an increase in the initial modulus, and of ultimate strength, and also determines a decrease of the elongation to break.

An interconnected morphology developing upon increasing the HS content has been also evidenced by use of AFM. At *low HS contents*, the deformation mechanism involves extension of the SS chains. The HS are mixed within the continuous domain without significant aggregation so alignment and break-up of hard domain structure is not in operation, resulting in lower levels of hysteresis. As the *HS content is increased*, elongation of the more crystalline, interconnected hard domains produces plastic deformation and disruption of the hydrogen-bonded and/or crystalline hard block structure, which is reflected in increasing amounts of hysteresis [56, 176]. This is also consistent with our recent observations made on PUs based on HS crystallizing or non-crystallizing [135].

Qi and Boyce conducted a systematic experimental study on the large strain stress-strain behaviour of polyetheric segmented PUs based on the diisocyanate MDI and BDO [301]. The PUs composition was changed from sample to sample to provide *different HS ratios*. Materials were examined by uniaxial compression tests in both the quasi-static and *high strain rate regimes, from* 10^3–10^4 s^{-1}. The highly non-linear large observed stress-strain behaviour demonstrated a strong hysteresis and rate dependent character. Cyclic softening was observed for the materials with crystalline HS content and was absent for the material with an amorphous well-dispersed HS content [302].

Similar conclusions were drawn by us [185]. As reported, the strain energy input increased with increasing the concentration of the HS urethane groups, reflecting the increase in concentration of hydrogen bonds. In contrast, the fractional recovered energy decreased, revealing the fact that the increase in the strain energy input is associated with increased plastic dissipation of energy, resulting from stress-induced breakdown of the hydrogen bonds.

Another example is a recent study which we made on materials obtained with the diisocyanates MDI (non-crystallizing) or DBDI (crystallizing), where we aimed to investigate how aspects of the mechanical responses vary with the increase of the HS percent, from 20% to 80%. A series of 10 materials (as shown in Tables 4.8 and 4.9) based on the chain extenders DEG or EG was investigated. *DEG was used to follow the PUs mechanical response with increasing the HS fraction whereas the influence of the degree of crosslinking (isocyanic index 100 or 110) was followed for EG based materials of ca 40% HS fraction.* The soft segment PTMO of molar mass 2 000 g/mol was used in the synthesis of all materials.

Table 4.8 Compositions of the family of thermoplastic polyurethane elastomers [185]

DI	CE	PTHF
MDI	EG	PU1
MDI	DEG	PU2
DBDI	EG	PU3
DBDI	DEG	PU4

Table 4.9 Molar compositions of the PU1–6 materials (with isocyanic index I = 100 or 110) [185]

PU material	PU1	PU2	PU3	PU4	PU5	PU6
Moles of polyol	1	1	1	1	1	1
Moles of MDI	4	4	0	0	0	4
Moles of DBDI	0	0	4	4	4	0
Moles of EG	3	0	3	0	2.64	2.64
Moles of DEG	0	3	0	3	0	0

(A) To produce *thermoplastic PUs of variable HS content (I = 100)*, the chain extender DEG was used in the syntheses. Obviously, the molar composition was changed from sample to sample in order to provide different HS ratios. The HS fractions ranged between 20 and 80%, as shown in Tables 4.10 and 4.11. The synthesis steps were performed as described elsewhere [185].

The tests included load-unload cycles to a nominal strain of 3, at constant rate of extension, with measurement of hysteresis and strain recovery. They were performed

Table 4.10 Compositions of the family of DBDI based polyurethane elastomers (D), with increasing the hard segment content from 20% to 80% [185]

PU material	D20	D30	D40	D50	D80
Moles of ($PTHF_{2000}$)	1	1	1	1	1
Moles of DEG	0.5	1.27	2.03	2.78	5.06
Moles of MDI	0	0	0	0	0
Moles of DBDI	1.5	2.27	3.03	3.78	6.06

Table 4.11 Compositions of the family of MDI based polyurethane elastomers (M), with increasing the hard segment content from 20% to 80% [185]

PU material	M20	M30	M40	M50	M80
Moles of $PTHF_{2000}$	1	1	1	1	1
Moles of DEG	0.6	1.4	2.2	3	5.4
Moles of MDI	1.6	2.4	3.2	4	6.4
Moles of DBDI	0	0	0	0	0

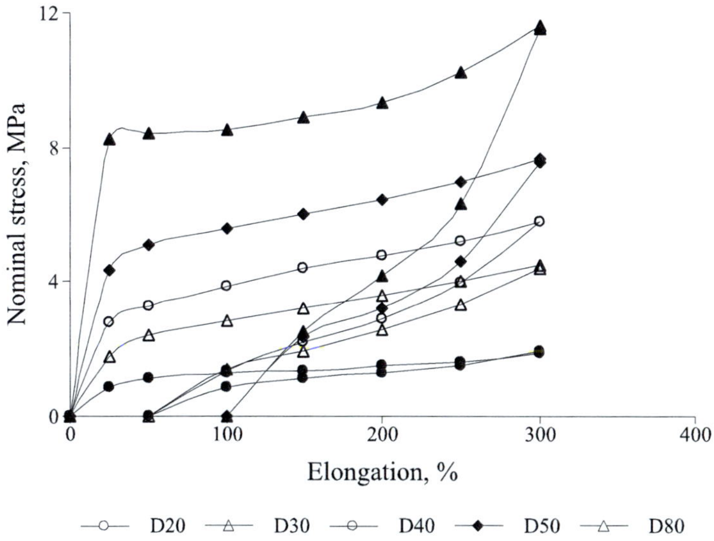

Fig. 4.20 Tensile load/unload cycles for material based on DBDI (D): (DEG-PTHF-DBDI, I = 100), at strain rate 0.03 s^{-1}, as the HS content was increased from 20% to 80% [185]

on two systems of materials with analogous structures, where only the type of diisocyanate (MDI or DBDI) differed (Fig. 4.20 and 4.21).

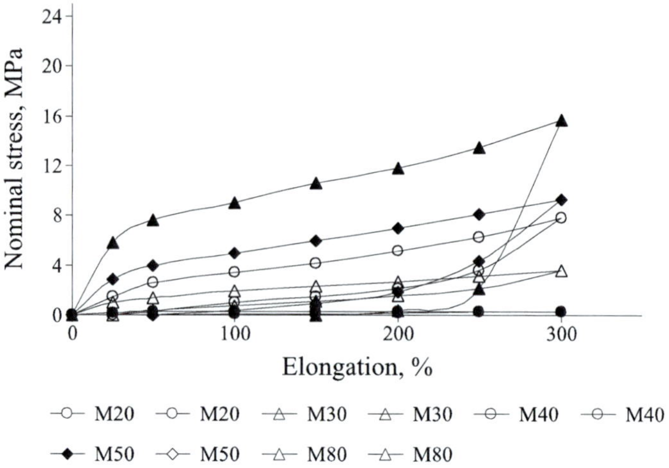

Fig. 4.21 Tensile load/unload cycles for a thermoplastic material based on MDI (M): (DEG-PTHF-MDI, I = 100), at strain rate 0.03 s^{-1}, as the hard segment percent was increased from 20% to 80% [185]

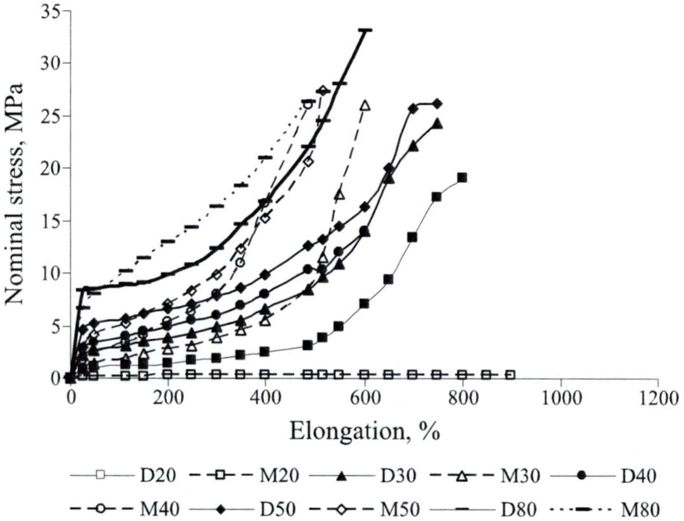

Fig. 4.22 Comparison of the stress-strain curves for polymers based on DBDI (D) and MDI (M) with HS increasing percentage from 20% to 80%, during loading at nominal strain-rate 0.03 s^{-1} [185]

Also, shown in Fig 4.22 is the influence of increasing the HS percentage from 20 to 80% on the tensile stress-strain data of PUs based on the chain extender DEG.

Increasing the HS percentage led systematically to an increase in the input strain energy elongation [185]. The materials based on DBDI hard segments displayed higher stiffness and strength than did the conventional MDI based PUs. As seen in Fig. 4.20 and 4.21 the recoverability of the MDI based materials was higher than that of the DBDI based PUs. Such features of the response were attributed to differences in hard phase plastic flow stress, resulting from crystallinity in the DBDI phase. As the HS percentage increased from 20 to 80%, a lower strain recovery and strain energy recovery on cycling were observed in the case of PUs derived from dibenzyl structures. When comparing successive load-unload cycles, it was found that irreversible changes to the stress-strain response were confined essentially to the first loading cycle. In subsequent cycles the load-unload stress-strain curves remained almost unchanged. The Young modulus increased proportionally with increasing the HS content and was observed to be higher for materials obtained with DBDI [185]. In the first deformation steps the materials with HS based on DBDI displayed a higher strength, followed by values of strains which required smaller stress increments. In contrast, similar materials but based on the MDI-DEG couple, where the HS crystallinity was practically absent, exhibited proportionality between the increase of nominal stress and stretching degree [185].

Overall, both for the MDI and DBDI materials, the hysteresis percentage increased with increasing the HS content, which is a direct reflection of the two-phase morphology displayed in segmented copolymers [176, 303].

(B) In the second part of the study we also aimed to investigate how aspects of the mechanical responses of selected PUs vary with the value of the isocyanate index (I). Two series of similar *PUs were investigated, achieved with excess (I = 110) and with no excess (I = 100) of isocyanate NCO groups.* The HS fraction was kept constant at ca 40%.The materials studied are listed in Tables 4.9 and 4.12.

The range of polymers was designed to reveal *the role of the crosslinking degree* on the PUs mechanical response. The stress-strain data of the two series of materials were compared. The stress-strain data are collected in Table 4.12 for four of the materials studied, chain extended with EG and based on DBDI or MDI respectively.

Table 4.12 Comparison between the stress-strain data of similar PUs achieved as crosslinked and thermoplastic materials, based on the chain extender EG, and diisocyanates DBDI or MDI[185]

PU	Type of PU structure* and moles	Isocyanic index %	Tensile stress 100% [MPa]	Tensile stress 300% [MPa]	Strength stress [MPa]	Elongation at break (%)	Residual elongation (%)
PU3	EG-PTHF-DBDI 3: 1: 4	100	9.7	13.3	34.4	550	189
PU5	EG-PTHF-DBDI 2.64: 1: 4	110	13.6	14.8	35.6	525	160
PU1	EG-PTHF-MDI 3: 1: 4	100	4.3	8.7	29.0	400	21
PU6	EG-PTHF-MDI 2.64: 1: 4	110	8.7	10.4	31.4	300	10

* The structure is shortly represented by the nature and the molar ratios of the monomers used in synthesis.

As seen, the stress-strain data of the postcured PUs obtained with an isocyanic index I = 110 were somewhat higher than those of similar but thermoplastic materials when no excess of isocyanate was used. However, in the case of PU5 (I = 110) as compared to its analogous PU3 (I = 100) the crosslinks, if any, were very small, for example, PU3 has HS% = 38.33% whereas PU5 has HS% = 37.91% [185]. The crosslinks were in fact the inevitable interchain physical association by hydrogen bonds between urethane groups.

The thermoplastic character of PUs obtained using I = 100 was mainly evidenced by an increase of the residual elongations. Similar conclusions were drawn in the case of materials achieved with the chain extender DEG which inhibited crystallization, but where the residual elongations were significantly lower irrespective of the type of diisocyanate employed in the material synthesis.

Regardless the crosslinking degree, in the case of polymers based on EG the 300% modulus, *tensile strength and residual elongation values were significantly higher for the materials based on DBDI than for the PUs derived from MDI.* This was due to the HS having a higher flow stress in the presence of DBDI relative to MDI, associated with increased hydrogen bonding in DBDI-based polymers,which was enhanced by HS crystallinity.

4.4.3 Cycling tensile responses of polyurethane elastomers achieved with no excess of isocyanate NCO groups

The overall PUs mechanical properties depend upon the relative volume fractions of soft and hard phases, and on the intrinsic properties of each of the phases. These in turn depend on details of molecular packing of the constituents within the phases, including the density of hydrogen bonds. Crystallinity has been observed in the soft phase when the macrodiol chain is long enough [4, 43, 126], and it is also sometimes present in the hard phase. The latter is usually limited to only a few percent for most HS structures when solidified from the melt, but as previousy shown, there is one particular diisocyanate, DBDI that, in the presence of a suitable chain extenders, gives rise to significant degrees of crystallinity [60, 61, 67, 127, 175, 304, 305] and this is included in the present work.

There have been several previous studies of the effects of varying HS fraction on the mechanical properties of thermoplastic PUs. The purpose of our present study has been to investigate the role of other important structural features, as summarized above. This was achieved by us, by a systematic variation of the three chemical constituents: diisocyanate, macrodiol and chain extender. Two diisocyanates were considered: the frequently employed MDI, and its close relation DBDI, that is of special interest because of its tendency to crystallize on cooling from the melt in the presence of some chain extenders [60, 61, 67, 127, 135]. Three macrodiols were considered: PEA, PTHF PBA. These were chosen because polyethers such as PTHF are well-known to promote phase-separation from the diisocyanates, while polyesters such as PEA and PBA have a greater affinity for the diisocyanate through hydrogen bonding to their ester groups and hence are more miscible with the diisocyanate, and phase segregation is expected to be less pronounced [4, 263]. Three chain extenders were considered: ethylene glycol (EG), di-ethylene glycol (DEG) and 1,4 butylene glycol (BG). These were selected so as to include two chain extenders that promote crystallinity with DBDI (EG and BG), and DEG that has been shown previously to inhibit crystallinity in DBDI [61].

Results are reported here for a set of 14 model thermoplastic materials prepared with differing combinations of diisocyanates, macrodiols and chain extenders, but all having approximately *the same mass fraction of* HS *(ca 40%)*, and all having SS of the same molar mass (2 000 g/mol). As expected, the differing chemical compositions of the thermoplastic PUs caused variation in the degree of phase separation and crystallinity.

Note that the series of materials investigated in this study is previously characterized in Chapter 2, by small-angle X-ray scattering (SAXS) and wide-angle X-ray scattering (WAXS). SAXS and WAXS results of these materials are previously detailed in section 2.3.1, Table 2.1. Example SAXS patterns are shown in Fig. 2.15, and WAXS patterns are illustrated in Fig. 2.15, also in section 2.3.1.

The materials were subject to a series of cyclic uniaxial tensile tests at room temperature and ambient humidity, designed to characterize features of their constitutive response relevant to their performance as thermoplastic elastomers, especially focusing on their stiffness and their deviations from purely elastic behaviour.

The three components were always mixed in the molar proportions hard segment diisocyanate (DI):macrodiol(MD):chain extender(CE) = 4:1:3, giving hard segment mass fractions in the region of 40%, and isocyanic index I = 100. Synthesis was carried out by the pre-polymer route described previously by Prisacariu et al [61, 135]. The DI and MD components were reacted together with vigorous mixing under vacuum at 100°C, to give prepolymer consisting of MD terminated at each end by DI. This was then thoroughly mixed with the CE at 90°C, and cast into closed sheet moulds for curing at 110°C over 24 hours. The final result was polymer with Mw in the range 60–120 kg mol^{-1}, in the form of sheets with thickness in the range 0.3–0.6 mm. The sheets were stored at room temperature for at least one month before testing. They were labeled PU1-PU14 according to their combination of HS, macrodiol and chain extendder, as indicated in Table 4.13. It should be noted that the stoichiometric proportions used in these polymers (I = 100) means that are truly thermoplastic. They do not have the potential for further reaction with ambient humidity to produce chain lengthening and allophanate cross-linking, seen in similar polymers but with excess isocyanate groups (e.g. I = 110) [17, 60, 135].

The materials are numbered similarly as in Table 2.1, section 2.3.1. as long as the PUs family investigated in this work is the same.

The sheet materials of thickness 0.3–0.6 mm were cut into rectangular strips of width 6 mm, and tested in uniaxial tension at ambient temperature (23±1°C) and humidity (41±7% RH) using an Instron model 4204 testing machine at Oxford, with 50 mm between the grips. Extension was measured in the centre of the specimen using an Instron "elastomer" model extensometer, with a gauge length of 20 mm. No strain localization was detected within the guage length. The experimental

Table 4.13 Composition of the materials studied in this work

Material	DI	MD	CE
PU1	MDI	PEA	EG
PU2	MDI	PEA	DEG
PU3	MDI	PEA	BG
PU4	DBDI	PEA	EG
PU5	DBDI	PEA	DEG
PU6	DBDI	PEA	BG
PU7	MDI	PTHF	EG
PU8	MDI	PTHF	DEG
PU9	DBDI	PTHF	EG
PU10	DBDI	PTHF	DEG
PU11	MDI	PBA	EG
PU12	MDI	PBA	BG
PU13	DBDI	PBA	DEG
PU14	DBDI	PBA	BG

programme was divided into two phases, with a different experimental protocol in each phase. The resulting data are expressed below in terms of nominal stress s and nominal strain e.

Phase 1 consisted of three load-unload cycles to a maximum nominal strain $e_{\max} = 3$, using a constant nominal strain-rate of magnitude 0.03 s^{-1} [135]. No dwell time was allowed when the straining changed direction. From the stress-strain data obtained, several properties of interest were determined. The tensile modulus E was calculated as grad$\frac{ds}{de}$ at $e = 0.01$, determined by fitting a quadratic function to the stress-strain curve over the range 0.005–0.015 and differentiating analytically. The values quoted below are means over an average of 4.5 samples for each material, and in each case the standard error of the mean was 2.4 MPa or less. Assuming linear viscoelasticity, this value of E corresponds to the isochronal 0.3 s tensile stress relaxation modulus. Also determined was the residual strain on first unloading e_{r}, expressed as relative residual strain $e_{\mathrm{r}}^* = e_{\mathrm{r}}/e_{\max}$. In addition, the data were used to compute various measures of specific work input W, obtained by the integration $W = \int s\, de$. The first cycle work input W_1 was obtained by integration over the first loading up to $e_{\max}$; first cycle hysteresis ΔW_1 by integration over the whole first load-unload sequence; second cycle work input W_2 by integration over the second loading, from e_{r} to $e_{\max}$; and second cycle hysteresis ΔW_2 by integration over the whole second load-unload sequence [135]. The reproducibility of these work calculations was better than 8% in all cases. In comparing the materials, it was convenient to express ΔW_1 as the relative first cycle hysteresis $\Delta W_1^* = \Delta W_1 / W_2$.

Phase 2 consisted of a series of load-unload cycles conducted at the same rate as in Phase 1, but taken to progressively higher nominal strains $e_{\max}$ on each cycle: 1, 2, 3.... Between 3 and 5 cycles were possible before failure with most materials. These may be described as 'pseudo-cyclic' experiments, to distinguish them from those in Phase 1 where the maximum strain was constant. As in Phase 1, for each value of e_{max}, the second cycle work input W_2 was obtained by integrating over the second loading from e_r to e_{max}. In this phase, however, the second cycle to any particular e_{max} was left incomplete, since straining continued to the next value of e_{max}. Therefore the second cycle hysteresis ΔW_2 was *approximated* by integrating over the first unloading from e_{max} and the second loading to e_{max}, and we shall distinguish it in the following with the notation $\Delta W_2'$. In this way, for each material, a series of values of W_2 and $\Delta W_2'$ was obtained, for various values of e_{max} [135].

Phase 1: Example stress-strain curves obtained are shown in Fig. 4.23, where the directions of straining are indicated by arrows. The three curves shown exemplify the response seen with non-crystalline MDI-based PUs (PU2), with semicrystalline DBDI-based PUs (PU6), and with non-crystalline DBDI-based PUs (PU10). Several features are apparent immediately.

All three materials show pronounced hysteresis, and all three cyclic responses exhibit the well-known Mullins effect, whereby in second loading to a given strain a material follows a stress-strain path closer to the previous *unloading* path than to the first *loading* path. Comparing the curves in Fig. 4.23, we may also see consequences of differing chemical structures, that were replicated across the whole set of materials. Firstly it is clear that material PU2 (MDI-based) has a discernibly lower initial stiffness than the other two materials (DBDI-based). Secondly the hysteresis and residual strain in the case of PU6 (semicrystalline and DBDI-based) are significantly greater than in the case of both the other two materials (noncrystalline and either MDI or DBDI-based). Thirdly, material PU2 shows significantly more strain hardening than the two DBDI-based materials.

The four parameters used to characterize the first and second load/unload cycles are collected in columns 2–5 of Table 4.14. Materials PU9 and PU11 are missing (except for the measurement of modulus), since they were insufficiently ductile to withstand cyclic straining to $e = 3$.

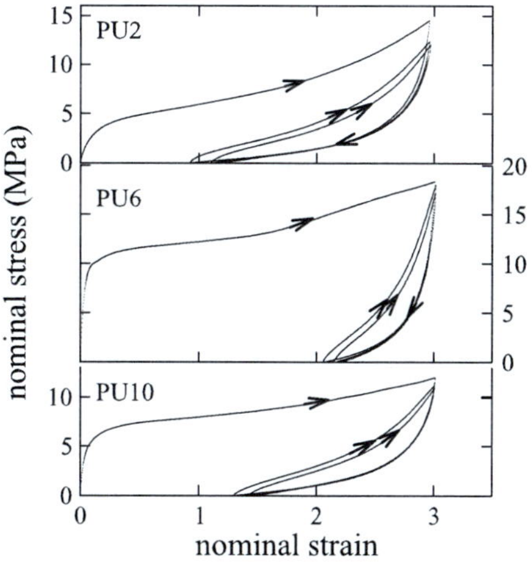

Fig. 4.23 Examples of stress-strain curves obtained in the 3-cycle load-unload tensile tests of Phase 1. Arrows indicate the direction of straining. Nominal strain-rate = 0.03 s^{-1} [135]

Table 4.14 Tensile test results for first and second loading/unloading cycles to nominal strain $e = 3$. Test temperature = 23±1°C, relative humidity = 41±7% RH, and nominal strain-rate = 0.03 s^{-1} [135]

Material	1st cycle tensile modulus E (MPa)	1st cycle work input W_1 (MJm^{-3})	1st cycle relative hysteresis ΔW_1^*	1st cycle relative residual nominal strain e_r^*	2nd cycle work input W_2 (MJm-3)	2nd cycle hysteresis ΔW_2 (MJm^{-3})
PU1	85.7	28.2	0.834	0.463	9.61	5.14
PU2	45.0	22.5	0.788	0.323	9.51	5.05
PU3	47.3	28.2	0.829	0.416	10.3	5.57
PU4	324	50.1	0.910	0.586	9.64	5.32
PU5	138	19.2	0.889	0.516	4.11	2.08
PU6	185	40.8	0.924	0.690	6.34	3.35
PU7	70.5	29.8	0.757	0.246	11.7	5.35
PU8	39.2	26.7	0.744	0.228	11.8	5.52
PU9	305	-	-	-	-	-
PU10	125	26.3	0.855	0.435	7.31	3.61
PU11	74.3	-	-	-	-	-
PU12	72.9	34.8	0.857	0.524	11.0	6.28
PU13	122	24.8	0.904	0.681	5.09	2.82
PU14	250	45.6	0.939	0.747	5.72	3.15

4.4.3.1 Tensile modulus

Consider first the isochronal tensile modulus E (0.3 s). It varies widely between 39 and 324 MPa. But within this range there is a clear pattern. In every case, the modulus of MDI-based materials (maximum value 86 MPa) is lower than that of DBDI-based materials (minimum value 122 MPa). Moreover, the modulus of those DBDI-based materials without crystallinity (maximum value 138 MPa) is always lower than that of DBDI-based materials with crystallinity (minimum value 185 MPa). It is informative also to consider how the modulus depends on physical structure of the polymers. In Fig. 4.24 the modulus is plotted versus Q, quantifying the degree of phase separation, for all the materials. It shows clearly that the modulus is independent of Q when the degree of phase separation varies through change of macrodiol, while DI and CE are unchanged. Fig. 4.25, however, shows the modulus plotted versus degree of crystallinity [135]. This shows clearly that the modulus does increase with increasing degree of crystallinity in the DBDI-based materials, and that in addition there is the purely chemical effect mentioned above: DBDI always gives a polymer with higher modulus than MDI, irrespective of choice of CE and MD [135].

How can such large variations in tensile modulus be explained within a family of thermoplastic PUs where the hard segment concentration is almost constant and the soft segment chain length is constant? They are two-phase materials: HS-rich hard domains are surrounded by a continuous SS-rich soft matrix. But there is likely to

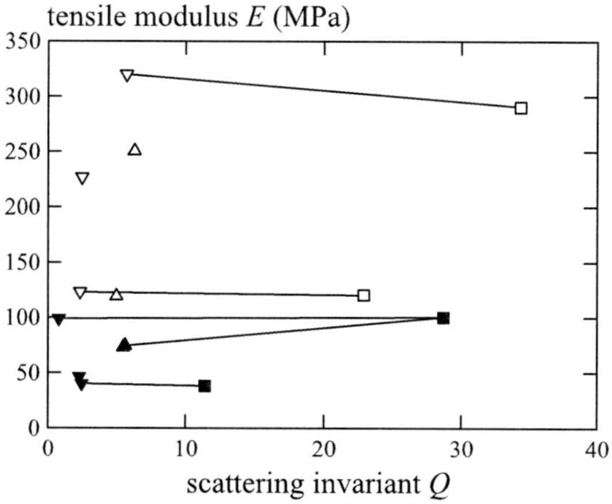

Fig. 4.24 Tensile modulus versus SAXS invariant Q (in arbitrary units). Filled symbols = MDI-based polymers; open symbols = DBDI-based polymers. Symbol shapes indicate the macrodiol: ▽ PEA; □ PTHF; △ PBA. Lines are only to guide the eye: they link materials differing only in their macrodiol [135]

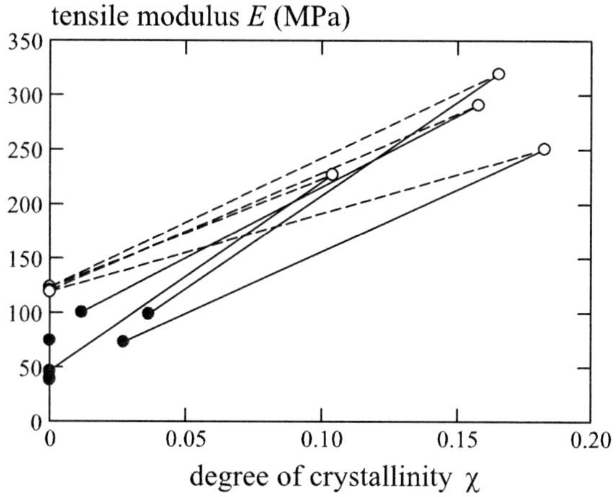

Fig. 4.25 Tensile modulus versus degree of crystallinity deduced from WAXS. Filled symbols = MDI-based polymers; open symbols = DBDI-based polymers. Lines are only to guide the eye: they link materials differing only in diisocyanate (full lines) or in chain extender (dashed lines) [135]

be, in general, some phase mixing (HS occurring in the matrix and SS occurring in the hard domains). In interpreting physical properties of such a system, some insight can be gained by considering it as a particulate composite material, comprising two phases each of which has *effective* continuum properties such as elastic constants. Applying this approach to the present TPU systems, there are three possible sources of variation in elastic constants: variation of the hard phase volume fraction φ_h, variation of elastic constants of the soft matrix, and variation in elastic constants of the hard particles. Here, to make the problem tractable we invoke two, physically reasonable, approximations: (1) both phases are isotropic elastic continua; (2) the elastomers are incompressible—i.e. Poisson's ratio $\nu = 0.5$.

To understand the variation of tensile modulus of such a system requires a solution to the well-known problem of predicting elastic constants of a 2-phase material. In general there is no analytical solution to this problem. The only rigorous predictions for arbitrary particle shape are bounds on the bulk and shear modulus—the tightest bounds are those proposed by Hashin and Shtrikman [306]: see for example the review by Christensen [307]. For the present materials, however (one phase rubbery and the other glassy or crystalline) the bounds are likely to be rather too far apart to be useful. A convenient empirical description that lies approximately midway between the bounds is the log-law used by Gray and McCrum in interpreting the dependence of shear modulus on degree of crystallinity in polyethylene [308]. Applied to the present TPUs, this gives the shear modulus G as follows

$$\ln G = \varphi_h \ln G_H + (1 - \varphi_h) \ln G_S \tag{4.6}$$

where G_H and G_S are shear moduli of hard and soft phases respectively. Hence the tensile modulus of the composite can be obtained from $E = 2G(1 + \nu)$. Two interesting results follow.

Firstly, suppose there is some phase mixing. Let a fraction f_{hs} of the hard segments mix into the soft matrix and a fraction f_{sh} of the soft segments mix into the hard particles. Suppose also that equations analogous to equation (2.4) section 2.3.1. apply also to the shear moduli of the individual phases, in terms of shear moduli G_H and G_S intrinsic to the hard segments and soft segments respectively. Then there are two additional equations giving the hard and soft phase shear moduli in terms of the hard segment volume fraction φ_{H}:

$$\begin{aligned} \ln G_H &= \varphi_h^{-1} [\varphi_h (1 - f_{hs}) \ln G_H + (1 - \varphi_h) f_{sh} \ln G_S]; \\ \ln G_S &= (1 - \varphi_h)^{-1} [\varphi_h f_{sh} \ln G_H + (1 - \varphi_h)(1 - f_{sh}) \ln G_S + \varphi_h f_{hs} \ln G_H] \end{aligned} \tag{4.7}$$

Combining equations (2.4) section 2.3.1. and (4.6), gives the shear modulus in terms of intrinsic hard and soft segment shear moduli:

$$\ln G = \varphi_h \ln G_H + (1 - \varphi_h) \ln G_S \tag{4.8}$$

We see that the shear modulus is predicted to be independent of the degree of phase mixing if G_H and G_S remain constant. This explains the results in Fig. 4.24 [135]. There is no significant change in tensile modulus when the degree of phase mixing varies through change only in MD, while diisocyanate and chain extender remain the same. There is no a priori reason to expect G_H or G_S to vary when only the MD changes. Note that all three soft segments have the same molar mass MS and hence would have the same rubber elastic modulus if crosslinked at their ends.

Secondly, predictions from equation (4.7) may be compared with measurements of tensile modulus to estimate changes that occur in the hard and soft phase properties, as the chemical and physical structures of the TPUs vary. Fig. 4.26 shows contours of calculated tensile modulus E plotted versus the hard and soft phase shear moduli, according to equation (3.1), with $\nu = 0.5$ and $\varphi_H = 0.354$ (the mean hard segment volume fraction) [135]. It is possible to place some bounds on the ranges of hard and soft phase moduli. For example, all three macrodiols are above their glass transition at room temperature. Since they all have molar mass $M_S = 2\,000$, if the hard segments at their ends acted as simple crosslinks, this would lead to an intrinsic soft segment shear modulus $G_S = \rho RT/M_S = 1.2$ MPa at room temperature for all the polymers, according to the affine chain theory of rubber elasticity (we neglect here the small differences in density). But this assumes the chains are completely free of constraint except from crosslinks. In the TPUs this must represent a *lower* bound on G_S, since the chains are subject to additional constraint from the excluded volume of, and connectivity to, the hard domains *as noted previously in section 2.3.1*. Thus a higher value is expected for G_S.

Similarly, in the absence of any phase mixing, the hard phase is a hydrogen-bonded glassy or semicrystalline polymer. Experience of other polymers suggests that $G_H \approx 1\,000$ MPa at room temperature [135]. But this is a typical value for a macroscopic sample of polymer. In the thermoplastic PUs it must represent an *upper* bound on G_H, since in this case the hard segments are confined to such small domains that a large fraction of them reside at the particle surfaces, adjacent to the more mobile matrix, again *as shown previously in Section 2.3.1*. So a lower value is expected for G_H.

Fig. 4.26 provides a chart on which we may place the various materials. Consider the MDI-based polymers, with E in the range 39 to 86 MPa. It may be seen that these values could be achieved with a range of combinations of G_H and G_S, varying from a glassy value of G_H combined with almost no increase of G_S above its lower bound, to a substantially reduced G_H (indicating an unusually compliant glass) combined with significant increase in G_S. Consider now the DBDI-based, but noncrystalline polymers with $E = 122 - 138$ MPa. Fig. 4.26 shows that in this case there is definitely substantial soft-phase stiffening, by a factor of at least 6. It can be explained by the closer packing and hydrogen bonding of DBDI-based hard segments in hard domains, causing higher constraint at the ends of soft segments. Finally, the DBDI-based semicrystalline polymers with $E = 185 - 320$ MPa must have even higher soft-phase stiffening. In the most extreme case G_S must be at least ca 20 MPa. Clearly, crystallization in the hard domains causes severe constraint on the soft phase segments because of molecular connectivity between the phases.

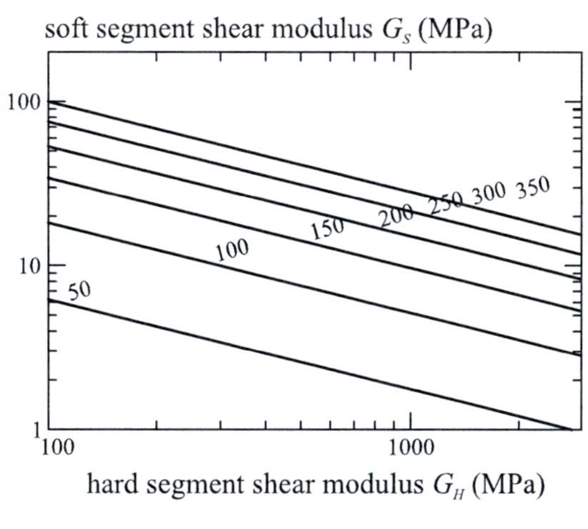

Fig. 4.26 Contours of tensile modulus (MPa) for a PU with hard segment volume fraction $\varphi_H = 0.354$ and Poisson's ratio $\nu = 0.5$, according to the empirical log law for shear modulus of a two-phase composite material (equation (4.7)) [135]

4.4.3.2 First cycle work input, unrecovered strain and hysteresis

Fig. 4.27 and 4.28 show the 1st cycle work input W_1 plotted versus scattering invariant Q and degree of crystallinity χ. W_1 is a measure of the mean flow stress during first extension to $e = 3$. If the soft phase is rubberlike with mobile soft segments (i.e. with minimal phase-mixing), the origin of the flow stress and hence W_1 is expected to be plastic flow in only the hard domains. With phase-mixing, the hard phase flow stress will decrease as more mobile soft segments penetrate the hard domains, but at the same time the soft matrix will develop a resistance to flow, as relatively low mobility hard segments are now interspersed among the soft segments. These two effects will act in opposition, and Fig. 4.27 shows that they approximately cancel: there is no trend of varying W_1 with change in degree of phase separation (indicated by Q). As with the modulus, however, Fig. 4.28 reveals a significant increase in W_1 with increasing degree of crystallinity. This is readily explicable, since more efficient molecular packing is expected to cause an increase in the activation barrier, and hence the driving stress and hence W_1, associated with the flow process. Interestingly, unlike the case of the modulus, there is no evidence in Fig. 4.28 for DBDI-based hard domains requiring a higher flow stress than MDI domains in the absence of crystallinity.

Since the materials are candidates for use as thermoplastic elastomers, the degree of strain recovery (or its converse, permanent set) is of great practical interest. The examples in Fig. 4.23 illustrate the fact that strain recovery varies widely among these materials. It is apparent that residual strain is a consequence of the unloading stress-strain curve deviating from the loading curve, in other words it is a consequence of hysteresis. For this reason, the data in columns 4 and 5 of Table 4.14 are closely related. In Fig. 4.29 relative residual strain e_r^* is plotted versus relative first cycle hysteresis ΔW_1^*, and they can be seen to be well correlated across the range of materials. This figure also reveals two important trends with respect to the effects of chemical composition on first cycle hysteresis. Firstly, MDI-based polymers show

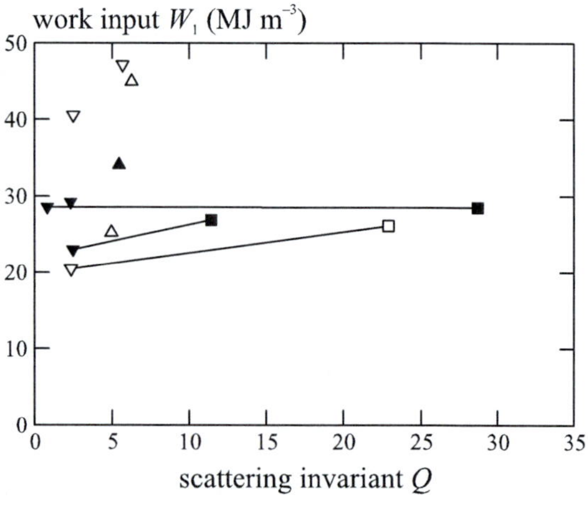

Fig. 4.27 First cycle work input W_1 on loading to $e = 3$, plotted versus SAXS invariant Q. Symbols as in Fig. 4.24. Lines are only to guide the eye: they link materials differing only in their macrodiol [135]

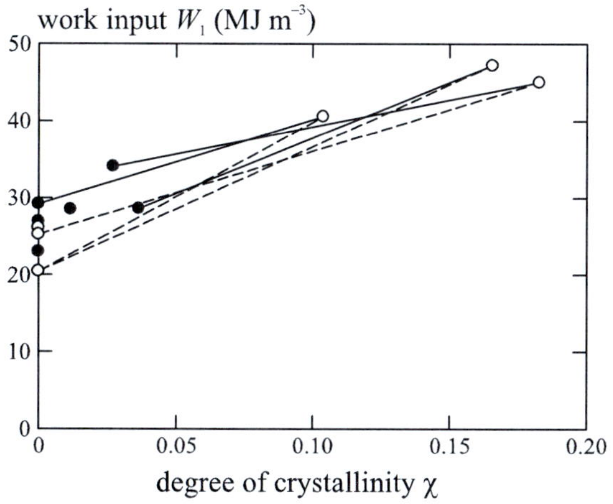

Fig. 4.28 First cycle work input W_1 on loading to $e = 3$, plotted versus degree of crystallinity. Filled symbols = MDI-based polymers; open symbols = DBDI-based polymers. Lines are only to guide the eye: they link materials differing only in diisocyanate (full lines) or in chain extender (dashed lines) [135]

substantially lower residual strain and relative hysteresis than the corresponding DBDI-based polymers. Secondly, for a given DI, polymers with PTHF as MD show lower residual strain and hysteresis than the corresponding polymers with PEA or PBA as MD.

The relation between ΔW_1^* and physical structure is revealed in Fig. 4.30 and 4.31. Fig. 4.30 shows that it tends to fall when phase separation increases, and this is why the PTHF-based polymers show the lowest hysteresis. Conversely, ΔW_1^* is found to increase with degree of crystallinity, as shown in Fig. 4.31. To explain these trends in the response to loading followed by *unloading* requires an understanding of the structural damage taking place during first loading.

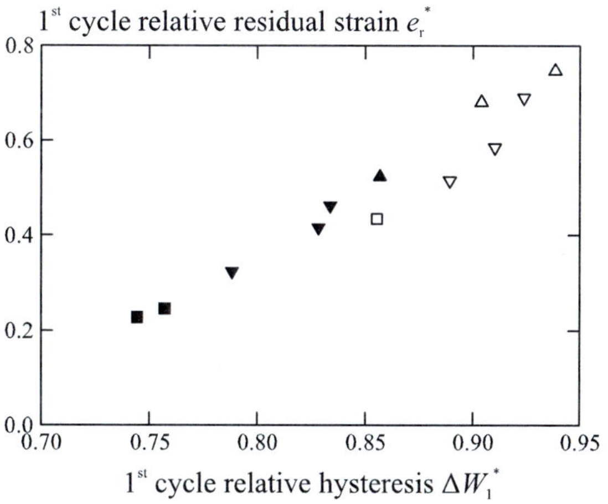

Fig. 4.29 First cycle relative residual strain on loading/unloading to $e = 3$, plotted versus first cycle relative hysteresis, showing correlation and trends with respect to choice of DI and MD. Symbols as in Fig. 4.24 [135]

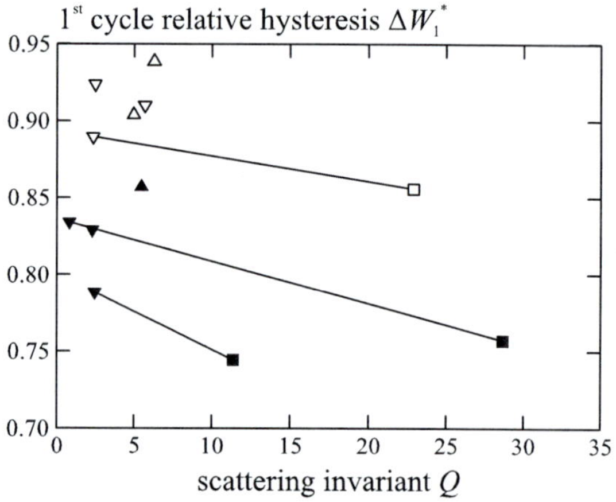

Fig. 4.30 First cycle relative hysteresis on loading/unloading to $e = 3$, plotted versus SAXS invariant Q. Symbols as in Fig. 4.24. Lines are only to guide the eye: they link materials differing only in their macrodiol [135]

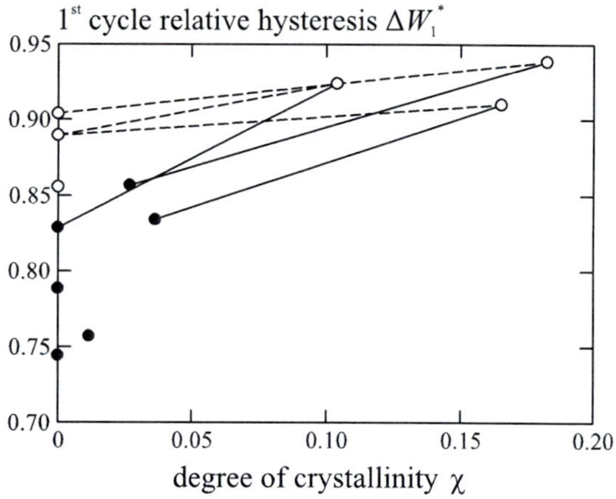

Fig. 4.31 First cycle relative hysteresis on loading to $e = 3$, plotted versus degree of crystallinity. Filled symbols = MDI-based polymers; open symbols = DBDI-based polymers. Lines are only to guide the eye: they link materials varying only in diisocyanate (full lines) or in chain extender (dashed lines) [135]

4.4.3.3 Second cycle work input and hysteresis

The example stress-strain curves in Fig. 4.23 illustrate the dramatic difference in response of these materials in second and subsequent load/unload cycles to a given strain level, as compared to the first cycle. Clearly, damage occurs during the first cycle. The second cycle response is effectively that of a new material. The work input W_2 and hysteresis ΔW_2 on the second cycle are included as columns 6 and 7 in Table 4.14, as obtained from the cyclic experiments of Phase 1. They are plotted one versus the other in Fig. 4.32, for all the 12 materials that could withstand cycling to $e_{max} = 3$. The remarkable feature of this graph is that there appears to be a single linear relationship between ΔW_2 and W_2 encompassing all the materials *except* PU7 and PU8: the MDI-based polymers with highest phase separation. The dashed line shown is a linear regression through all the points except these two, with $R_2 = 0.986$, as follows (uncertainties quoted are the standard errors of the coefficients)[135]:

Fig. 4.32 Hysteresis versus work input for the second load/unload cycle to $e = 3$. Symbols as in Fig. 4.24. The dashed line is a linear regression through all points except those of PU7 and PU8 [135]

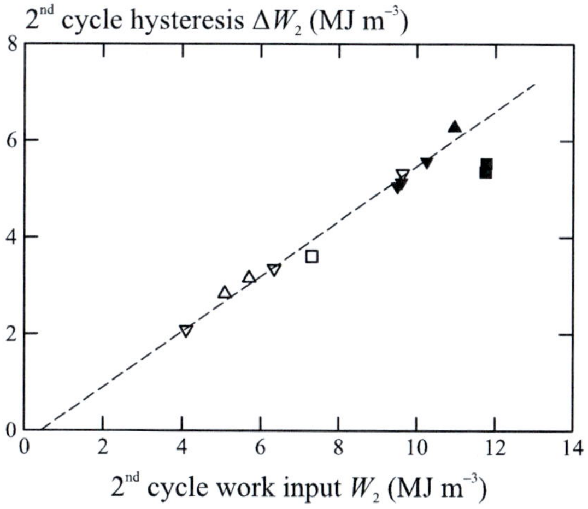

$$\Delta W_2 = (0.57 \pm 0.02) W_2 - (0.25 \pm 0.20) \quad (4.9)$$

However, the various materials appear at widely differing positions along this line. It is interesting to note that W_2 (and therefore ΔW_2) tends to be lowest for the DBDI-based materials, irrespective of their crystallinity. The reason for the highly phase-separated materials PU7 and PU8 having a slightly lower relative 2nd cycle hysteresis $\Delta W_2/W_2$ than all the other polymers is unknown at this stage, but it is interesting to note that these materials also showed the lowest relative hysteresis in the first cycle (see Fig. 4.30 and Table 4.14).

Phase 2: Examples of stress-strain curves obtained in the more complex pseudo-cyclic experiments of Phase 2 are shown in Fig. 4.33–4.35, where pairs of materials are compared. Fig. 4.33 compares PU1 and PU2. These materials are closely related, both being based on MDI and PEA, and differing only in chain extender (*see Table 4.13 in this section and also Table 2.1 in previous section 2.3.1*). However PU1 has a small degree of hard phase crystallinity (4%), while PU2 is amorphous. It is interesting to note the differences in the curves: the slight degree of crystallinity in PU1 clearly increases the stress levels reached throughout the test, and gives a much higher modulus (see also Table 4.14). Fig. 4.34 [135] compares the two DBDI-based polymers PU4 and PU5. These both have PEA as MD, but differ in CE. PU4 has EG as chain extender and therefore has significant crystallinity (17%), while PU5 has DEG as chain extender and therefore has no detectable crystallinity. The effect of crystallinity in PU4 is clearly to cause a large increase in stress throughout, and to increase the residual strain on each cycle. Fig. 4.35 compares PU8 and PU9, that are both based on PTHF and DEG, but differ in the diisocyanate. They are both amorphous, and both highly phase-separated. But the presence of DBDI in PU10 can be seen to increase residual strain in each cycle, and to reduce the rate of strain-hardening compared to PU8. An equivalent comparison can be made between PU2 in Fig. 4.33 and PU5 in Fig. 4.34, where the same pattern of differences is visible [135].

Fig. 4.33–4.35 illustrate the fact that stress-strain paths followed in the pseudo-cyclic tests show significant quantitative differences between the materials, although all exhibit the characteristic Mullins effect and varying degrees of residual strain, increasing with $e_{\max}$. It is interesting, therefore, to discover whether the second cycle responses at each strain level reflect the remarkable degree of commonality seen in Phase 1 experiments. As examples, Fig. 4.36 shows second cycle hysteresis and work input for two materials: PU2 (MDI + DEG + PEA) and PU13 (DBDI + DEG + PBA). For each material, $\Delta W_2'$ is plotted versus W_2 in successive cycles. Arrows indicate the direction of increasing $e_{\max}$. This graph reveals two surprising features. First, for each material the points appear to fall on a straight line, with a single direction of travel away from the origin with increasing $e_{\max}$. Second, data for both materials (with differing the diisocyanate and macrodiol) appear to lie on the *same* straight line.

It is of interest to know whether this commonality extends to *all* the cycles for *all* the materials. Fig. 4.37 therefore shows all the measurements of W_2 and $\Delta W_2'$: data for all the ten materials included in Phase 2 (materials 9, 11, 12 and 14 failed before completion of at least three cycles), and all cycles in their tests.

Fig. 4.37 confirms the trend seen already in Fig. 4.32, but with a much larger data set. The data for *all* the materials except PU7 and PU8, for *all* loading cycles, lie on a common straight line. Moreover, in every case, the sequence of points was found to be the same as in Fig. 4.36: towards higher W_2 with increasing maximum strain. Materials PU7 and PU8 behave similarly, but lie on a different line [135]. These observations may be quantified by fitting linear regressions through the data points.

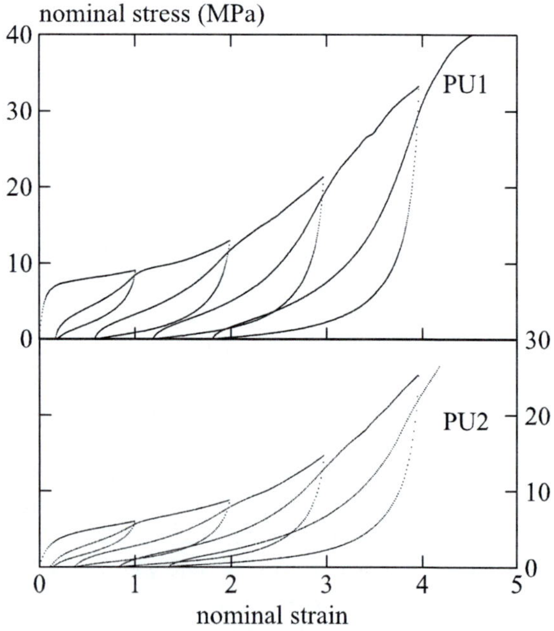

Fig. 4.33 Stress versus strain for "pseudo-cyclic" loading of two materials at nominal strain-rate 0.03 s^{-1}. PU1 and PU2 are both based on MDI and PEA but differ in their chain extender. PU1 has slight crystallinity (4%) while PU2 is amorphous [135]

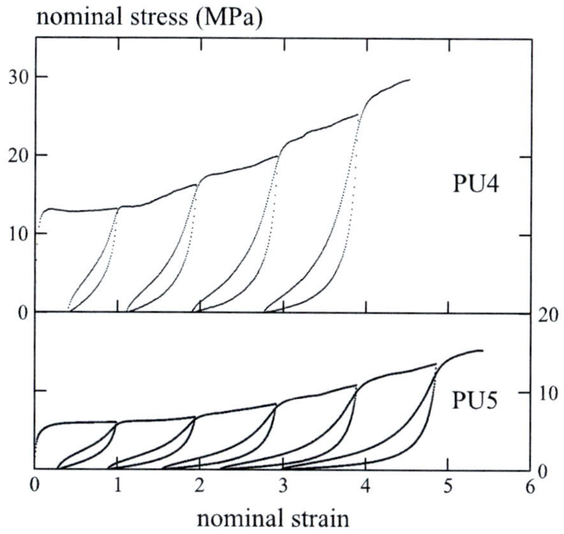

Fig. 4.34 Stress versus strain for "pseudo-cyclic" loading of two materials at nominal strain-rate 0.03 s^{-1}. PU4 and PU5 are both based on DBDI and PEA, but differ in their chain extender. PU4 has significant crystallinity (17%) while PU5 is amorphous [135]

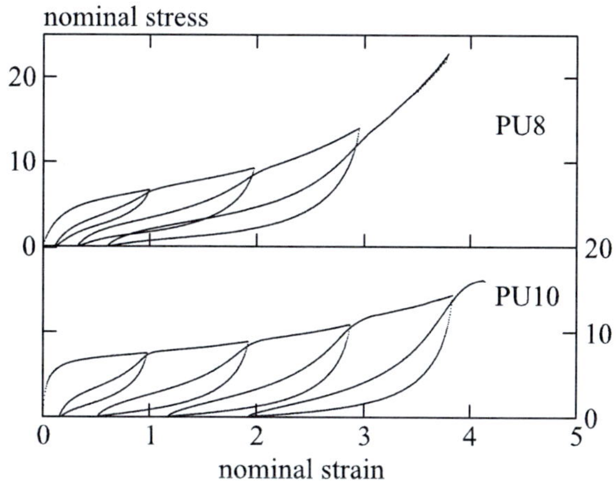

Fig. 4.35 Stress versus strain for "pseudo-cyclic" loading of two materials at nominal strain-rate 0.03 s^{-1}. PU8 is PTHF+DEG+MDI, PU10 is PTHF+DEG+DBDI. Neither has detectable crystallinity. Both have significant phase separation [135]

Thus the data for eight materials PU1, PU2, PU3, PU4, PU5, PU6, PU10 and PU13 (34 data points in total) fitted the following relation with $R_2 = 0.995$ (line A in Fig. 4.37):

$$\Delta W_2' = (0.565 \pm 0.007) W_2 - (0.48 \pm 0.05)\ \text{MJm}^{-3} \tag{4.10}$$

The data for materials PU7 and PU8 (7 points only) fitted the different relation with $R_2 = 0.988$ (line B in Fig. 2.7):

$$\Delta W_2' = (0.44 \pm 0.02) W_2 - (0.50 \pm 0.19)\ \text{MJm}^{-3} \tag{4.11}$$

Thus we see the pattern reported in Fig. 4.32 is confirmed and greatly extended in its reach. For most of the materials in the present study there is a common linear relationship between second cycle hysteresis and second cycle work input, and this

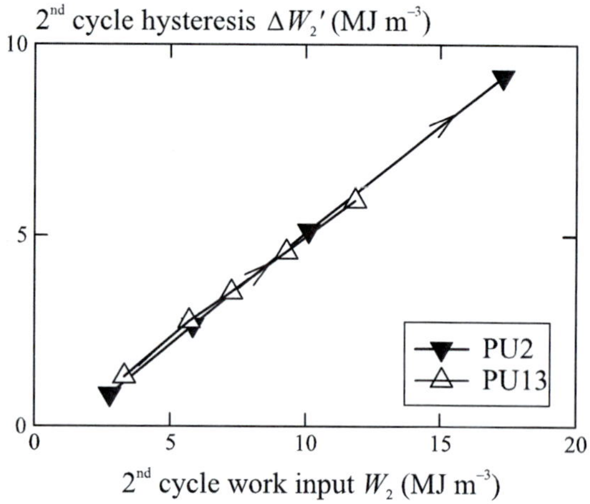

Fig. 4.36 Hysteresis versus work input for load/unload cycles to increasing e_{max}, in the pseudo-cyclic experiments of Phase 2, for materials PU2 and PU13. Arrows indicate the direction of increasing e_{max} [135]

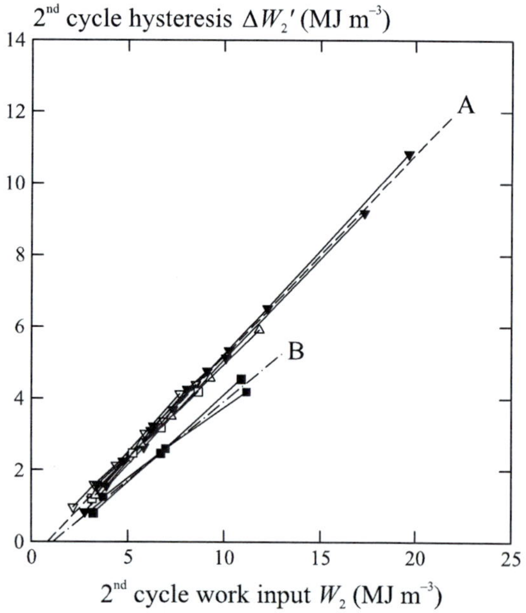

Fig. 4.37 Hysteresis versus work input for second cycle load/unload to increasing e_{max}, in the pseudo-cyclic experiments of Phase 2. Symbols are as in Fig. 4.24. The two dashed lines are linear regressions through data for sub-sets of the materials: A (all materials except PU7 and PU8), and B (PU7 and PU8)

relationship is independent of e_{max} reached on the first cycle. Note that the linear relations described by equations (4.8) and (4.9) are identical to within the standard errors of the coefficients. However, a sub-set of the present materials (PU7 and PU8) follow a different relation, characterized by lower relative second cycle hysteresis.

The results presented above reveal a dramatic difference in response between first loading to a given deformation, and second and subsequent loadings. The model thermoplastic PUs studied in this work showed large, systematic variations in *first* loading response—as quantified by tensile modulus, work input, hysteresis and unrecovered strain—depending on their chemical structure and molecular packing.

But on *second* loading to the same strain they showed a remarkable degree of uniformity in their relative hysteresis. The clear conclusion is that first loading causes significant damage to the initial structure, such that second loading corresponds to deformation of a new structure. Moreover, whereas the initial physical structure varies greatly between the materials, the new structure dominant in second loading must be similar between the materials.

To understand this result we may consult the evidence concerning structural changes during deformation of similar materials, provided by in-situ mechano-calorimetry [294, 298], infra-red (IR) spectroscopy [51, 309], WAXS and SAXS [141, 284, 310]. Although the details differ with the precise chemical composition, some features seem to be common to thermoplastic PU elastomers.

Several studies have followed deformation of the two-phase structure by means of in-situ SAXS. On first loading, small deformations are accompanied by affine deformation of the arrangement of hard domains, but this is followed after stretching to $\lambda \approx 2$ by a deviation towards reduction of long period d relative to affine deformation [284, 310], and possibly even to an absolute reduction in d [310]. In some cases the original isotropic scattering is replaced by a four-point pattern [284]. These observations are all explained in terms of breakage of the original hard domains into smaller units during plastic deformation [284, 310]. Simultaneously, calorimetric measurements of heat flux, combined with work input, indicate a significant increase in internal energy [294] associated with the less efficient molecular packing of hard segments, once the original two-phase structure starts to break up. This might be thought to suggest a reduction in H-bonding during deformation, but in-situ IR spectroscopy suggests otherwise: the density of H-bonds appears to be independent of strain [141]. Thus, if H-bonds are broken during deformation, new bonds must be formed rapidly on the time-scale of straining. Another interesting observation is the progress of orientation of the various bonds identifiable in IR spectra obtained during straining. Initially the chain orientation as indicated by >N–H and urethane or urea >C=O groups (i.e. the HS) orients perpendicular to direction of extension, while the chain axes of soft segments align parallel to the direction of extension. But beyond $\lambda \approx 2-3$, the hard segments rotate and their chain direction follows that of soft segments, parallel to the direction of extension [51]. This may also be interpreted plausibly in terms of breakage and re-reformation of H-bonds.

On subsequent cycling, within the envelope of previous strains, a totally different pattern has been observed. In-situ SAXS simultaneous with cyclic deformations showed reversible movements of unchanged four-point patterns, although the movements were in opposite directions for two variants on the same TPU, with monodisperse or polydisperse hard segments [284]. Meanwhile, in-situ IR studies simultaneous with pseudo-cyclic experiments of the type in Phase 2 of this work, showed that the orientation pattern of hard segment and soft segment portions of chains was also reversible [51]. Calorimetry during subsequent loading of TPUs with hard segment fractions similar to the polymers of the present study showed that there was no increase in internal energy [298], consistent with no further structural breakdown.

Combining such structural information with results from the present work, a plausible picture of deformation in thermoplastic PU elastomers emerges. In small deformations below the yield stress, the elastic response is that of a two-phase composite material. The soft matrix responds with a stiffness higher than that of a homogeneous crosslinked elastomer with the soft segment composition, because of constraint from the excluded volume of, and molecular connectivity to, the hard domains, especially so when the hard domains are semicrystalline, and possibly because of some phase mixing with hard segments. At strains beyond yield, break-up of the two-phase structure commences. Bonart suggested how this could proceed by sliding of hard segments relative to their neighbors within the hard domains [311]. There are two likely consequences. (I) the sliding could lead to irreversible deformation and hence residual strain, and (II) the sliding could lead to stripping of segments from the hard domains to provide new soft matrix in series with the hard domains, as suggested by Kilian and co-workers [312]. Both these processes would be facilitated by the small sizes of the particles. *As noted in previously in section 2.3.1, a large proportion of hard segment monomers lie at the surfaces of the hard domains, and therefore must be relatively weakly bound.* On unloading and subsequent re-loading, while the stress remains below the flow stress of the hard domains, the structure is expected to remain constant and reversible deformation to occur. Since the newly-created soft phase is much more compliant than the hard domains and is coupled in series with them, the reversible deformation occurs essentially entirely within this soft phase, characterized by much lower stiffness and lower hysteresis than the original structure. If the new (series-coupled) structure is loaded up to the flow stress, structural break-up resumes.

This description accounts qualitatively for the characteristic constitutive response seen in the present work, and provides a basis for understanding its dependence on chemical and physical structure. But the new results also add an unexpected twist. The reproducibility of relative hysteresis on second and subsequent loading indicates that the new series-coupled soft phase, formed by stripping segments from the hard domains, has almost constant relative hysteresis, independent of the strain level at which it is formed. Moreover it is (with two minor deviations among the range of polymers studied) independent of chemical composition of DI, CE and MD, when the MDs are of equal length and HS fractions are equal. Why the small sub-set PU7 and PU8, based on MDI but with highest phase segregation should have a slightly lower relative hysteresis than the others remains unclear. However, one could speculate that it results from a lower degree of intermolecular H-bonding being achieved in the new soft phase (because of the combination of (a) absence of ester >C=O groups and (b) the relatively less mobile MDI).

To understand the relative first cycle hysteresis ΔW_1^* we need to consider the processes of both first loading and unloading. In terms of the description above, its value is expected to be sensitive to the relative importance of the two damage processes identified: (I) and (II). Thus its reduction with replacement of DBDI by MDI, or with increasing phase separation (Fig. 4.29 and 4.30) could be caused by increase in contribution from process II, relative to process I, in these cases, presumably because of weaker binding of segments to the hard phase.

The explanation offered above for the Mullins effect in these materials differs in an important respect from the model proposed by Qi and Boyce [301]. That model invokes parallel coupling between the hard domains and (strain amplified) soft matrix, and explain the Mullins effect in terms of a reduction of strain amplification in the matrix with increasing strain. In such a parallel-coupled structure, second and subsequent cycles of loading would involve cyclic deformation of the hard phase in addition to the rubbery soft phase. This appears to be inconsistent with the structural studies reported above, and not to be reconcilable with the new results, since the two hard phases show greatly differing relative hysteresis (as seen on the first cycle) while there is remarkable commonality of relative hysteresis in second and subsequent cycles [135].

Overall the following is a plausible picture of events during strain cycling beyond yield, consistent with in-situ structural evidence and present results. During first loading, break-up of the hard domains occurs, producing new soft phase coupled in series with the remaining hard phase, as suggested by Kilian and co-workers. During unloading and subsequent reloading, it is this more compliant phase that dominates the deformation. Such a description explains why second and subsequent loading/unloading inside the envelope of previous maximum strain gave responses totally different from first loading. They were characterized by much lower stiffness and hysteresis (the Mullins effect). It also sheds light on the most remarkable discovery from the present work: the apparently almost unique linear relation between second cycle hysteresis and work input. This relation is independent of the previous maximum strain reached, and it is *almost* independent of choice of DI, MD and CE. Curiously, two polymers—the best phase-separated polymers based on MDI—showed slightly lower second cycle hysteresis. In terms of the Kilian hypothesis, these results indicate the mechanical response of the new soft phase is remarkably independent of the chemical structure of the thermoplastic PU if, as in the present work, the hard segment fraction and soft segment chain length are held constant.

One aspect of the present results—relative first cycle hysteresis—depends upon the responses of both the initial structure and of the new soft phase, and is also sensitive to the precise mechanism of flow and break-up of the hard domains. It was observed to reduce with increasing phase segregation and with replacing DBDI with MDI as diisocyanate. In terms of the picture of structural break-up invoked above, this suggests that in such cases there is relatively more stripping of segments from hard domains, and less plastic deformation of them. A possible explanation is weaker binding of segments at the hard domain boundaries.

4.4.4 Cycling tensile responses of polyurethane elastomers achieved with excess of isocyanate NCO groups

A range of PUs has been synthesized by us, in which the HS was the conventional MDI, the more flexible DBDI. The materials are listed in Table 4.15 below.

The cycling responses were investigated. The effects of varying HS were studied in the context of a variety of combinations of chain extender, macrodiol and their proportions: a total of 19 formulations in total.

The family of PU elastomers depicted in Table 4.15 was based on the two diisocyanates DBDI or MDI. Samples were prepared with systematic variations in diisocyanate (DI), macrodiol (MD) and chain extender (CE). Molar ratios were as given in column 6. All polymers (except PU5, PU 18 and PU19) were synthesized with *I = 110 (10% excess of NCO groups), corresponding to the molar ratios diisocyanate (DI):macrodiol (MD):chain extender (CE) = 4:1:2.64.*

All the materials except two (PU18 and PU19) were based on a single DI which was either MDI or DBDI according to the structures shown in Table 4.15.

For comparison reasons, in the series of materials it was included a polymer achieved with no excess of NCO groups (I = 100, PU5) with the molar ratio DI:MD:CE = 4:1:3.

The two copolyurethane elastomers of type C (PU18 (I = 110) and 19 (I = 110) in Table 4.15), containing mixtures of the diisocyanates MDI and DBDI randomly distributed, were synthesized by using a one step prepolymer synthesis where a melt mixture of DBDI and MDI was used in equal molar proportions, which was then introduced into the anhydrous macrodiol [60, 61].

Note that these 2 materials are identical with the materials denoted PU_{C1} in Tables 2.2 and 2.3, sections 2.3.2.1 and 2.3.2.2. where the materials were characterized by WAXS and SEM. A detailed description of the PUs synthesis is detailed elsehere [60, 61].

The performance as elastomers, was assessed by us by means of a series of cyclic tensile tests, at a constant or variable nominal strain-rate of $3.1 \times 10^{-2}\ s^{-1}$, cycling between 300% elongation and zero load. From these tests, the initial input strain energy density E_{1C} was computed as the area under the first cycle loading curve of nominal stress versus nominal strain up to the elongation limit of 300%. The calculation was repeated for the corresponding quantity for the second load cycle: E_{2C}. Similarly, recovered energy densities E_{1R} and E_{2R} were computed as the areas under the unloading curves for first and second cycles respectively. The areas were measured on paper copies of the curves, using a digital planimeter. For *PUs systems achieved with excess NCO groups* E_{2C}, E_{1R} and E_{2R} were expressed as fractions of the initial input energy density E_{1C}, and denoted as such by an asterisk. The fractional energy dissipated on each cycle was expressed as the fractional hysteresis energy density on first and second cycles, obtained from the expressions:

$$E^*_{1H} = 1 - E^*_{1R} \tag{4.12}$$

$$E^*_{2H} = E^*_{1C} - E^*_{2R} \tag{4.13}$$

Comparing successive load-unload cycles, it was found that irreversible changes to the stress-strain response were confined essentially to the first loading cycle. In subsequent cycles the load-unload stress-strain curves remained almost unchanged. Example materials are shown in Table 4.15, where it may be seen that the recovered energies in first and second cycles are equal in most cases, to a good approximation. Similarly, increasing the time interval between the end of one unloading cycle and the start of the next, from a few seconds to several hours (up to 24 hours), had no sensible effect on the measured recoverable strain energy. This was in contrast to the input energy, which was clearly dependent on both the previous strain history and the relaxation time between loading cycles [61].

Table 4.15 Example values of the input and recovered energies of the 1st and 2nd cycle for selected PUs [61]

PU #	PU Type	PU structure DI	MD	CE	DI:MD:CE	I%	E_{1C} MJ m^{-3}	E^*_{2C} %	E^*_{1R} %	E^*_{2R} %
1	D	DBDI	PTHF	DEG	1.5:1:0.38	110	8.00	77.8	50.3	48.8
2	D	DBDI	PTHF	DEG	2.27:1:1.065	110	14.3	62.5	31.1	31.0
3	D	DBDI	PTHF	DEG	3:1:1.73	110	23.1	52.2	18.9	18.9
4	D	DBDI	PTHF	DEG	4:1:2.64	110	25.3	48.0	16.6	17.0
5	D	*DBDI*	*PTHF*	*DEG*	*3:1:2*	*100*	*20.9*	*51.3*	*17.7*	*18.0*
6	D	DBDI	PEA	DEG	3:1:1.73	110	24.0	53.6	15.6	15.0
7	M	MDI	PTHF	DEG	3:1:1.73	110	19.0	63.8	39.1	39.0
8	M	MDI	PEA	DEG	3:1:1.73	110	16.3	72.8	41.7	38.1
9	D	DBDI	PEA	BG	3:1:1.73	110	41.7	32.3	16.7	11.0
10	D	DBDI	PTHF	BG	3:1:1.73	110	25.3	40.9	14.2	14.0
11	M	MDI	PEA	BG	3:1:1.73	110	27.2	60.9	25.3	26.1
12	D	DBDI	PEA	EG	3:1:1.73	110	41.8	33.3	10.2	13.8
13	M	MDI	PEA	EG	3:1:1.73	110	21.7	62.7	33.3	33.0
14	D	DBDI	PEA	EG	4:1:2.64	110	46.6	40.9	11.5	12.2
15	M	MDI	PEA	EG	4:1:2.64	110	30.4	64.6	17.6	20.2
16	D	DBDI	PTHF	EG	4:1:2.64	110	41.1	36.2	10.2	10.8
17	M	MDI	PTHF	EG	4:1:2.64	110	30.6	63.2	38.9	35.2
18	C	*DBDI+MDI*	*PEA*	*EG*	*2:2:1:2.64*	*110*	*40.1*	*55.3*	*15.7*	*15.7*
19	C	*DBDI+MDI*	*PTHF*	*EG*	*2:2:1:2.64*	*110*	*35.7*	*50.2*	*27.9*	*25.5*

Strain energy input (at 300% elongation) and strain energy recovery (on unloading from 300% elongation) were measured in cyclic tensile tests at a nominal strain-rate of 3.1×10^{-2} s^{-1}. Asterisks denote that input energy on the second cycle E^*_{2C} and recovered energies on first and second cycles E^*_{1R} and E^*_{2R} are expressed as percentages of the input energy on the first cycle E1C.

Fig. 4.38 shows the dependence of E_{1C} and E^*_{1R} on the equivalent hard segment urethane group concentration in PUs of structure DEG-PTHF-DBDI (PU1-PU4 in Table 4.15). As may be seen, E_{1C} increases with increase in concentration of the HS urethane groups, reflecting the increase in concentration of hydrogen bonds. In contrast, the fractional recovered energy E^*_{1R} decreases, revealing that the increase in E_{1C} is associated with increased plastic dissipation of energy, presumably resulting from stress-induced breakdown of the hydrogen bonds [61].

Comparing the DBDI and MDI based polyurethanes of different recipes in Table 4.15, it is clear that, in recipes otherwise identical (same MD and CE, and equal hard segment molar concentration) the DBDI based PUs display a substantially *higher* input energy density E_{1C} and *lower* fractional energy recovery E^*_{1R} than the corresponding MDI based PUs [61]. This applied for all MD/CE combinations studied: PTHF-DEG (compare PU3 with PU7); PEA-DEG (compare PU6 with

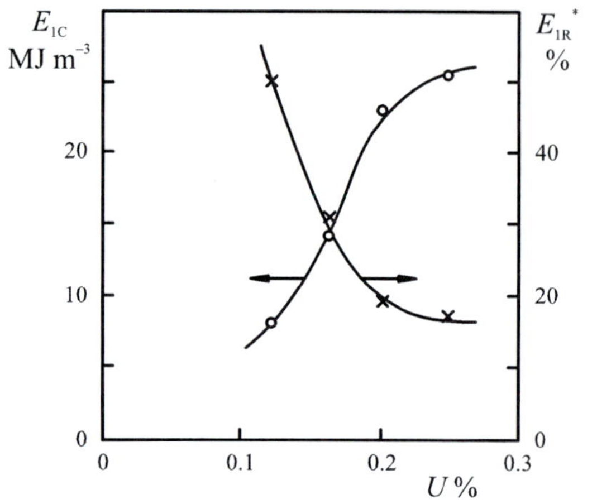

Fig. 4.38 Variations of input strain energy at 300% elongation E_{1C}, and first cycle recovered strain energy E^*_{1R}, with concentration U of urethane groups (expressed as equivalent urethanic groups/100g PU), in the case of a PU system based on DBDI-PTHF$_{2000}$-DEG (polymers PU1-PU4) [61]

PU8); PEA-BG (compare PU9 with PU11); PEA-EG (compare PU12 with PU13 and PU14 with PU15); PTHF-EG (compare PU16 with PU17). These findings are summarized in Table 4.16, where the percentage excess of E_{1C} of the DBDI-based PUs over the MDI-based PUs is shown.

Table 4.16 The increase of strain energy input E_{1C} for DBDI-based PUs over corresponding MDI-based PUs, for various combinations of macrodiol and chain extender. From the data in Table 4.15 [61]

Polymers compared	MD	CE	Relative E_{1C} increase,%
PU7 (MDI) and PU3 (DBDI)	PTHF	DEG	21.6
PU8 (MDI) and PU6 (DBDI)	PEA	DEG	47.2
PU11 (MDI) and PU9 (DBDI)	PEA	BG	53.3
PU13 (MDI) and PU12 (DBDI)	PEA	EG	92.6
PU15 (MDI) and PU14 (DBDI)	PEA	EG	53.2
PU17 (MDI) and PU16 (DBDI)	PTHF	EG	34.3

A further interesting observation from Table 4.15 is that, when an equimolar mixture of DBDI+MDI was used for the hard segments, the values of E_{1C} and E^*_{1R} obtained were intermediate between those obtained for the corresponding recipes with either DBDI or MDI alone (consider the families of materials PU14/PU15/PU18 and PU16/PU17/PU19). Finally, comparison of results for PU5 (DBDI-PTHF-DEG, I = 100) and PU3 (DBDI-PTHF-DEG, I = 110) revealed the effects of PU chain structure modification, with the appearance in PU3 of new urea groups and possible formation of some allophanate branching, in the presence of humidity when there is an isocyanate excess. PU5 has a larger input strain energy E_{1C} but also a higher recovery of energy E^*_{1R} than PU3 [61].

The influence of HS type on recoverability of energy and strain was examined more closely by means of single-cycle load-unload tensile tests conducted to a range of maximum elongations at a nominal strain-rate of 4.2×10^{-3} s^{-1}, on a group of three PU materials with equivalent recipes, differing only in the choice of hard segment: PU17 (MDI), PU16 (DBDI) and PU19 (DBDI+MDI). In every case the macrodiol was PTHF and the chain extender was EG, and molar ratios were 4:1:2.64. Results are shown in Fig. 4.39-4.41. The influences of HS structure on E_{1C} and hysteresis, already referred to, can clearly be seen by eye by comparing Fig. 4.39 to 4.41.

It is interesting to note that, while polymer C occupies an intermediate position relative to the MDI and DBDI materials in terms of E_{1C} and E^*_{1R}, it gives the lowest residual elongation. The dominant feature is a hysteresis loop with a width that varied between the three polymers. Polymer with MDI hard segments (Fig. 4.39) gives the lowest stresses and least hysteresis, whereas polymer with DBDI hard segments gives the highest stresses and most hysteresis, Fig. 4.40). It is interesting to note that polymer with mixed MDI/DBDI hard segments (Fig. 4.41) occupies an intermediate position.

4.4.5 Interpretation of mechanical response

Prediction is very difficult, especially about the future. (Niels Bohr)

The stress-strain curves in Fig. 4.39 to 4.41 have the characteristic shape for elastomers containing reinforcing particles. There is a rate-dependent flow stress associated with stress-activated segmental diffusion at the surface or interior of reinforcing particles, superimposed on the strain-stiffening hyperelastic stress response of the elastomeric matrix. Such a system is amenable to a quantitative description in terms

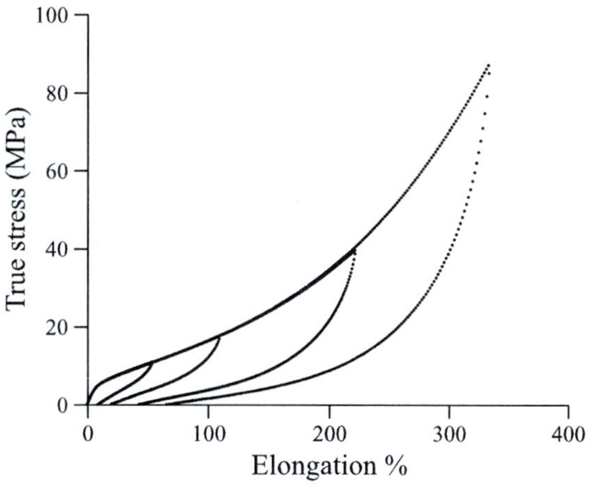

Fig. 4.39 Tensile load/unload cycles for a type M polymer: PU17 (MDI-PTHF-EG, I = 110). Nominal extension rate was 0.0042 s^{-1}. Each cycle was obtained with a different specimen [61]

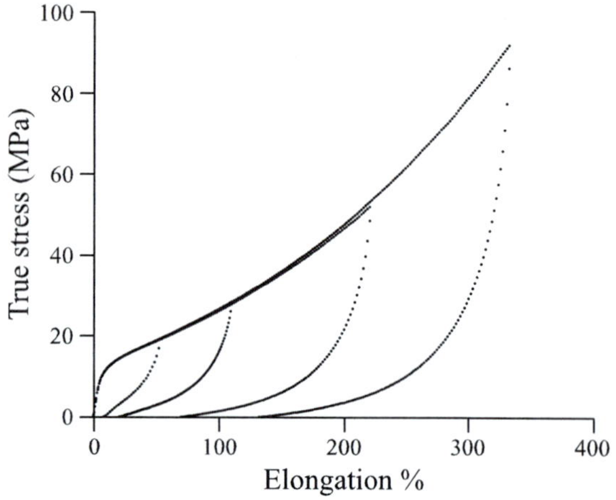

Fig. 4.40 As Fig. 4.39, but for type D polymer: PU16 (DBDI-PTHF-EG, I = 110) [61]

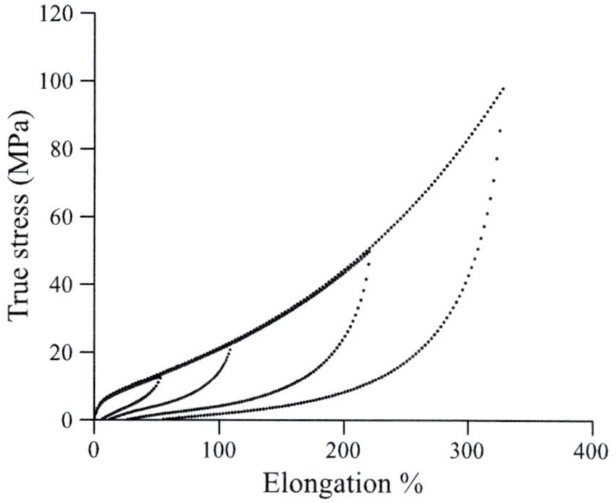

Fig. 4.41 As Fig. 4.39, but for type C polymer: PU19 (DBDI+MDI-PTHF-EG, I = 110) [61]

of the Glass-Rubber (GR) constitutive model framework [313] that has been applied previously to *homogeneous* amorphous polymers. The (GR) constitutive model for amorphous polymers [61, 313] provides a convenient means to interpret and explain the variations in mechanical properties reported above. Even though PU elastomers are well-known to be heterogeneous on a length scale of order 10 nm, they remain subject to the general truth that, under an imposed deformation, polymers possess only two major free energy sinks: (1) perturbation of inter-atom potentials between and within molecules, i.e. what might be called the "bond-stretching" free energy density A^b, and (2) perturbation of conformational entropy, i.e. the "conformational" free energy density A^c. To proceed further with the GR model, we now invoke two important assumptions. Firstly, *at the continuum level* the state of deformation throughout the material is uniform. Secondly, within a certain volume fraction v_s of the elastomer (defined as the "soft" phase) the part of the free energy contribution A^b associated with shape change is fully relaxed. Thus the soft phase is as an amorphous polymer above its glass transition. Combining these two assumptions, it follows that the deviatory Cauchy stress tensor s at any instant may be expressed as the sum of two terms: a conformational stress s^c pervading the whole material and a bond-stretching stress s^b acting only within the "hard" phase, with volume fraction $\nu_h = 1 - \nu_s$. Thus the deviatoric stress is given by

$$s = \nu_h s^b + s^c \tag{4.14}$$

and the total Cauchy stress is given by

$$\rho = \nu_h s^b + s^c - p\mathbf{I} \tag{4.15}$$

where the mean stress is $-p$.

Consider a uniaxial tensile test, parallel to axis 1, say, in which at any instant the stretch is λ and the rate of stretching is $\dot{\lambda}$: hence $\sigma_{11} = \sigma$, $\sigma_{22} = 0$. It follows from equation (4.15) that the applied true stress σ may be expressed unambiguously as the sum of two distinct terms:

$$\sigma = \nu_h \sigma^b + \sigma^c \tag{4.16}$$

where $\sigma^b = s^b_{11} - s^b_{22}$ and $\sigma^c = s^c_{11} - s^c_{22}$. The contribution σ^b is expected to be elastic-viscoplastic in character, while the contribution σ^c is assumed here to be hyperelastic in character. Entanglement slippage is neglected in the present materials prepared with I = 110, since the molecular length is believed to be sufficient for entanglements to act as physical crosslinks, together with any chemical allophanate crosslinks [61].

Under these conditions, the contribution $\sigma^c(\lambda)$ will be path-independent, while the elastic-viscoplastic contribution σ^b will vary with strain history. Under steady-state flow, however, after completion of the start-up transient, the GR model framework predicts σ^b will be a function of only the instantaneous values of $\dot{\lambda}$, λ and, in general, the mean stress $-p$. For the present purpose we now neglect the effect of varying mean stress, as the contribution is minor in a uniaxial test [61, 313]. Now consider a load-unload strain cycle, where the magnitude of the rate of stretch is equal and opposite in load and unload portions of the cycle. At a given value of elongation during loading we have $\sigma^b = \sigma^b(+\dot{\lambda}, \lambda)$, whereas at the same elongation during unloading we have $\sigma^b = \sigma^b(-\dot{\lambda}, \lambda)$. But, in view of the associative form of flow rule employed in the model [313], reversing the sign of the rate of stretch will reverse the sign of the deviatoric stresses s^b_{11} and s^b_{22}, from which it follows that

$$\sigma^b(-\dot{\lambda}, \lambda) = -\sigma^b(+\dot{\lambda}, \lambda) \tag{4.17}$$

This is a useful result, as it means that stress-strain data for load-unload cycles may be used to separate out bond-stretching and conformational contributions to the stress [61]. Thus we may evaluate them as follows, applying equation (4.16) to loading and unloading, and making use of equation (4.17), for any value of elongation where the above restrictions apply (particularly *where stress transients can*

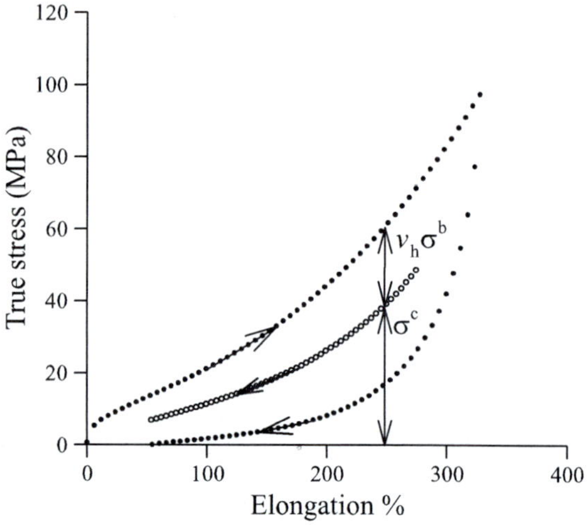

Fig. 4.42 Apportioning of true stress into contributions from bond-stretching and conformational entropy for polymer PU19, according to equations (4.18) [61]

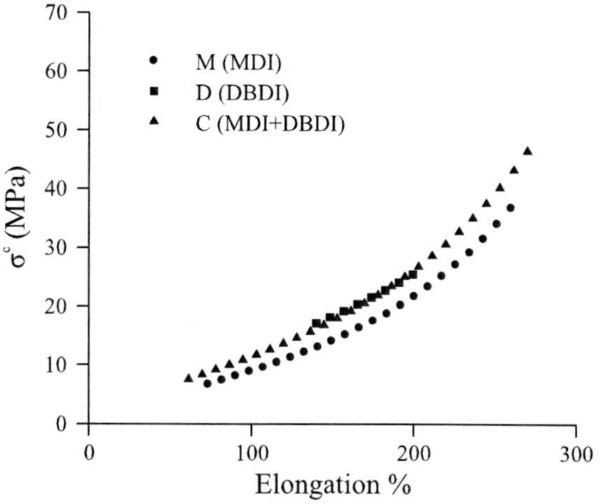

Fig. 4.43 Conformational entropy contribution to the true stress, σ^c, as deduced from equations (4.18) from the data shown in Fig. 4.39-4.41, for the polymers of type M, D and C [61].

reasonably be neglected):

$$\sigma^c = \frac{1}{2}\left[\sigma(+\dot{\lambda},\lambda)\right], \quad \nu_h\sigma^b = \frac{1}{2}\left[\sigma(+\dot{\lambda},\lambda) - \sigma(-\dot{\lambda},\lambda)\right] \tag{4.18}$$

This separation is illustrated in Fig. 4.42, for a single load-unload cycle on PU19 to ca 300% elongation, and results of the analysis for materials PU16, PU17 and PU19 are presented in Fig. 4.43 and 4.44.

Fig. 4.43 shows that the apparent conformational entropy contribution to stress deduced in this way differs little between the three materials. This result is expected, since the soft segment macrodiol employed in all three materials is the same: PTHF. The main difference visible is a slightly smaller stress contribution from the MDI-based polymer, possibly originating in a lower entanglement density in this polymer

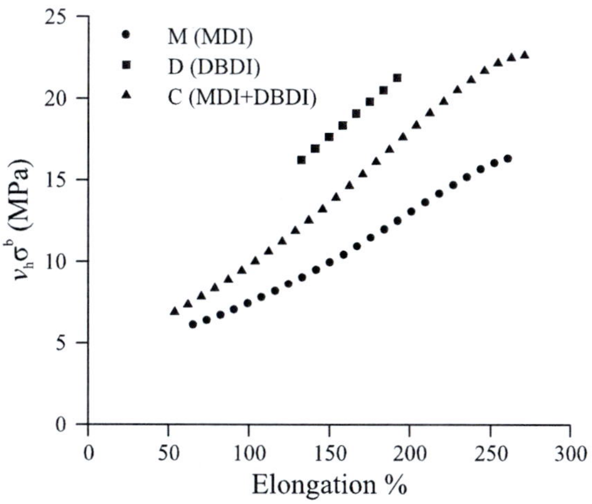

Fig. 4.44 As Fig. 4.43, but for the bond-stretching contribution to the stress, $\nu_h \sigma^b$ [61]

in view of the lower flexibility of its chains relative to the polymers containing DBDI. By contrast, Fig. 4.44 shows large differences in the magnitude of the viscoplastic flow stress σ^b in the three polymers. The DBDI-based polymer PU16 shows the greatest flow stress, the MDI-based polymer PU17 the lowest, and the DBDI+MDI-based polymer PU19 an intermediate value. These differences may be understood in terms of the crystallinity and more pronounced hydrogen bonding seen in the materials containing DBDI. They provide a common explanation for most of the trends in mechanical response discussed above—see below.

Notwithstanding the success of this approach, it is interesting to note that the results in Fig. 4.44 reveal an inconsistency with the original form of the GR constitutive model. It would predict *negative* gradients for $\sigma^b(\lambda)$, instead of the positive gradients shown, since the true strain-rate falls with increasing elongation in tests at constant nominal strain-rate. A possible explanation for this discrepancy is that thermally activated flow events in the hard phase are intrinsically anisotropic, whereas the model as previously described assumes they are isotropic [313]. Extension of the GR model to accommodate intrinsic anisotropy, has been discussed in outline [61, 314] and a preliminary discussion of the application to PU elastomers was given before [280]. Further details will be given elsewhere.

The primary concern of this study was the question of how the choice of hard segment composition in PUs elastomers affects their mechanical response, especially when the relatively rigid MDI segment is replaced partially or entirely by the relatively more flexible, and crystallisable, DBDI segment. It revealed a consistent pattern of behaviour. Other parameters being equal, replacement of MDI by DBDI was seen to lead to a response that is substantially stiffer (e.g. higher E_{1C}), shows higher hysteresis (e.g. lower E^*_{1R}) and lower strain recovery, and higher stress relaxation. The separation of stress into a viscoplastic component, dominated by the hard phase, and a hyperelastic component, dominated by the soft phase, described in the previous section, allows all these results to be explained on a common basis. Applying equation (4.16) to cyclic uniaxial extension at constant elongation rate $\dot{\lambda}$ to a maximum elongation $\lambda_{max} - 1$ allowed the input strain energy density and recovered energy density to be evaluated:

$$E_{1C} = \int_1^{\lambda_{max}} \left[\sigma^c(\lambda) + \nu_h \sigma^b(\dot{\lambda}, \lambda)\right] d\lambda; \; E_{1R} = \int_{\lambda_r}^{\lambda_{max}} \left[\sigma^c(\lambda) - \nu_h \sigma^b(\dot{\lambda}, \lambda)\right] d\lambda \quad (4.19)$$

where $\lambda_r - 1$ is the residual elongation. Suppose that the pattern of behaviour seen in Fig. 4.43 and 4.44 were replicated across the range of combinations of PU recipes. It was immediately clear from the dependence of σ^b on hard segment content, combined with the relative insensitivity of σ^c to the hard segment, that the observed trends in the results for E_{1C} and E^*_{1R} would then be expected from equation (4.19). Similarly, of the two contributions, only σ^b is history-dependent. Thus stress relaxation must be attributed to this portion of the stress, and the differences in stress relaxation observed, for example in Fig. 4.7 and 4.8 (see previous section 3.4.), can also be explained in terms of differences in σb similar to those seen in Fig. 4.44 (for a different PU recipe).

Thus it was clear that the important effect of inclusion of the DBDI hard segment is to increase the flow stress of the hard phase. This might be considered anomalous in view of the greater conformational mobility of DBDI provided by the $-CH_2-CH_2-$ bridge. The explanation is provided by the structural evidence, revealing more complete phase segregation in the presence of DBDI (other features of the recipe and preparation remaining constant) [60], and in some cases hard phase crystallinity in addition. Closer interactions between the isocyanate groups provided by these structural differences are expected to be associated with a higher density of hard segment hydrogen bonds, leading to a higher stress required to overcome these bonds and achieve plastic flow of the hard phase.

The performance of these materials as elastomers was shown to vary greatly depending on the composition. The most extreme difference between the effects of MDI and DBDI hard segments is seen in the cases where the chain extender-diisocyanate couple is EG-DBDI or BG-DBDI, where the hard segments are observed to crystallise. In such materials the stiffness, hysteresis, residual elongation and stress relaxation are all increased relative to the corresponding MDI based PU. When the couple is DEG-DBDI the same trend is observed, although the differences are less pronounced because the flexible DEG inhibits crystallization. These differences between DBDI and MDI may be attributed to the greater flexibility of DBDI allowing a higher tendency to self-associate by hydrogen bonding. A particularly good balance of elastomeric mechanical properties is observed in the case of PU derived from equimolar mixtures of MDI and DBDI. These polymers are stiffer than those based on pure MDI but show the highest strain recovery of all the systems studied, a primary consideration for elastomers.

Results of the mechanical tests can be understood in terms of a physical model in which the stress is separated objectively into two distinct contributions: an elastic-viscoplastic stress σ^b rising from bond-stretching in the hard phase, and a hyperelastic stress σ^c. This separation has been carried out in the case of the EG-PEA combination and varying DBDI content. The primary effect of the presence of DBDI hard segments was shown to be a substantial increase in σ^b, for any given elongation, presumably as a result of the closer self-association of hard segments by hydrogen bonding. This single effect of the DBDI is sufficient to explain most of the DBDI-induced differences in the observed mechanical response, across the range of materials studied.

4.4.6 Interrupted tests

Further tensile load-unload stress relaxation tests were conducted by us by interrupting constant strain-rate tests (again 3.1×10^{-2} s^{-1}) with 100 s waiting times during loading and reloading, to observe the relaxation of stress at constant strain, before resuming extension [69, 127]. The interesting points to emerge from the interrupted tests concerned the magnitude and sign of stress relaxation over the 100 s interruptions during loading and unloading. Stress relaxation in interrupted tests increased with hysteretic energy loss. The observed effects of varying HS could be explained by the hard domains having a higher flow stress in the presence of DBDI relative to MDI, associated with increased hydrogen bonding in DBDI-based polymers, which is enhanced in by hard segment crystallinity [69, 127].

Shown in Fig. 4.45 and 4.46 are the load/unload cycles of two analogous materials. Only the type of diisocyanate differs, (DBDI—Fig. 4.45 or MDI—Fig. 4.46).

It observed that: *(a)* the magnitude of the mechanical stress relaxation was higher in the case of PUs with DBDI than with MDI; this was again consistent with the larger contribution of hard phase stress in the case of the DBDI based PUs [60, 61, 69, 127]. The magnitude of stress relaxation increased with strain during loading. This indicated that the stress contribution from the hard phase increased with strain, i.e. the strain-stiffening was not only from the soft (rubber-like) matrix [69, 127]. This was consistent with the arguments used in our earlier works, [61, 135] that the flow process needs to be considered as intrinsically anisotropic: with deformation the relaxing units tend to align with the direction of straining; *(b)* the sign of stress changed and was different between loading and unloading of the material (see Fig.

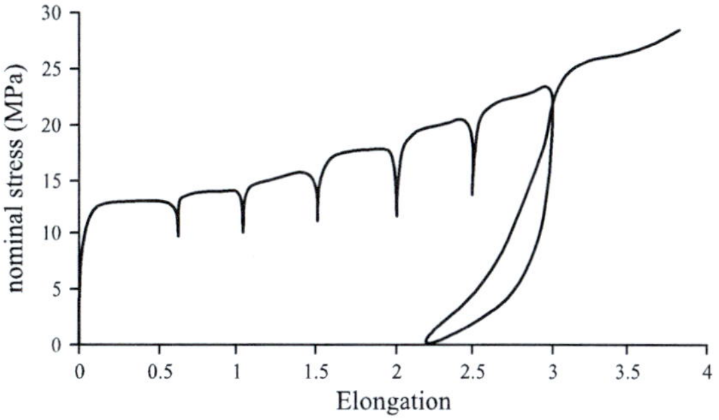

Fig. 4.45 Loading and unloading of a DBDI material (DBDI:EG:PEA). Interrupted test with 100 s waiting time during loading [127]

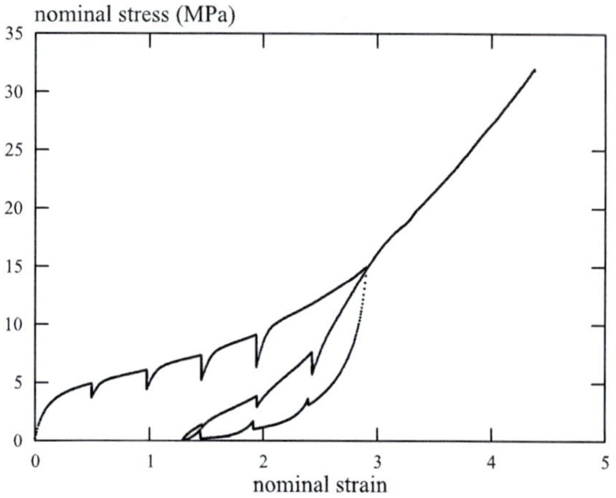

Fig. 4.46 Interrupted test with 100 s waiting time during loading and unloading of a MDI material (MDI:EG:PEA) [69]

4.46). This was consistent with a two phase microstructure where there are two contributions to the stress: an elastic component and a viscous component [61]. The sign of the viscous stress changed sign on reversal of straining, and therefore relaxation of this stress during unloading produced an increase in the net stress.

On loading, a common pattern of behaviour for the MDI and DBDI materials could be seen. When straining was resumed after stress relaxation for 100 s, the stress raised rapidly before continuing a steady increase with strain, resuming approximately the original curve. But, for the material based on DBDI there were observed higher E modulus and higher stiffness. Also lower strain recovery and strain energy recovery and higher residual strain, were retained with DBDI. It should be mentioned that these are examples of the preliminary research regarding the PUs mechanical behaviour when subjected to interrupted tests. Further research will be reported soon.

4.5 Mechanical and thermal aspects of polyurethane elastomers extended with diamines

Much of outcomes research is a systematic attempt to exploit what is known and make it better (Kevin Kelly)

A series of polyurethane-ureas (PUUs) has been synthesized by us, based on the aromatic diisocyanates DBDI or MDI. The materials were synthesized with two SS macrodiols, the polyether macrodiol PTMO or the polyester SS, PEA.

In the first stage of our study, three aromatic diamines were used as chain extenders: a non-symmetrical chain extender, 2,4-toluene diamine (TDA); two symmetric aromatic diamines: 4,4'-methylene dianiline (MDA) and 4,4'-diamino-dibenzyl (DAB). TDA was used in order to improve the mechanical properties and thermal stability. The materials were synthesized by a two steps polyaddition reaction as described elsewhere [315, 316]. In the second step the prepolymer was extended by a solution polyaddition reaction using a diamine as chain extender: DAPO, DAB, or MDA (Fig. 4.47). The materials were synthesized with a molar ratio between diisocyanate:macrodiol:chain extender = 2:1:1. PUUs films with thickness in the range 0.3-0.6 mm were obtained. The PUUs mechanical properties and thermal behaviour as revealed by TGA and DSC measurements were followed. Examples of HS schemes of MDI/ MDA and DBDI/DAB based materials are shown in Fig. 4.48 and 4.49.

The PUUs structures are shown in Table 4.17.

Table 4.17 Structure of polyurethane ureas obtained with diamines TDA, DAB and MDA

Polymer	Composition			Hard segment (%)
	DI	MD	CE	
PUU1	DBDI	PTMO	TDA	24.56
PUU2	MDI	PTMO	TDA	23.72
PUU3	DBDI	PEA	TDA	24.56
PUU4	MDI	PEA	TDA	23.72
PUDB	DBDI	PTMO	DAB	27.03
PUMD	MDI	PTMO	MDA	25.89

DI–diisocyanate; MD–macrodiol; CE–chain extender

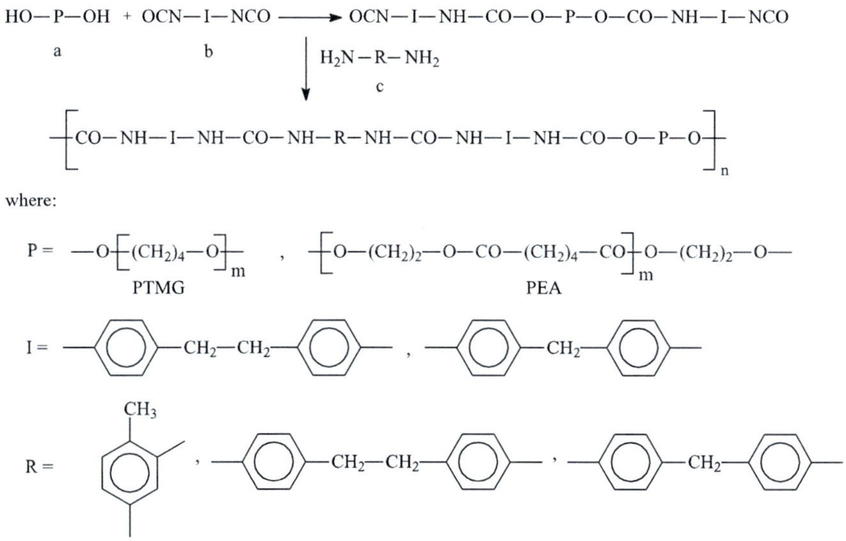

Fig. 4.47 Schematic of PUUs: P-polyol; I -isocyanate

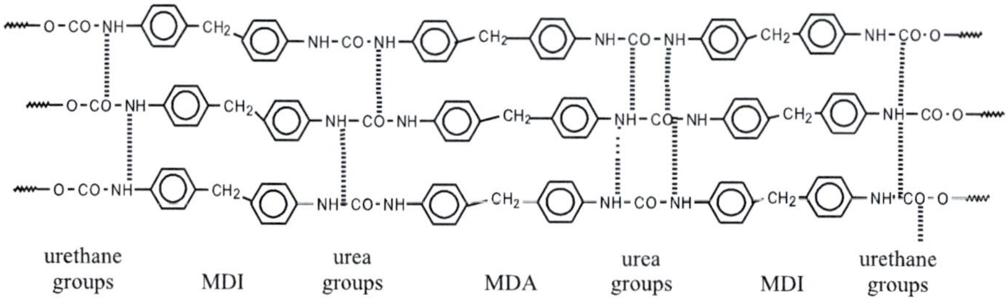

Fig. 4.48 Schematic of a hard segment of a MDI based PUU extended with aromatic diamine MDA

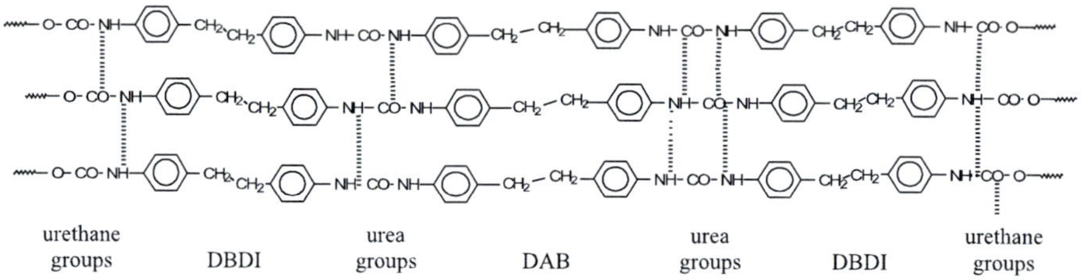

Fig. 4.49 Hard segment of a DBDI based PUU extended with aromatic diamine DAB

The type of diisocyanate (MDI or DBDI) strongly influenced the course of polymer synthesis and the polymer properties. As the HS contained urea and urethane groups, some polymers showed a tendency of precipitation (polymers PUDB based on DAB and DBDI) [17]. On the other hand, the HS ordered structure allowed physical crosslinks by hydrogen bonds which resulted in an increase of the thermal stability and of tensile properties data, as shown in Tables 4.18 and 4.19. In spite of the lack of symmetry induced by TDA, the PUs series chain extended with this diamine displayed a higher thermal stability. As in the case of polymer PUMD obtained with DBDI/DAB, all the materials based on TDA were soluble and viscous polymers without any precipitation tendency, unlike material PUDB, obtained with DBDI and extended with DAB. The PUMD material had higher viscosity values due to the MDI+MDA content. All the materials PUU_{1-4} (extended with TDA) had a good solubility because the non-symmetrical TDA caused an interruption of the succession of para-substituted aromatic rings of HS (DBDI, MDI) and cancelled the tendency of ordering induced by the diisocyanate DBDI [316]. The polymers PUU_{1-4} showed a good solubility into the reaction solvent and produced transparent and flexible films. Similar observations were made for the material PUMD based on MDI+MDA.

The viscosity values (η_{inh}) of the PUU_{1-4} series were in the range of 0.43-0.76 dL/g. The highest viscosity (0.99 dL/g) was found for the material PUMD. A similar observation was made for the polymer PUDB which had a surprisingly good viscosity (0.94 dL/g) although this material contained HS based exclusively on dibenzyl rests. This was explained by strong interactions due to intermolecular hydrogen bonds that can act even in solution. While PUDB containing DBDI resulted in opaque films that shrank during drying (up to break), in the case of PUUs based on TDA flexible and transparent films were obtained [316].

The resulting linear polymers were characterized by TGA, DSC, inherent viscosity measurements, and mechanical properties as shown in Tables 4.18 and 4.19.

Table 4.18 Characterization of polyurethane-ureas (PUU_{1-4})

PUU_i	η_{inh} (dL/g)	TGA (°C)	
		T_5^*	Max DTG
PUU_1	0.67	300	440
PUU_2	0.76	310	435
PUU_3	0.63	315	375
PUU_4	0.43	285	405
PUDB	0.94	191	348
PUMD	0.99	265	355

$^*T_5$—5% weight loss temperatures

The effect of the urea linkage (in the HS), on the extent of phase separation and other polymer properties can be revealed by the study of polymers thermal behaviour (TGA and DSC). As seen in Table 4.18, the chemical structure influenced the PUUs thermal stability. The materials PUU_{1-4} had 5% weight loss temperatures (T5) ranging between 285 – 315°C, while those of materials PUDB and PUMD were of 191 – 265°C. Polymers PUU_{1-4} showed DTG maxima between 375 – 440°C, the highest value belonging to PUU^1 obtained with DBDI and PTMO.

PUUs based on TDA showed higher thermal stabilities due to the proximity of urea and urethane groups in the HS as compared to other PUUs. In the case of PUDB and PUMD that contained diamines with two aromatic rings, the number of hydrogen bonds was not sufficient, (the distance between successive urea and urethane groups being too large). On the other hand, it is well known that the polyester SS can contribute to greater phase mixing than polyether, in spite of the tendency of the polyesters to crystallize more readily than polyethers. This behaviour implied that the hydrogen bonding of urethane to polyester is stronger than that to polyether. As shown elsewhere unlike in the case of PUDB and PUMD, the DSC tests revealed that PUU_{1-4} were amorphous polymers [316]. Thus, in the case of $PUU_{1,2}$ which contained PTMO, the SS phase was amorphous, since the melting endotherm was absent over −27°C. However, PUU_2 presented a small endotherm at 232°C which was attributed to the glass transition temperature of HS domains. The polyester materials $PUU_{3,4}$ based on PEA had melting endotherms of the crystalline SS at 42.5°C and 44.7°C, and small exotherms were also observed, which was due to the HS crystallization that occurred at 197.7°C and 196.6°C, respectively [316].

The tensile properties of the PUUs series listed in Table 4.17 were determined (Table 4.19).

Table 4.19 Tensile properties of the PUUs series labelled in Table 4.17

PUU_i	Stress at 100% and 300% nominal strain [MPa] 100%	300%	Tensile strength [MPa]	Elongation at break [%]	Residual elongation [%]
PUU_1	4.3	7.3	25	800	120
PUU_2	4.4	9.8	39	750	16.6
PUU_3	4.2	10.3	36	750	77.7
PUU_4	3.4	6.9	23	700	127.7
PUDB	5.9	10.6	36	850	389
PUMD	6.3	11.4	33.5	600	50

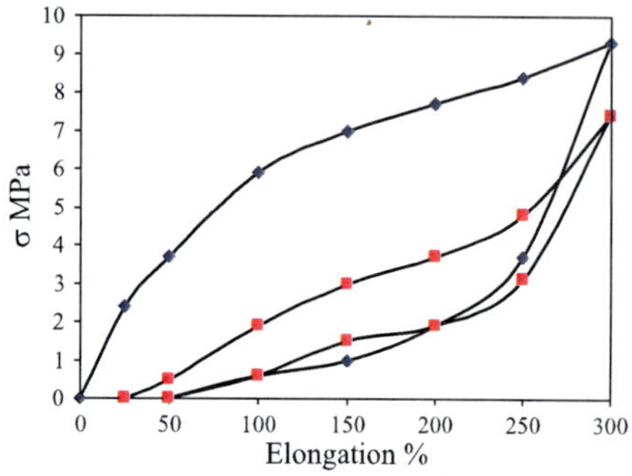

Fig. 4.50 Tensile load/unload cycles for polyurethane-urea based on DBDI: PUDB (PTHF/DBDI/DAB)

The PUUs tensile properties depended on the molecular weights (expressed as the inherent viscosity) and were strongly influenced by the chemical structure. Higher stress-strain data of the series PUU_{1-4} were observed for the materials PUU_2 and PUU_3, comparable with the data of PUDB based on DBDI. The best results were found for the material PUU_2 with the highest tensile strength value (39 MPa) and lowest residual elongation (16.6%).

The materials were also subjected to cyclic tensile experiments. Example *tensile load/unload cycles* are shown in Fig. 4.50 and 4.51 for the materials PUDB and PUMD. Polymers based on DBDI hard segments, displayed higher stiffness and strength than did the conventional MDI based PUs. Polymer with MDI hard segments gave lower stresses and least hysteresis, whereas polymer with DBDI hard segments gave higher stresses and most hysteresis.

Similar conclusions could be drawn as in the case of dibenzyl polyurethanes chain extended with diols like BDO or EG as shown in previous sections: that such features of the response can be attributed to differences in hard phase plastic flow stress, resulting from crystallinity in the DBDI phase.

In a second stage of our study we compared the DAB and MDA based materials (PUDB and PUDM from above) with other PUUs obtained by using the same macrodiols, (polyester PEA or polyether PTMO) *but the diamine chain extenders 2,5-bis-(4-amino-phenylene)-1,3,4-oxadiazole (DAPO) or 2,6-diaminopyridine (DAPy).* The materials were obtained similarly as in the case of the TDA and MDA

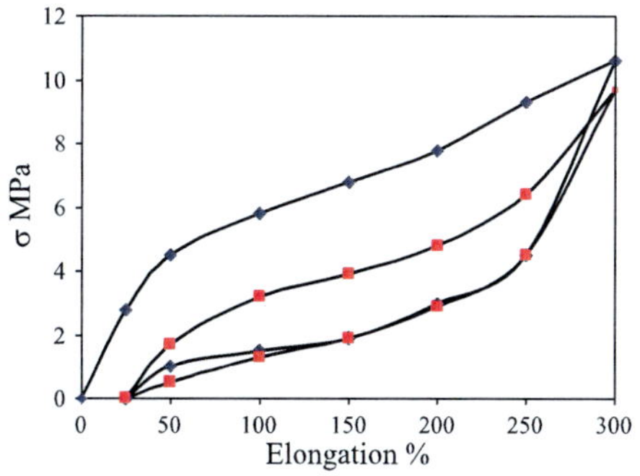

Fig. 4.51 Tensile load/unload cycles for polyurethane-urea based on MDI: PUMD (PTHF/MDI/MDA)

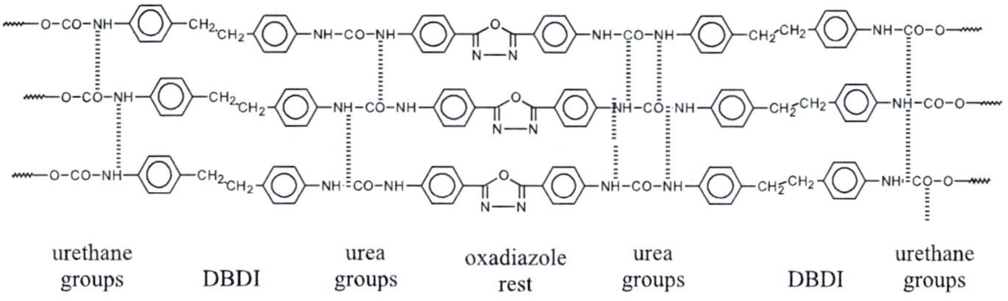

Fig. 4.52 Schematic of a PUs hard segment of DBDI extended with DAPO

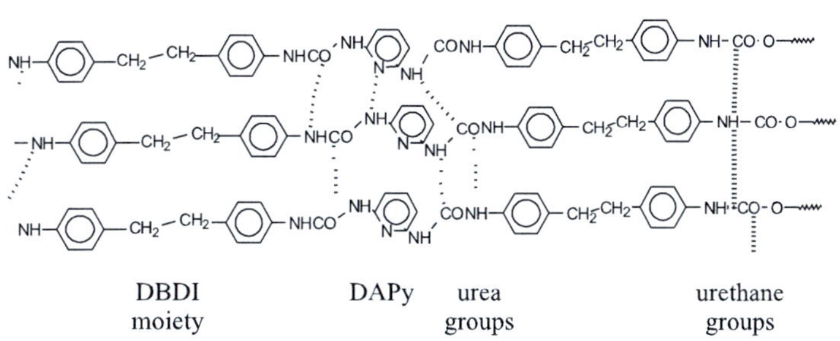

Fig. 4.53 Schematic of a PUs hard segment of DBDI extended with DAPy

polymers. Examples of HS schemes of MDI/ MDA and DBDI/DAB based materials are shown in Fig. 4.52 and 4.53.

Shown in Table 4.20 is the composition and the inherent viscosity (η_{inh}) of the materials studied:

The *viscosity values* of the $PUPY_{1-4}$ series were in the range of 0.37-0.55 dL/g. This was due to the worm-like configuration of the macromolecular chain even though relatively high molecular weights were revealed by the GPC data. Some increase of polydispersity (polymer $PUPY_4$) was observed. Moreover, a correlation has been observed between viscosity and polydispersity: the smaller the viscosity values, the higher the polydispersity and molecular weight. The relative small

Table 4.20 PUUs composition and properties: materials extended with DAPO, DAPy, DAB and MDA

PU	Composition	Molar ratio P/DI/DA	Hard segment (%)	η_{inh} (dL/g)	TGA (°C) T_5/T_{10}	TGA (°C) Max DTG	DSC (°C) Endo	DSC (°C) Exo
$PUOX_1$	PEA/DBDI/DAPO	1/2/1	27.02	0.62	248/275	309/344	38	95
$PUOX_2$	PEA/MDI/DAPO	1/2/1	26.30	0.88	253/285	313/358	43	95
$PUOX_3$	PTHF/DBDI/DAPO	1/2/1	28.08	0.61	234/261	288/361	49	183.5
$PUOX_4$	PTHF/DBDI/DAPO	1/3/2	39.35	0.45	224/255	278/364	45	93, 221
$PUPY_1$	PEA/DBDI/DAPy	1/2/1	24.09	0.55	268/288	305/361	35, 60	90
$PUPY_2$	PTHF/DBDI/DAPy	1/2/1	24.93	0.43	261/275	330/358	30, 64	115
$PUPY_3$	PTHF/DBDI/DAPy	1/3/2	34.44	0.50	287/298	361/396	27, 60	87, 190
$PUPY_4$	PEA/MDI/DAPy	1/2/1	23.13	0.37	285/309	351/388	40	-
PUDB	PTHF/DBDI/DAB	1/2/1	27.03	0.94	191/224	268/348	51, 167	70
PUMD	PTHF/MDI/MDA	1/2/1	25.89	0.99	265/285	292/355	56, 177	75

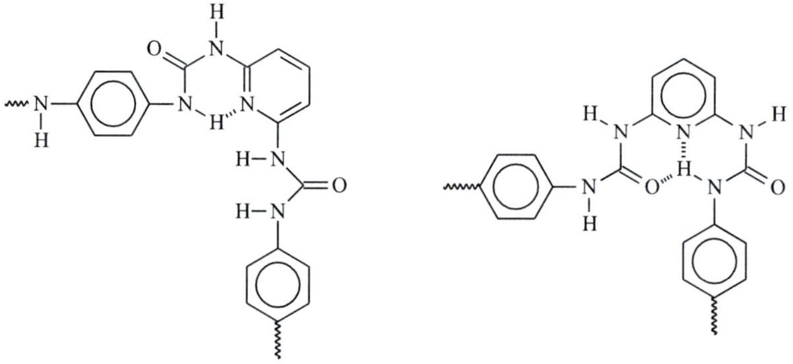

Fig. 4.54 Intramolecular hydrogen bonds attributed to polyurethanes with pyridine rings

viscosity could be explained by the meta-substitution of pyridine rings, the stronger interaction of polymer-solvent type and the increased tendency of 2,6-pyridine rest to form intramolecular hydrogen bonds with urea or urethane groups (Fig. 4.54).

Due to the presence of MDI and highest molecular weight, the material $PUOX_2$ showed a good viscosity value (0.62 dl/g). The $PUOX_{1-4}$ series (especially the DBDI materials) and polymer PUDB displayed a semi-crystalline structure. The crystallinity degree varied from 5.49% ($PUOX_3$) to 18.84% ($PUOX_4$).

The ordered structure of HS allowed physical crosslinks by hydrogen bonds which resulted in an increase of the thermal stability. The PUUs series extended with DAPO showed higher thermal stability due to the symmetry and rigidity of this heterocyclic diamine. The series $PUOX_{1-4}$ had the 5% loss weight temperatures (T_5) between 225 – 252°C, as compared to PUDB and PUMD (191 – 265°C).

Polymers $PUPY_{1-4}$ decomposed above 260°C. The highest value was observed for material $PUPY_3$ with the longest HS (298°C).

The materials $PUOX_2$ and PUMD (both of them obtained with MDI), showed higher viscosity values. PUs extended with DAPy had a good solubility since the meta-substituted pyridine ring caused an interruption of the succession of para-substituted aromatic rings of HS (DBDI or MDI) and cancelled the tendency of ordering induced by DBDI. For example, because the HS of $PUPY_3$ contained a urea-urethane group ratio of 2/1, the polymer $PUPY_3$ showed a good solubility into reaction solvent and produced transparent films. The viscosity values of the $PUPY_{1-4}$ series were in the range of 0.37 – 0.55 dL/g. The series of materials extended with DAPO showed a higher thermal stability due to the symmetry and rigidity of DAPO.

The thermal behaviour of the materials labbeled in Table 4.20 was also investigated by means of DSC measurements. Results are shown in Fig. 4.55 and 4.56 and in Table 4.21.

Table 4.21 DSC thermal behaviour of the materials listed in Table X

ENDOTHERMS	PEA melting peaks:	38°C ($PUOX_1$), 43°C ($PUOX_2$) and 40°C ($PUPY_4$)
	PTMO melting peaks:	35°C ($PUPY_1$), 30°C ($PUPY_2$) and 27°C ($PUPY_3$)
	T_G of HS and/or dissociation of HS hydrogen bondings:	49°C ($PUOX_3$), 45°C ($PUOX_4$), 60°C ($PUPY_{1,3}$), 64°C ($PUPY_2$), 51°C (PUDB) and 56°C (PUMD)
	Conformational changes in the soft segments:	160°C ($PUOX_3$), 165°C ($PUPY_3$), 167°C (PUDB), 177°C (PUMD), 185°C ($PUPY_1$) and 190°C ($PUOX_4$)
	Endotherms that represent the beginning of decomposition process:	240°C ($PUOX_1$) and 230°C ($PUPY_4$)
EXOTHERMS	Conformational changes in mixed phases:	70°C (PUDB), 75°C (PUMD), 87°C ($PUPY_3$), 90°C ($PUPY_1$), 93°C ($PUOX_4$), 95°C ($PUOX_1$, $PUOX_2$), and 115°C ($PUPY_2$):
	Conformational changes during of the crystallization process of HS:	155°C ($PUOX_2$), 190°C ($PUPY_3$) and 221°C ($PUOX_4$)

The HS crystallization processes and conformational modifications were revealed by DSC exotherms especially in the case of PUUs with the longest HS based on DBDI ($PUOX_4$ and $PUPY_3$).Conformational changes occurring in the SS were observed mostly in PUUs with longer HS displaying the smallest SS melting temperatures e.g. better phase separation ($PUOX_3$, $PUOX_4$, $PUPY_3$, and PUDB or PUMD).

In a third stage of this study, the tensile properties of the series of materials labeled in Table 4.20 (extended with the diamines DAPO, DAPy, DAB and MDA)

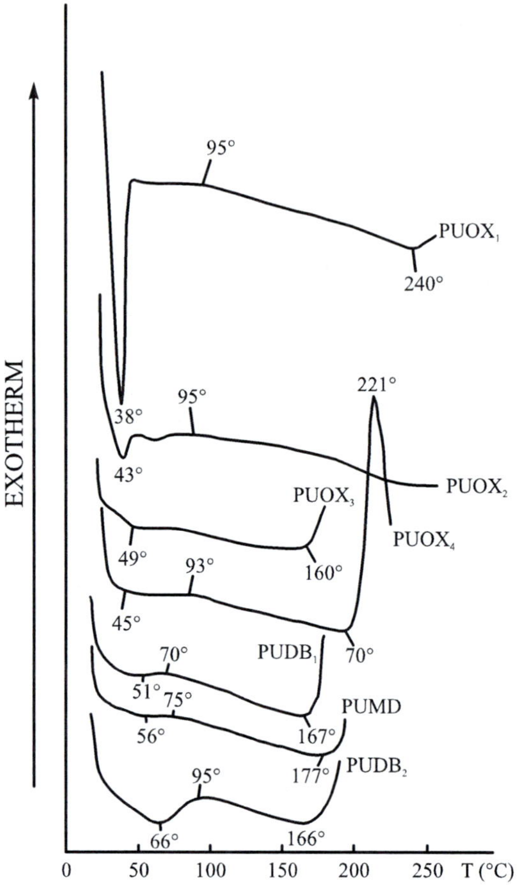

Fig. 4.55 DSC curves for the series of materials $PUOX_{1-4}$ extended with DAPO

were compared by us, with two PUs extended with the low molecular weight diol EG, (see series $PUEG_{1-2}$ in Table 4.22).

Table 4.22 Composition of diol-extended polyurethanes

PU	Structure	Molar ratio	Hard segment (%)
PUEG1	EG-PEA-DBDI	3/1/4	38.31
PUEG2	EG-PEA-MDI	3/1/4	37.23

Shown in Table 4.23 are the tensile properties of the materials listed in Tables 4.20 (extended with diamines) and 4.22 (extended with the diol EG).

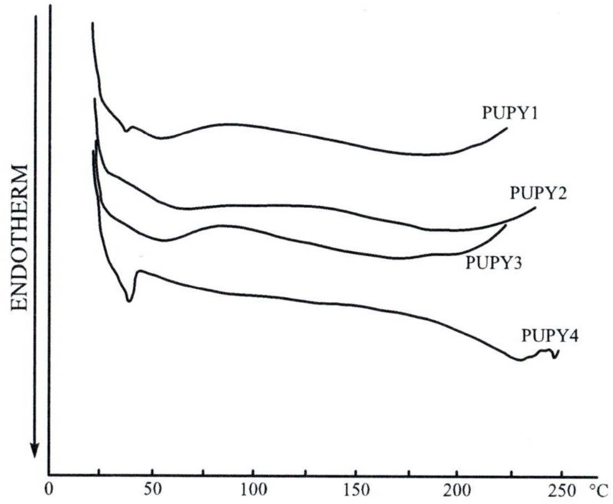

Fig. 4.56 DSC curves for the series of materials $PUPY_{1-4}$ extended with DAPy

Of all PUUs, the DBDI based material PUDB had the highest tensile strength (33.5 MPa), followed by material $PUPY_3$ (also based on DBDI) with HS = 34.44% (31.5 MPa). In such materials, the HS adopt a more compact packing arrangement (given by the presence of DBDI) by hydrogen bond interactions. The number of hydrogen bonds is not sufficient if the distance between successive urea and urethane groups is too large.

Table 4.23 Tensile properties of the materials listed in Tables 4.20 and 4.22

PU	Modulus [MPa] 100%	Modulus [MPa] 300%	Strength stress [MPa]	Elongation at break [%]	Residual elongation [%]
PUOX1	1.4	-	1.8	250	39
PUOX2	1.9	3.4	15.7	800	216
PUOX3	3.5	5.9	17.3	850	339
PUOX4	3.3	-	6.7	200	39
PUPY1	1.9	-	1.9	100	0
PUPY2	2.0	3.0	5.7	700	72
PUPY3	4.8	7.9	31.5	1000	67
PUPY4	1.52	2.8	5.3	750	23
PUDB	5.9	10.6	33.5	850	389
PUMD	6.3	11.4	28.5	600	50
PUEG1	13.7	17.0	40.7	350	145
PUEG2	8.42	20.6	37.5	450	75

Of all the materials included in Table 4.23, the highest tensile strength (40.7 MPa), was found for the material $PUEG_1$ based on the couple DBDI-EG, followed by the material $PUEG_2$ (37.5 MPa) obtained with MDI, followed by material PUDB.

The polymers extended with EG had the highest HS content ((37.23% (MDI) and 38.31% (DBDI)), with the shortest distance between urethane groups belonging to the same HS, as built during the synthesis (molar ratio:diisocyanate:macrodiol:chain extender = 4:1:3). As previously shown, in PUs based on the chain extender-diisocyanate couples EG-DBDI (or BDO-DBDI), the HS were observed to crystallize. PUUs with HS containing para-substituted aromatic rings showed the weakest mechanical properties ($PUOX_{1,4}$). Similar observations were made for the PUUs with HS containing at least one meta-substituted aromatic ring ($PUPY_{2,4}$).

The smaller values observed for the DAPO based materials ($PUOX_{1-4}$ series) was explained by the lower interaction between aromatic rings than in the case of the hydrogen bonding type.

4.6 Mechanical properties of dibenzyl based polyurethane elastomers containing a molecularly dispersed UV absorber

No one can live under degradation (Walid Jumblatt)

It is true. As for things, some of them resist to degradation much better than others ☺ Like other polymers, PUs are subjected to *thermal and photo-oxidative degradations*, even during their synthesis. In order to reduce these effects polymers can be treated with suitable antioxidants and light-stabilizers [64, 317].

The photo-oxidation mechanism of polyurethanes is generally less known and the products are often complex and difficult to separate and identify [317]. This mechanism is considered to be similar to that of polyamides due to the similitude of their chemical structure. The mechanism of photo-oxidation of polyurethanes was studied on model systems [318]. In the presence of oxygen, the degradation starts by the attack of oxygen atom on α-methane groups (proton abduction) and formation of the hydroperoxydes. The rate of oxygen absorption depends on the chemical constitution of diisocyanate and of the hydroxyl-containing partner used in the synthesis [319]. With regard to the initial rate of oxygen absorption, PUs based on different isocyanates may be placed in the following order: toluene diisocyanate (TDI) > diphenylmethane diisocyanate (MDI) > hexamethylene diisocyanate (HDI). Photo-rearranged intermediates are suggested as precursors of colored products in the photo-oxidation of polymers. PUs based on MDI exhibit intense yellowing tendency during photo-oxidation due to formation of quinoid groups as shown in Fig. 4.57.

The PUs photo-degradation processes are accelerated by water and ammonia [319].

In this work, a series of PUs based on DBDI were prepared. They presented some specific features due to the diisocyanate structure [320]. Thus, when subjected to the photo-degradation process, the PUs based on DBDI showed a lower yellowing tendency than those based on MDI. This could be explained by the particular behaviour of the ethylene bridge between the two aromatic rings against the photo-degradation

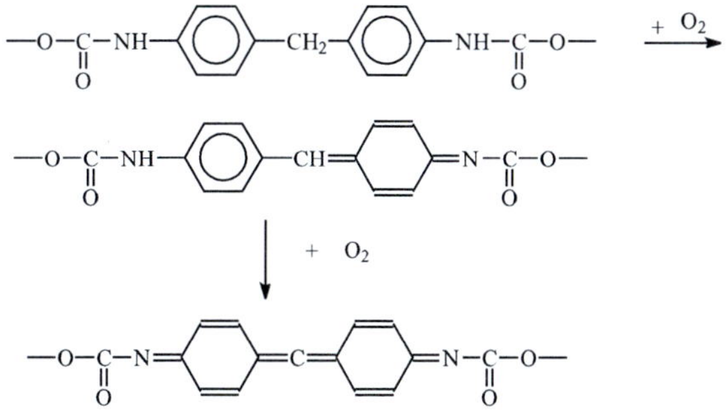

Fig. 4.57 The photo-oxidation process of the polyurethanes based on MDI [64]

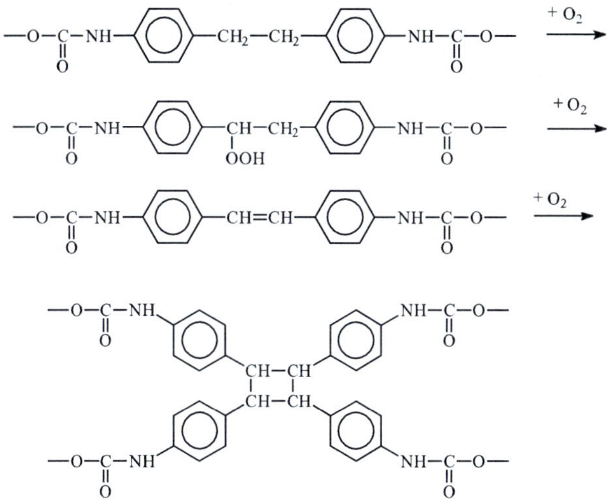

Fig. 4.58 The photo-oxidation process of the polyurethanes based on DBDI [64]

process. Whereas the methylene bridge of MDI is simultaneously activated by the two aromatic rings (Fig. 4.57), in the case of DBDI, each methylene group is activated by only one aromatic ring. The photo-oxidation reaction of PUs based on DBDI for each degradation step could be described as in Fig. 4.58.

As shown in Fig. 4.58, the hydroperoxydes, which result in the initial stage of the oxidation process, can eliminate water leading to the stilbene structures. The stilben moieties can further suffer a 2 + 2 cycloaddition reaction under UV radiation leading to crosslinked colorless structures of cyclobutane type [59, 321]. These new structures of cyclobutane hinder the formation of the systems containing quinoid groups [59]. This is the explanation of the *improved light stability of all PUs based on DBDI as compared to those based on MDI.* As a result the yellowing of PUs based on DBDI appear much slower. The photo-oxidative degradation of these materials can be improved by treating them with UV absorbers. Frequently, the addition of an

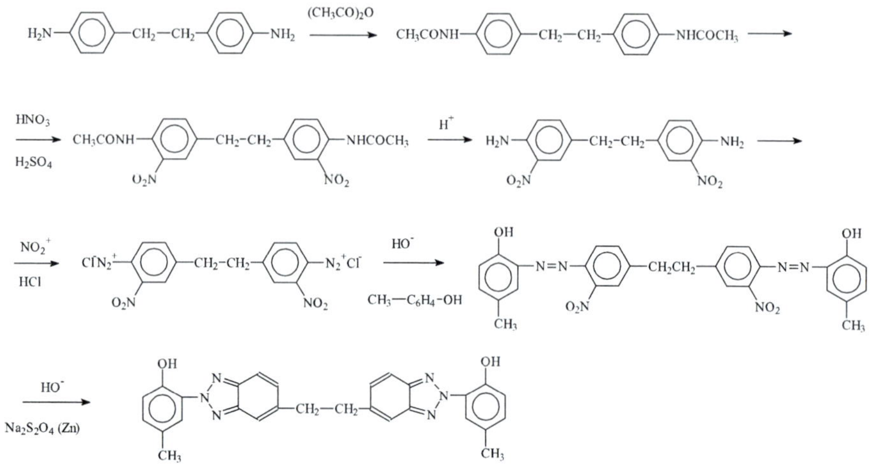

Fig. 4.59 Synthesis of the UV absorber: 1,2-bis-[2-(2-hydroxy-5-methylphenyl)-5-benzotriazolyl]-ethane (BHMBE) [64]

insoluble UV absorber to the polymers leads to a more or less fine solid dispersion in the polymer matrix. One of the principal conditions to enhance the effectiveness of the additive is to obtain as fine as possible particle dispersion of UV absorber in polymers [64].

In this work we have studied a series of blends made up by PUs based on DBDI and an UV absorber having a good behaviour when exposed to environment. The materials were obtained by two methods as described elsewhere [64]. The UV absorber was obtained as described earlier (Fig. 4.59) [322].

The materials were obtained with PEA of molar mass 2 000 g/mol and chain extended with DEG.

The synthesis of the UV absorber was performed by starting from 4,4'-diaminodibenzyl through a multistep reaction as follows: acetylation, nitration, hydrolysis, diazotizing, azo-coupling and, finally, reductive cyclization leading to the 1,2-bis-[2-(2-hydroxy-5-methylphenyl)-5-benzotriazolyl]-ethane (BHMBE) as shown in Fig. 4.59 [322]. This product was characterized by physical and chemical analyses and was used as UV absorber in blends with polyurethane [64, 322]. 1,2-bis-[2-(2-hydroxy-5-methylphenyl)-5-benzotriazolyl]-ethane (BHMBE) [322]: M. p. 208 – 210°C (toluene). Elemental analysis: Calculated for $C_{28}H_{24}N_6O_2$: C% = 70.60; H% = 5.04; N% = 17.65. Found: C% = 68.87; H% = 4.92; N% = 17.26. UV absorption at 304 and 344 nm. ^{1}H NMR ($CDCl_3$, ?): 11.15 (s, OH, 2H), 7.64 – 8.18 (m, aryl, 6H), 7.09 – 7.26 (m, benzotriazole, 6H), 4.45, 3.23 (d, CH_2, 4H), 2.4 (s, CH_3, 6H) [64, 322].

Two types of polyester PUs were synthesized:

(a) the first PU was achieved with a small degree of crosslinking: DBDI:PEA: DEG DBDI, and HS fraction at ca 40%. A 10% NCO excess (isocyanate I = 110) was obtained by the polyaddition reaction in bulk of 1 mol of the macrodiol PEA and 3 mol of DBDI that formed a prepolymer (P) having isocyanate (NCO) final groups. This prepolymer was chain-extended with 1.755 mol DEG leading to a reactive oligomer having final NCO groups. The crude material was cast as 2 mm thickness plates and was cured at 110°C for 24 h. The product was post-cured two weeks at room temperature. During this time the atmospheric humidity consumed the remaining small excess of isocyanate groups leading to urea bridges and rare allophanate groups too [17, 64].

(b) The second type of PU was linear: DBDI:PEA:DEG40 (I = 100) and it was obtained similarly as in the case (a), except the quantity of the chain-extender: 2 mol of DEG was used [64]. In this case, the product did not contain free isocyanate groups any longer and, consequently, the post-curing process by means of the atmospheric humidity did not take place any longer. Therefore, the urea and allophanate branches on the polymer chain were avoided.

The introduction of the UV-stabilizer in PUs was made by different ways. The first kind of stabilized polymer (a) was obtained by adding a solution of BHMBE in DMF (conc. = 4.34 g/100 ml) under stirring into the melted prepolymer. The solvent DMF was then removed by vacuum distillation, and the chain extender was added. A polyurethane containing 0.58% BHMBE was obtained (Fig. 4.60). A PU sample obtained by introducing the powdered UV-stabilizer in the macrodiol directly (forming a dispersed phase) presented poor mechanical properties because of their less homogeneous structure [64].

The second method used a solution of polyurethane of type (b) which was prepared by dissolving 16 g of polymer into 100 ml DMF under stirring at 70°C for 24 h. Based on this polyurethane solution, two blends were prepared each containing different proportions of BHMBE (0.5% and respectively 1%). Thin films (0.3 mm thickness) have been prepared from these solutions by casting on glass plates and drying at 60°C for 72 h. The samples prepared with 0.5 and respective 1% BHMBE concentration were exposed for 120 days at the sun light in atmospheric conditions. The mechanical properties were periodically monitored (Tables 4.24 and 4.25).

A practical route to reduce the effects of photo-degradation in PUs (as in the case of other polymers) is the addition of suitable light stabilizers by different ways. The most used procedure is the additive stabilization method. A more effective method to increase the maintenance of the stabilizer in the polymer is the chemical bonding of these compounds to the macromolecular chain even in the course of polyurethane synthesis. Some attempts to introduce in such way the stabilizers in polymers are known. Thus, a benzotriazole UV absorber was mixed with a prepolymer containing isocyanate groups and the chain extender directly into moulding machine (conc. 0.5%) [64, 323]. The stabilizer can be also mixed into polyol component before its reaction with the isocyanate component. The stabilizer that usually contains OH groups could form chemical bonds with the macromolecular chain in some extent [324].

4.6.1 Polyurethane-UV absorber blends

In this study, BHMBE containing two OH groups was added into the prepolymer leading to the formation of urethane adducts between phenolic OH groups and isocyanate groups as presented in Fig. 4.60. Such a type of reaction can have a high probability to occur because it is well known that isocyanates and phenols can react easily leading to blocked isocyanates [325]. A similar reaction was reported in the case of adducts between isocyanates and antioxidants that contain reactive hydroxylic groups also [324]. The isocyanates blocked with phenols are unstable at temperatures of 150 – 200°C, when they decompose by regenerating (releasing) the initial isocyanate structures, that makes them useful for many applications, like one-pack polyurethane systems [326]. Based on these considerations we presumed that the UV absorber does not consume in fact the isocyanate groups because the initially formed adducts suffer a thermal dissociation during the final cure of polyurethane objects (temperatures above 100°C). Moreover, because of the temporary binding to the PU chain, the UV absorber has much more chances to reform, but in a more dispersed state in the bulk of polymer instead to migrate to amorphous zones (soft segments) as in usual cases [64].

The second method for the addition of BHMBE in polyurethane consisted of a simple mixing of the stabilizer with the solution of polymer following the classic route. From these solutions test-pieces were prepared and subjected to the sun light in the condition of atmospheric ageing process for 120 days and were tested periodically from the view-point of the mechanical properties (Tables 4.24 and 4.25) [64].

4.6.2 Effect of the UV stabilization on mechanical properties

Shown in Table 4.24 is the behaviour of specimens of the PU type DBDI:PEA: DEG40 Index 110 (PU 110) with or without BHMBE component, exposed to the same environmental conditions. The introduction of BHMBE into the polyurethanes modified in a small extent their mechanical properties. It observed that during the ageing testing conditions the tensile strength of the simple PU decreased with 17.5% (from 50.54 to 41.7 MPa), while the tensile strength of the protected PU remained practically unchanged. Small variations of 100% and 300% tensile stress was observed for the stabilized material while the value of tensile stress of the polymer without stabilizer a little increased. These results had to be considered cautiously because the ageing phenomena in the case of 2 mm thickness plates are relatively insignificant due to the relative thickness of test-pieces [64]. In these conditions the degradation processes (occurring mainly at the surface) required longer time to migrate in the bulk of specimens.

These considerations enabled us to study the effect of the stabilizer addition on samples of smaller thickness (0.3 mm). The data on the behaviour of these specimens from the viewpoint of mechanical properties are shown in Table 4.25.

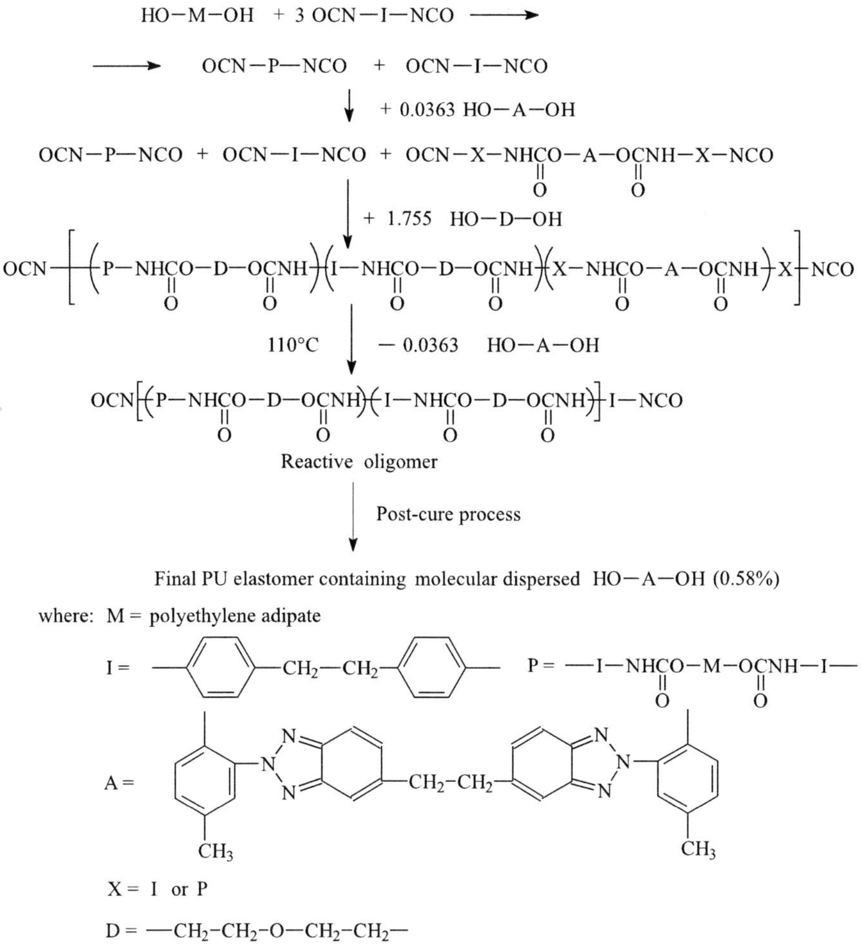

Fig. 4.60 Synthesis of the type (a) polyurethane blended with UV absorber [64]

A more significant variation of the properties could be observed during 120 days of ageing process. After the first 30 days, almost all samples presented a significant decrease of the tensile strength. Thus, the PU sample without stabilizer showed a decrease of the tensile strength of 58% (from 27.60 to 16.05 MPa), while in the case of the sample with 0.5% stabilizer the strength was reduced to 56.5% (from 27.27 to 15.40 MPa) and it was reduced to only 32.3% (from 27.12 to 8.75 MPa) for the sample with 1% stabilizer [64]. In the following period (until 120 days) all the mechanical properties presented smaller variations, the observed differences being favorable in the samples with higher content of stabilizer. For example, the tensile strength of the untreated sample diminished with 65.4% (from 27.60 to 18.06 MPa) whereas in the case of the sample with 0.5% stabilizer the reduced value was 59%

Table 4.24 Tensile properties of *plates* of polyurethane DBDI:PEA:DEG_{40} (I = 110) with or without UV absorber [64]

UV Absorber	PU composition	Mechanical measurements	Time (days) 7	30	60	90	120
BHMBE Conc. = 0.58%	DEG_{40} PEA_{2000} DBDI I = 110	100% tensile stress (MPa)	3.6	3.57	3.7	3.8	3.7
		300% tensile stress (MPa)	10.4	6.86	7.7	7.7	6.9
		Tensile Strength (MPa)	42.2	42.3	41.7	39.1	44
		Elongation at break (%)	605	600	550	550	625
		Residual elongation (%)	20	20	20	20	20
–	DEG_{40} PEA_{2000} DBDI I = 110	100% tensile stress (MPa)	3.9	4	3.9	3.9	4.1
		300% tensile stress (MPa)	7.8	8.3	8.4	7.8	8.2
		Tensile Strength (MPa)	50.5	45.2	42.6	38.2	41.7
		Elongation at break (%)	623	600	550	550	550
		Residual elongation (%)	10	10	10	20	10

Table 4.25 Tensile properties of *films* of polyurethane DBDI:PEA:DEG_{40} (I = 110) with or without UV absorber [64]

UV Absorber BHMBE	PU composition	Mechanical measurements	Time (days) Initial	30	60	90	120
0	DEG_{40} PEA_{2000} DBDI I = 110	100% tensile stress (MPa)	3.34	3.3	4.1	4.8	5.2
		300% tensile stress (MPa)	5.9	5.4	6.7	6.9	6.9
		Tensile Strength (MPa)	27.6	11.5	8.3	10.4	9.5
		Elongation at break (%)	724	650	400	550	500
		Residual elongation (%)	85	105	50	70	60
0.5	DEG_{40} PEA_{2000} DBDI I = 110	100% tensile stress (MPa)	2.6	2.7	4.1	4.2	4.7
		300% tensile stress (MPa)	6.1	5.0	6.7	5.8	6.5
		Tensile Strength (MPa)	27.3	11.9	12.0	8.4	11.2
		Elongation at break (%)	600	700	550	500	600
		Residual elongation (%)	90	110	70	70	100
1	DEG_{40} PEA_{2000} DBDI I = 110	100% tensile stress (MPa)	2.9	3.4	3.5	4	4.2
		300% tensile stress (MPa)	5.86	5.2	6.3	6	6.8
		Tensile Strength (MPa)	27.2	18.4	17.4	13.3	14.4
		Elongation at break (%)	640	700	675	700	675
		Residual elongation (%)	60	110	80	90	70

(from 27.27 to 16.06 MPa) and it was 47% (from 27.12 to 12.78 MPa) for the sample with 1% stabilizer [64].

Overall, it could be concluded that the addition of UV stabilizer 1,2-bis[2-(2-hydroxy-5-methylphenyl)-5-benzotriazolyl]ethane (BHMBE) to the prepolymer step of the studied PUs can be used successfully [64]. The UV absorber first formed a temporary chemical binding as a result of the addition reaction between OH phenolic groups and the isocyanate groups belonging to the prepolymer. During the post-cure, the new urethane groups broke releasing the absorber that remained trapped in the polymer matrix as a molecular level dispersion.

This method of mixing the UV absorber with PUs was used without damage or significant decrease of mechanical properties. The presence of BHMBE determined temperate stabilization effects. In this first stage of our research the influence of the stabilizer of a concentration range up to 1% in polymer was studied [64].

This work will be continued by us with the investigation of the stabilization effects on PUs containing higher concentrations of UV absorber.

4.7 Influence of hydrogen bonding on the mechanical response of polyurethane elastomers

If you have a difficult task, give it to a lazy person; they will find an easier way to do it.
(Hlade's Law)

We did not do so, even if this study has not been quite an easy task for us ☺. The topic has been challenging ☺.

As seen in previous chapters, it is clear that the mechanical properties of PUs are mainly determined by their microphase morphologies. They are extensively hydrogen bonded. The hydrogen bonding is the strongest secondary chemical bond, with a strength estimated to about 20-50 kJ/mol [4].

The effect of hydrogen bonding is frequently used to explain various anomalies or improved properties. However it is difficult to isolate this effect from the effects of physical and chemical structure. Some studies have concluded that molecular mobility is not controlled by hydrogen bonding. Instead it is believed that a rapid increase in molecular mobility accompanying glass transition allows the hydrogen bonding to dissociate [2–4]. Therefore the hydrogen bonding does not necessarily enhance mechanical properties, although there is insufficient published data that quantitatively demonstrates the effect of hydrogen bonding on the mechanical properties.

As shown [4], a clear effect of hydrogen bonding could be observed only if mechanical tests are carried out on PUs of analogous structures with and without hydrogen bonding.

Therefore in our studies this problem was addressed by investigating a series of PUs with analogous structures but achieved with or without hydrogen bonding. The effect of hydrogen bonding on the PUs mechanical behaviour was undertaken. *To obtain polymers without hydrogen bondings by deuteration*, the following preparation procedures were adopted: (a) the direct substitution of hydrogen with deuterium of D_2O, in the case of 0.02 mm thin PUs films; (b) the deuteration by synthesis by using deuterated chain extenders; (c) the deuteration by synthesis using deuterated SS [327, 328]. *The hydrogen substitution with inert ($-CH_3$) groups* was also undertaken. It was carried out using the following synthesis routes: (a) chemical modification of usual PUs by substitution after synthesis, using lithium hydride and methyl chloride; (b) using dichloroformates based on PEA and DEG reacted with

–NN,– dimethyl diaminodiphenyl methane. In the study presented here, the materials were synthesized with the diisocyanate DBDI, and the polyesters PEA or PBA. To avoid crystallization, DEG has been used as a chain extender. The molar ratio of the components was diisocyanate:macrodiol:chain extender: 4:1:3. The stoichiometric proportions used in these polymers (I = 100) means that were truly thermoplastic. They did not have the potential for further reaction with ambient humidity to produce chain lengthening and allophanate cross-linking, seen in similar polymers but with excess isocyanate groups (e.g. I = 110) [17, 60, 61].

4.7.1 Deuterated polyurethane elastomers

The polyaddition procedures to obtain deuterated materials according to the methods (a)–(c) have been described elsewhere [127, 327, 328].

The polymers in the form of films were subjected to uniaxial tensile tests at constant nominal strain-rate (0.03 s^{-1}). In addition, to investigate mechanical hysteresis, cyclic tensile tests were carried out, cycling between a fixed strain limit (300%) and zero load, with the same magnitude of strain-rate for loading and unloading.

On employing the IR measurements the degree of isotope exchange was estimated by the decrease of the ν_{NH} stretching vibration and concomitantly the increase of ν_{ND} corresponding to the step by step appearance of the ND groups as a result of the hydrogen replacement by deuterium. *Quantitative IR measurements confirmed a 59% deuteration degree but only for 0.2 mm thin PUs films.* The position of the urethane hydrogen replacement was strongly influenced by the adopted polyaddition procedure [327]. While in the case of direct deuteration by isotope exchange using D_2O, the hydrogen replacement with deuterium occurred preferentially at the joint points between HS and polyesters, when using a deuterated glycol the exchange of hydrogen with deuterium proceeded especially within the middle of the HS urethane groups [328]. After deuteration with D_2O vapours, from the initial large 3 300–3 400 cm^{-1} N–H stretching band, it remained only a band centered at 3 305 cm^{-1} corresponding to the most associated state. In the case of the deuteration by synthesis of the initial large 3 300–3 400 cm^{-1} N–H stretching band, it resulted only a narrow band centered at 3 350 cm^{-1} corresponding to the lowest associated state [328].

In the case of *PUs thicker films*, the direct replacement of hydrogen with deuterium by using D_2O failed. This was due to the difficulty of D_2O macromolecules to penetrate through thicker PUs macromolecular networks. However, concomitantly with the experiments achieved on thicker sheets immersed in D_2O, similar experiments were done simultaneously on similar PU films but immersed in H_2O. The stress-strain data of the two series of polymers were achieved. We observed that while the samples immersed in H_2O presented significant hydrolytic degradation, the properties of the series of films immersed in D_2O were not altered. Not any degradation effects were evidenced. Apparently unexpectedly, for the similar samples subjected to the same treatment with H_2O, the mechanical properties decreased

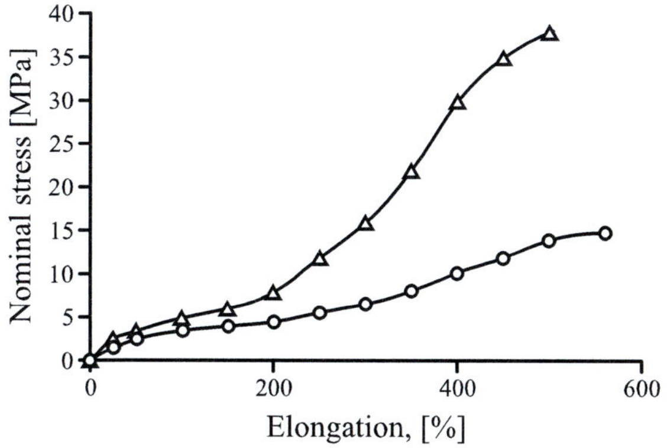

Fig. 4.61 Stress-strain diagrams of uniaxial extension: (▲) — PU immersed in D_2O; (•) — PU immersed in H_2O [327]

appreciably as a result of the appearance of a significant hydrolytic degradation (Fig. 4.61).

That has suggested that D_2O does not penetrate the macromolecular network because of its higher volume as compared to that of normal H_2O.

To enable the replacement of hydrogen with deuterium in the case of thicker PUs, we proceeded to the deuteration by synthesis. This was achieved by using a deuterated chain extender DO–CH_2–CH_2–O–CH_2CH_2–OD (*DEGd*) (isotopic purity 90%) instead of DEG. *DEGd* was obtained from a mixture of D_2O:DEG (25:6 vol) by fractionated distillation after 24 hrs of isotopic exchange at 60°C. Quantitative measurements confirmed a 66.7% deuteration degree. The stress-strain data of uniaxial tension (Table 4.26) showed that for the PUs derived from the deuterated chain extender *DEGd* it obtained better elastomeric properties. This was revealed by an increase of the tensile strength from 41.3 MPa (normal DEG) to 56.4 MPa (*DEGd*), and by a decrease of about 50% of the residual elongation (50%—normal DEG to 22%—*DEGd*). Also a higher strain energy (17.2 MPa) and a lower 300% permanent hysteresis (40.2 MPa were found in the PU obtained with *DEGd*, as compared to 14.7 MPa and 45.5 MPa respectively (normal DEG).

4.7.2 Polyurethane elastomers obtained by hydrogen substitution with inert (–CH_3) groups

When hydrogen was not substituted with deuterium but with inert (–CH_3) groups (Fig. 4.62) the polymer inelasticity increased dramatically, which resulted in a constant degree of elongation at break, and an increase of the residual elongation showing a pronounced tendency of polymer plasticization, as seen in Fig. 4.63.

Table 4.26 Tensile properties of a DBDI based PU obtained with PBA and normal/deuterated chain extender DEG [327]

The Tensile Property	Normal DEG	Deuterated DEG
100% Tensile Stress [MPa]	3.85	5.74
300% Tensile Stress, [MPa]	7.67	10.6
Tensile Strength [MPa]	41.2	56.3
Residual Elongation [%]	50.0	22.0
Hardness, Shore A	90	94
Strain Energy 300% [MPa]	14.7	17,2
300% Permanent Hysteresis,	45.5	40.2

Fig. 4.62 Hydrogen substitution with inert ($-CH_3$) groups

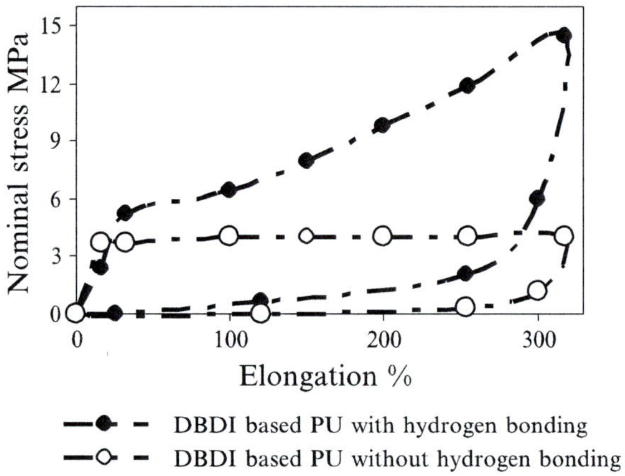

Fig. 4.63 Tensile load/unload cycles for a material with PBA and DBDI, at strain rate 0.03 s^{-1}, where hydrogen was substituted with inert ($-CH_3$) groups

Inelasticity (residual strain, hysteresis) increased when hydrogen was replaced by inert ($-CH_3$) groups (Table 4.27). The tensile properties data decreased dramatically: the tensile strength dropped from 66.7 MPa (PU with hydrogen) to 7.4 MPa (PU without hydrogen) whereas the residual elongation (one of the most important parameters to describe the material elasticity) increased from 20% (PU with hydrogen) to 220% (PU with no hydrogen). The molecular weight expressed by the inherent viscosity also decreased from 1.18 dl/g (PU with hydrogen) to 0.37 dl/g (PU with no hydrogen).

Table 4.27 Stress-strain data of a PU based on the SS macrodiol PBA with/without hydrogen bonding, obtained by the prepolymer technique

PU Synthesis Route	Inherent viscosity [dl/g]	Hardness, [Sh° A]	Stress at 300% strain, [MPa]
with hydrogen bonding	1.18	89	14.6
without hydrogen bonding	0.37	72	3.8
PU Synthesis Route	**Tensile strength, [MPa]**	**Elongation at break, [%]**	**Residual elongation, [%]**
with hydrogen bonding	66.7	550	20
without hydrogen bonding	7.4	780	220

Therefore while the hydrogen bonding did not necessarily enhance the PUs mechanical properties, it plays an essential role on the improvement of the materials performance as elastomers. Similar conclusions were obtained by us in other works, for materials obtained with the diisocyanate MDI and when using other macrodiols like the polyester PEA or the polyether PTMO [127, 327, 328].

4.8 Kinetics and postcure reactions mechanism in polyurethane elastomers achieved with excess of NCO groups

Time stays long enough for those who use it. (Leonardo da Vinci).

This also applies to PUs achieved with excess of NCO groups, as you will see in this section.☺

The postcure processes developed in solid state [3, 329, 330] play an important role in the polymer synthesis. Although an old subject, the postcuring mechanism of PUs remained so far incompletely elucidated.

In the casting PUs, which are traditionally synthesized using a small excess of up to 10% isocyanic groups (NCO) against the total amount of hydroxyl groups (OH) present in the system ("index number") [1, 3] I = 110, the mechanical properties enhance with time. Even though the atmosphere humidity has a vast access for thin films and porous materials, the preponderance of urea group formation has been unambiguously proven [1, 3, 59, 331–334]. The nature of the process, which performs in the depth of solid cast elastomers where water penetration takes place more difficultly, still remains open.

For the NCO excess group consumption many more reactions have been suggested. Also evidenced was the spontaneous isocyanate group dimerization at room temperature [322, 323]. In the hot cure reaction, the principal process, admitted by A. Awater, was the appearance of the allophanate and/or biuret linkages [335]. K.A.

Piggot [336] also presumed that the presence of some base or metal as a trace catalyst in the raw starting polyol can favor the biuret, allophanate of the trimerisation process. If considering the relatively long time of the postcure reaction, it is difficult to predict which of these reactions mainly continue to perform. Once the usual hot cure reaction has finished, it has been shown that when kept in time at room temperature, a cast PU continues to change its properties [331]. This change appears as a function of environmental exposure, extended use, or chemical reactions intrinsic to the material itself [1, 331, 332]. This phenomenon was previously observed by our Romanian colleagues on different types of PUs [320] with variable I values, and also using different reaction components.

In a first stage of our study we followed the evolution of the tensile properties in time for a series of polyether PUs obtained by using the two aromatic diisocyanates MDI or DBDI. The PUs composition is shown in Table 4.28.

Table 4.28 PUs composition

PU#	Diisocyanate	MD M = 2 000	CE	I
1	DBDI	PTHF	DEG	110
2	DBDI	PTHF	DEG	105
3	DBDI	PTHF	DEG	100
4	MDI	PTHF	DEG	110

where: MD — macrodiol; CE — chain extender.

The diisocyanate / macrodiol ratio (mol) was of 3.03. The amount of chain extender was calculated so to obtain a 0 to 10% excess of NCO groups [331]. The isocyanate index (I) was calculated according to eq. (4.20):

$$I = \frac{[NCO] \cdot 100}{[OH]_M + [OH]_{CE}} \tag{4.20}$$

where: [NCO] = equivalents of NCO groups from diisocyanate; $[OH]_M$ = equivalents of OH groups from macrodiol; $[OH]_{CE}$ = equivalents of OH groups from chain extender.

4.8.1 Postcure processes in polyurethane elastomers

The evolution of PUs postcure processes was compared by us by introducing and following a reduced 300% tensile stress function (α) which we calculated according to eq. (4.21) as follows:

$$\alpha = \left(1 - \frac{\sigma_{30}^{300} - \sigma_t^{300}}{\sigma_{30}^{300}}\right)100 \tag{4.21}$$

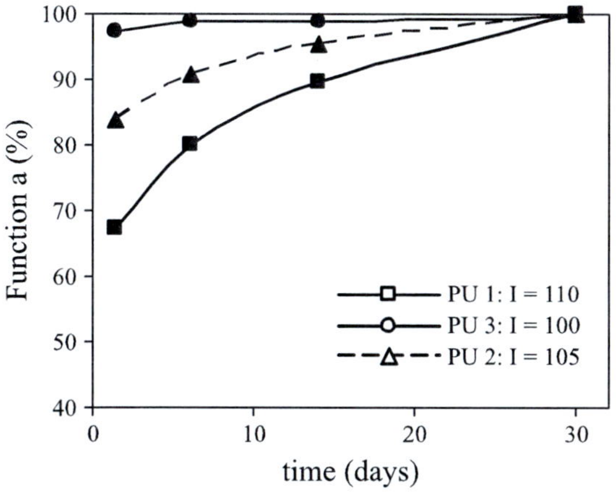

Fig. 4.64 α–t dependence as a function of time for the DBDI series of materials PU1, PU2 and PU3 of variable index I (100 to 110)(Table 4.28) [331]

The function reproduces the percent level of the 300% tensile stress (σ_t^{300}), achieved at a time (t) by comparing it to the final 300% tensile stress (σ_{30}^{300}), achieved 30 days after synthesis. The α—time dependence is a function of the chemical structure of the PUs hard constituent polymer blocks. The relationship between the function α and the isocyanate (NCO) excess was followed (Fig. 4.64). Decreasing the NCO groups excess (I) from the value of I = 110 to I = 105 and consequently to I = 100 resulted in a decrease of α from 33% to 13% and finally to 1.5%. Shown in Fig. 4.65 and 4.66 is the evolution in time of the tensile strength and residual elongation for the series of PUs listed in Table 4.28.

As seen in Fig. 4.65 and 4.66, regardless the type of diisocyanate (DBDI or MDI), the tensile strength and residual elongation developed towards stabilized values within specific periods of time, ranging from about 7 to 30 days. The higher the index values, the longer the postcure process[331].

The PUs groups of samples were kept under different conditions: (a) – ageing at 25°C and 50% RH; (b) – under vacuum of 0.1 mm Hg in sealed ampoules; (c) – maintained in closed P_2O_5 anhydrous atmosphere. PU postcuring reactions in vacuum and especially in P_2O_5 anhydrous atmosphere were found to be very poor as compared to the case when after synthesis, the materials were kept under atmospheric humidity at room temperature. This is shown in Fig. 4.67 for the polymer PU1 derived from DBDI and with the index I = 110.

As shown [331],these time-dependent maturation processes occurring in PUs were suggested to be caused mainly by the reaction of the excess of isocyanic groups with the atmospheric humidity which lasts until the total consumption of the excess of NCO groups.

Similar observations where made by us for polyester materials based on PEA and when using other chain extenders (BDO or EG) [331].

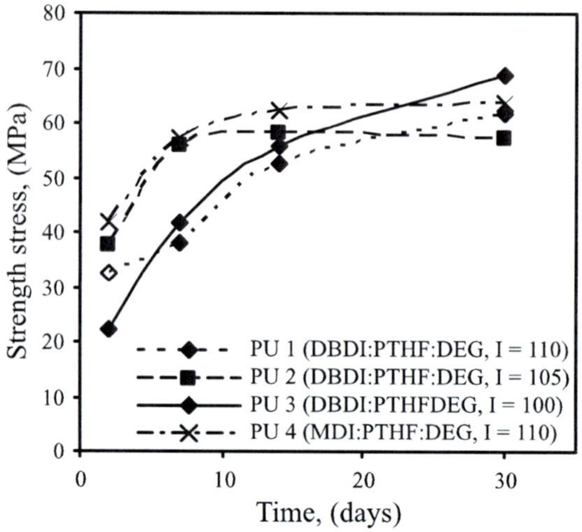

Fig. 4.65 Strength stress as a function of time for the series of PUs listed in Table 4.28 [331]

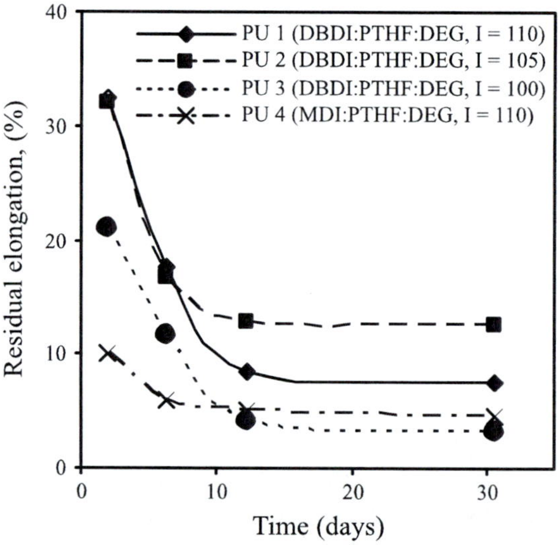

Fig. 4.66 Residual elongation as a function of time for the series of PUs listed in Table 4.28 [331]

In the second stage of our study we included only one of the most representative PU cases studied by us, a material based on PEA of molar mass 2 000 g/mol, the diisocyanate DBDI, and the chain extender DEG in a molar ratio 1:3.03:1.754.

From the various diisocyanates studied by us, we have chosen here DBDI, due to the fact that this isocyanate leads to the most stable PU toward environmental factors [59]. In this way, we intended to avoid the interference between the favorable

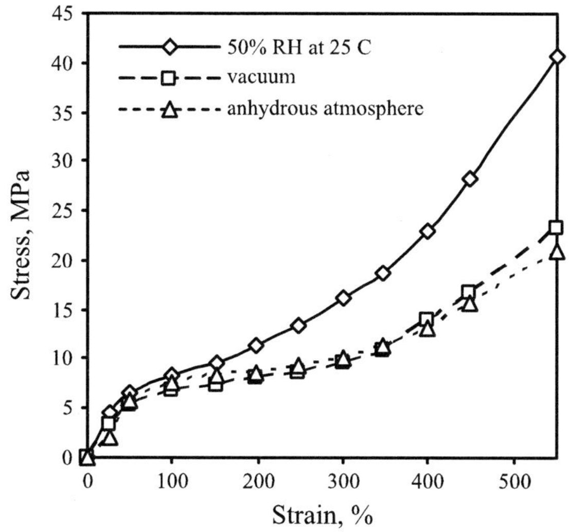

Fig. 4.67 Stress-strain curve of PU1 with DBDI (index I = 110) [331]

postcure processes and the eventual incipient parallel degradative phenomena as much as possible.

Recently, an excellent study presented by D.W. Duff and G.E. Maciel [337], proved that solid-state ^{15}N and ^{13}CCP/MAS NMR could be an extremely useful tool in pursuing the postcure reaction chemistry of residual isocyanate in *isocyanurate resins*. As reported by Duff and Maciel, the predominant postcure process should be due to the reaction of isocyanate group with the atmospheric humidity to form an amine which further condenses with an additional isocyanate from the immediate vicinity to form a urea linkage.

So far, this method was not extended to the post cure processes of PUs. However, if its use is intended, there are difficulties to be expected. The overlapping of the NMR nitrogen resonance signals as isocyanate with amine and urea with biuret mentioned by these authors [337] will be complicated here even more by overlapping with other additional nitrogen signals characteristic to a polyurethane system, like urethane and possible allophanate nitrogens [337–340].

Our own work was motivated by the hope that more detailed information on the postcure process of *polyurethane elastomers* could be obtained by combining the quantitative measurements like the enhancement in the mechanical properties in time with various dynamic processes which take place concomitantly, similarly as performed by Toffey et al in the case of *cellulose* crosslinking by urethane groups when the dynamic mechanical thermal analysis (DMTA) was used [341].

It is reasonable to suggest that the main process involved in the PUs maturation observed on the material obtained with a slight –NCO excess ($I \leq 110$) is one of a chemical nature. The presence of air humidity certainly plays a decisive role, but also, it cannot exclude other side reactions. These side reactions are well known, yet the results of the competition of these reactions in solid state during the long time postcuring and the correlation effects between the mechanical properties enhancement and also, such structure modifications are not clear.

Theoretically, several principal ways can be rationalized for the excess isocyanate consumption.

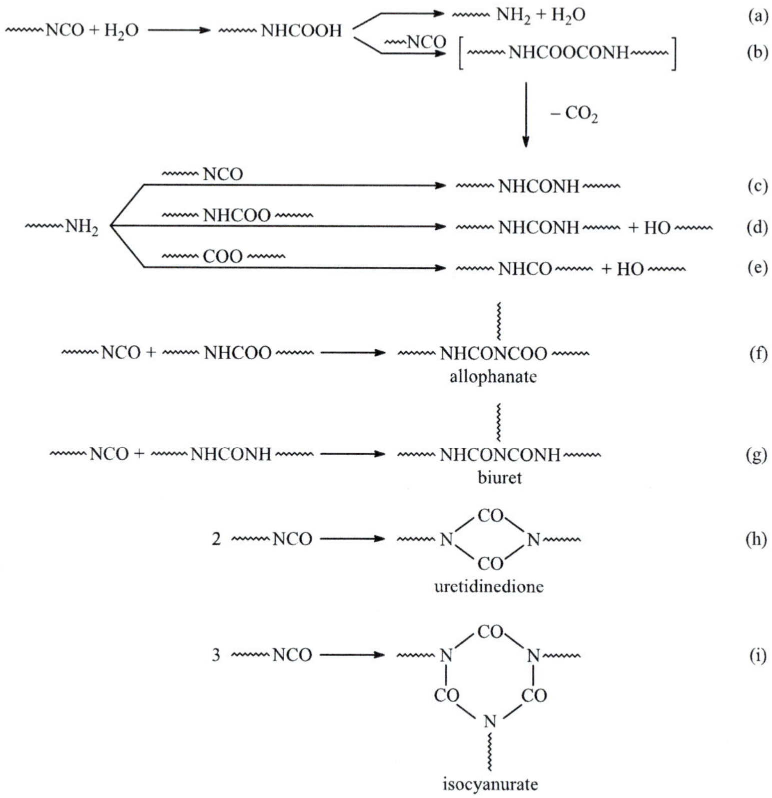

(4.22)

The intermediary appearance of the $-NH_2$ group in PUs as a result of water reaction (equation 4.22, (a)) was already confirmed quite a while ago [342]. Taking into account the necessity of the reactive end groups collision, the low concentration of these end groups and their quasi limited mobility, special consideration should be given to other possible $-NH_2$ group consumption. Under the circumstances, the much slower reactions with other groups like urethane and ester groups, which are present in high concentrations in PUs, should also be considered. As already mentioned, the aliphatic polyesters can be degraded by the amine groups yielding amide derivatives [17]. On the other hand, the interaction between the amine and urethane groups can also lead to the urea group formation [1]. Obviously, these reactions

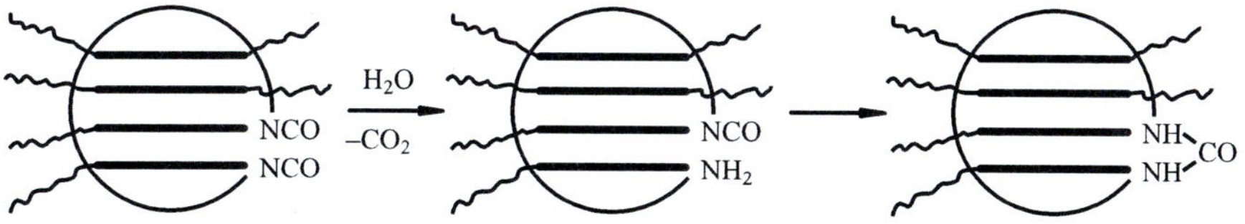

Fig. 4.68 Schematic representation of hard segment and chain end groups segregation [17]

normally perform at higher temperatures, yet it is necessary to determine if during the long period of the post cure phenomenon, such processes may or not perform to some extent, as in the case of the slow hydrolytic degradation process in time.

As observed in the first stage of our study, the considerable decrease of the rate of maturation process at room temperature in the absence of humidity shown in Fig. 4.67 from above [331], suggested that in our case the allophanate, biuret, uretidinedione and isocyanurate group formation equations (4.22 (d), (e) and (h) should play only a minor role in the PUs consolidation).

The aim of this research was to obtain new clarifications regarding the preponderant reactions responsible for the PU consolidation process and to study the evolution of the PUs mechanical performance during the post cure process in correlation with the dynamics of different diffusion processes like water absorption and CO_2 desorption.

From the data mentioned above in the first stage of this work, the first hypothesis regarding the modality of excess –NCO group consumption during the PU postcure process appeared to be related to the reactions leading to the formation of a urea group (equations 4.22 (a)–(c)). The development of reaction 4.22 (a) in solid state leading to the appearance of amine group does not seem to present problems due to the possibility of small water molecule able to diffuse relatively easy into the solid polymer. Regarding the further steps in the ureic groups formation according to equations 4.22 (b) and (c), although the reaction between –NCO and $-NH_2$ groups would be by all means the fastest one, the fact remains that in the solid state the probability of different macromolecule end groups colliding is low. In the alternative possibility, consideration should also be given to the fact that within the PUs matrix it occurs phase separation which segregate the HS in narrow areas. Taking into account that the free isocyanate groups are always bonded to the HS, these groups could be included in the hard block associates bringing the free isocyanate groups in a reciprocal proximity which would enhance the probability of the reaction 4.22 (b) taking place [17].

Taking into account the complexity of these problems, we first proposed to find a kinetic equation which could represent the change of the PUs mechanical properties with the time of maturation and to correlate it to the dynamics of other parallel processes related to these changes i.e., the decrease of free isocyanate group content, absorption of water from atmosphere and release of CO2 resulted from reaction.

4.8.2 Kinetics of maturation phenomenon

As mentioned above, the PUs investigated in the second part of this study were obtained by the prepolymer route with the SS macrodiol *PEA of molar mass 2 000 g/mol, the diisocyanate DBDI, and the chain extender DEG in a molar ratio 1:3.03:1.754.* The PUs synthesis is described elsewhere [17]. PUs sheets 1, 2, 3, 4, 5 and 6 mm thick and about 20μ PUs thin films were obtained. No release agent was used in order to avoid the disturbance of diffusion processes between air atmosphere and polymer during the postcure process. This procedure resulted in a PU with an isocyanate index I value of 110, which corresponded, in our case, to an excess of 0.0185 eq. NCO groups/100 g PUs. The density of this PU was 1.22 g/ml. The changes in mechanical properties were monitored starting immediately after demolding. In order to follow the postcure process, the samples were maintained in a tightly closed thermostat under controlled atmospheric conditions i.e., t = 25°C and 50±2% RH [17].

300% tensile stress (σ^{300}) of the stress-strain data, has proved to be a valuable criterion to follow the mechanical properties evolution in time of 1, 2, 3, 4, 5 and 6 mm thick PU sheets: the 300% stretching range was chosen because in this region of the $\sigma - \varepsilon$ curves the σ–time variation is sufficiently high so to ensure good monitoring and is sufficiently far from the polymer rupture region.

Fig. 4.69 illustratively presents the results obtained for the 1, 2, 4 and 6 mm thick PU sheets. We avoided to use the tensile strength value because, in this case, the reproducibility of the obtained data was lower. In order to express σ^{300} time-evolution, the maximum possible value reached by σ^{300} was used as a reference. In our case, this value was σ^{300}_{∞} = 14.1 MPa and was established as corresponding to a 1 mm thick PU, 60 days after synthesis. By using this reference, the simulated maturation phenomenon was found, expressed by $\sigma^{300}_{\text{t calc}}$ with the aid of the exponential equation:

$$\sigma^{300}_{\text{t calc}} = \sigma^{300}_{0} + (\sigma^{300}_{\infty} - \sigma^{300}_{0})(1 - e^{-k_m t}) \tag{4.23}$$

where the $\sigma^{300}_{0} = \sigma^{300}$ value found experimentally immediately after synthesis (in our case 5.15 MPa); t = maturation time (days); k_m = maturation process rate constant, leads to a fairly good agreement with the real σ^{300} time-evolution.

The k_m values corresponding to PU sheets of different thickness were determined experimentally as the respective straight line slopes derived from equation:

$$\ln(\sigma^{300}_{\infty} - \sigma^{300}_{\text{t exp}}) = -k_m t \tag{4.24}$$

where $\sigma^{300}_{\text{t exp}} = \sigma^{300}$ values found experimentally at the time t (days). Fig. 4.70 shows an exemplification for the 1 mm thick PU sheet case.

The corresponding k_m values to the six different sheet thickness are shown in Fig. 4.71.

For thicker PU films, the maturation proceeded slower and it was most likely that the weight of the slower reaction leading to allophanate bridges (equation 4.22(d)) should increase. The final σ^{300} values which could be reached, were all much lower

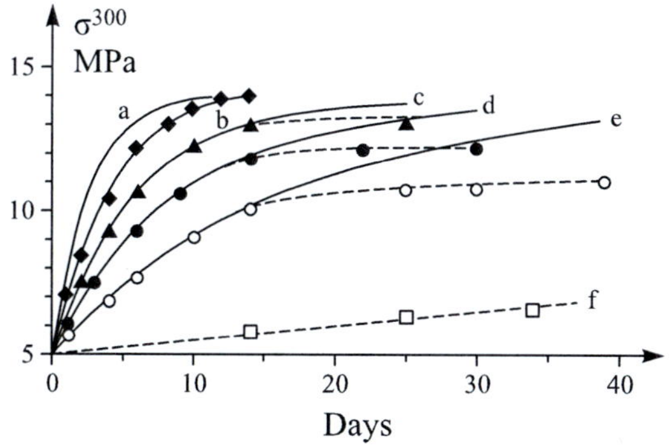

Fig. 4.69 Evolution of 300% tensile stress (σ^{300}) during postcure at 25°C and 50% relative humidity (RH) for different PU sheet thickness δ. Experimental (- - -): (+) – 1 mm; (Δ) – 2 mm; (•) – 4 mm and (o) – 6 mm. Calculated (——): (a) – virtual $\delta_{\lim\to 0}$; (b) – 1 mm; (c) – 2 mm; (d) – 4 mm; (e) – 6 mm. (f) – Experimental σ^{300} for 2 mm sheet in anhydrous atmosphere [17]

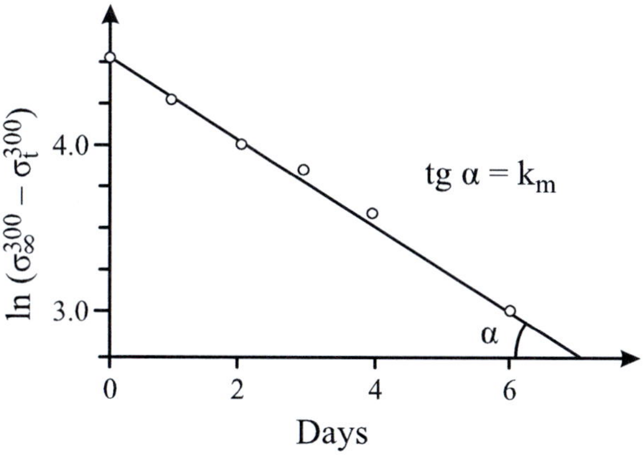

Fig. 4.70 Determination of maturation rate constant km using σ^{300} variation [17]

since the PU films thickness was higher. The existence of such a secondary process was supported by the fact that even in anhydrous atmosphere there exists however, a low increase of σ^{300} with time (see Fig. 4.67).

With regard to computer simulation of the maturation process, if in equation (4.23) the unique value $\sigma^{300}_{\infty} = 14.1$ MPa is replaced in every case by the particular σ^{300} maximum values reached with time for every sheet thickness, it is possible therefore, to obtain a much better agreement between the calculated $\sigma^{300}_{\text{t calc}}$ values and the $\sigma^{300}_{\text{t exp}}$ values experimentally found.

By calculating the product δk_m, we found that for PU films thicker than $\delta = 2$ mm, a relatively constant value of about 0.371±0.03 was obtained. This relation can be useful to estimate k_m values for other thickness and therefore, to approximate the length of the maturation process in other cases.

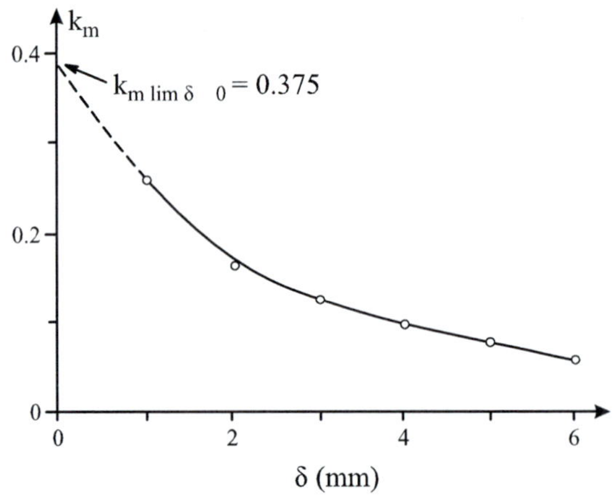

Fig. 4.71 Rate constant k_m versus thickness δ dependence [17]

Starting from the data in Fig. 4.71, we have extrapolated the k_m values to the zero thickness ($\delta \rightarrow 0$), characteristic to surface reaction, where neither water diffusion nor CO_2 desorption rates are involved; the value of $k_{m \lim \delta \rightarrow 0}$ was found to be 0.375. The virtual σ^{300} increase simulation, calculated with equation 4.23 for an idealized ultra thin PU film by using $k_m = 0.375$, is revealed in Fig. 4.69 by the solid line (a) [17].

4.8.3 *Kinetics of NCO excess consumption process*

The infrared method for the estimation of the NCO groups decrease in the PU postcure processes has been used by many authors [333, 334]. In the case of our PUs, we have determined the decrease in NCO group absorbance of the IR spectrum at 2 290 cm^{-1} on 20μ thin PU films cast from the reaction melt between two teflon sheets. The kinetic measurements started immediately after demolding. The films were maintained under the same maturation conditions as in the case of thicker sheets. These values have been converted into a NCO group equivalent concentrations per unit volume of polymer taking into account the measured absorbance and the polymer thickness. In every case, corrections for variation of thickness of PU films with time were made. The NCO group molar extinction coefficient was established on standard solutions of DBDI in benzene [17].

The kinetics of this process was tested by using an exponential equation similar to equation 4.23, e.g. equation 4.25:

$$[NCO]_t = [NCO]_0 e^{-k_{NCO} t} \tag{4.25}$$

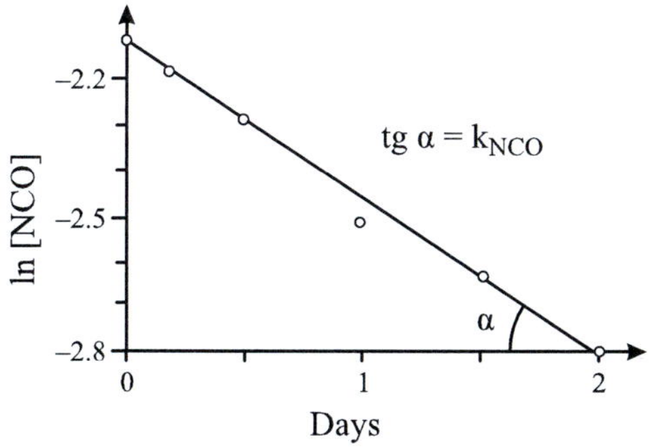

Fig. 4.72 Determination of NCO groups disappearance rate constant k_{NCO} [17]

where: $[NCO]_t$ = NCO group concentration at the time t (eq/l); $[NCO]_0$ = the initial NCO groups excess concentration in polymer calculated from the ratio between reactants i.e., for (I = 110) 0.227 eq/l polymer (before starting the postcure process); k_{NCO} = NCO consumption rate constant; t = time of exposure to atmosphere humidity (days).

Following the ln[NCO] variation as a function of time, an almost linear dependence could be observed for the initial stages characteristic to the surface process when the excess NCO consumption did not exceed more than 10% from its initial value. This conclusion allowed us to calculate the k_{NCO} value as shown in Fig. 4.72. As the reaction advanced inside the depth of PU film with time, the process became dependent also on the water diffusion rate in polymer [59].

The k_{NCO} value corresponding to a first stage of up to 10% NCO groups conversion, was found to be 0.35. In order to compare this value to the $k_{m\lim\delta\to 0}$ value, both of them being characteristic to the surface processes, we compared the time evolution degree of both processes expressed in percentages in relation to their respective final values. The virtual percentage of maturation was expressed by equation 4.26:

$$\text{Maturation }\% = \frac{\sigma^{300}_{\text{t calc}} - \sigma_0}{\sigma_\infty - \sigma_0} 100 \quad (4.26)$$

By combining equations 4.23 and 4.26, it obtained equation 4.27:

$$\text{Maturation }\% = (1 - e^{k_m t})100 \quad (4.27)$$

when for the PU thickness $\delta_{\lim\to 0}$ it obtained $k_{m\lim\delta\to 0} = 0.375$.

The percentage evolution of virtual surface NCO groups consumption $NCO_{cons\ \%}$ (representing the percent consumption from the initial $[NCO]_0$ content) was calculated by using a similar equation (4.28):

$$NCO_{cons\ \%} = \frac{[NCO]_0 - [NCO]_{calc}}{[NCO]_0} 100 \quad (4.28)$$

where $[NCO]_0$ is the initial NCO group excess concentration i.e., (for I = 110), 0.227 eq/l. By combining equations 4.28 and 4.25, it obtained equation 4.29, similar to equation 4.27:

$$NCO_{cons\ \%} = (1 - e^{-k_{NCO}t})100 \tag{4.29}$$

For the incipient NCO consumption process, we found $k_{NCO} = 0.35$ corresponding to the initial up to 10% NCO groups conversion (Fig. 4.71). We calculated the 300% tensile stress increase (maturation%) and the NCO group consumption for the for the virtual PU thickness $\delta_{\lim\to 0}$. The results are presented in Table 4.29. Both of the data series obtained are practically equal revealing that under these limit extrapolation conditions, both the surface processes, representing the surface NCO group consumption and the virtual surface σ increase. They proceed identically and are practically completed after 12 days (98.5%). When extending these findings to processes which perform in the depth of the polymer, it also should be taken into account the influence of some possible changes in the polymer morphology. The DSC measurements showed some peak changes when changing from a 20μ thin PU film to a 1 mm thick PU sheet [59]. However, further PU thickness increase over the values of 1, up to 6 mm, did not produce more significant changes in the DSC peaks [17, 59].

Table 4.29 The evolution in time of 300% tensile stress increase (maturation%) and NCO group consumption (NCOcons%) for a virtual thickness $\delta_{\lim\to 0}$ [17]

Time [Days]	Maturation, [%]	NCO_{cons}, [%]	Difference, $\Delta(a-b)$
0.5	17.10	16.05	1.05
1	31.28	29.53	1.75
2	52.77	50.34	2.43
4	77.70	75.34	2.36
8	95.04	93.92	1.12
12	98.90	98.50	0.40
16	99.77	99.63	0.14
32	100.0	99.90	0.01

4.8.4 Water absorption and desorption kinetics

Water, taken in moderation, cannot hurt anybody (Mark Twain).

This is true for polyurethanes ☺ as long as it has been already reported that PUs may absorb moderate quantities of water which increase with the HS concentration [343]. In our study we have followed the increase in weight of already maturated 2 mm thick PU sheets previously anhydrided and then submitted to a controlled

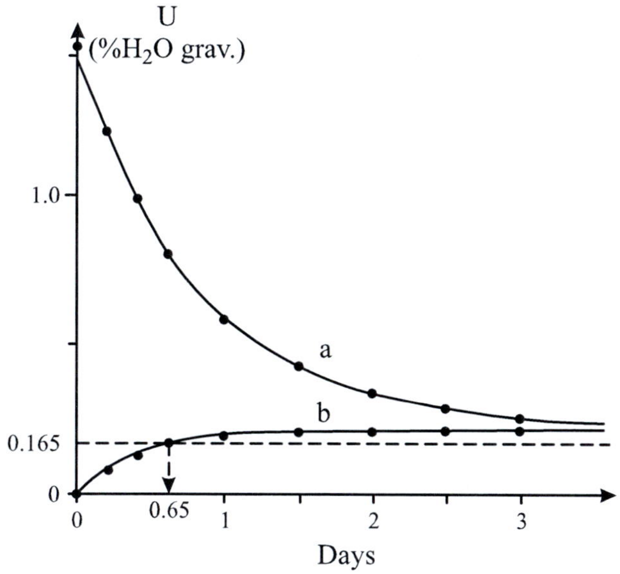

Fig. 4.73 Water desorption (a) and absorption (b) in a $\delta = 2$ mm PU sheet at 25°C and 50% RH; (– – – – –) the stoicheoimetric quantity of water necessary to react with 10% NCO excess groups [17]

atmosphere with 50% RH at 25°C i.e., preserving the same conditions as for maturation.

PUs sheets of different thickness, previously completely maturated, were anhydrided at 60°C under vacuum of 0.1 mm Hg in the presence of anhydrous $CaCl_2$ up to constant weights. On these films, the humidity absorption rate was measured by the increase of their weight under the same atmosphere conditions as those used in the maturation postcure process (e.g. 25°C and 50% RH).

To determine water desorption, PU sheets of different thickness previously maturated, were immersed for 7 days in water, then the surface humidity was wiped off and maintained at the same controlled atmosphere of 50% RH and $t = 25$°C. The decrease of weight due to gradual drying was followed [17].

The gravimetric increase i.e., the variation of humidity (U) with time for 2 mm thick PUs films is plotted as shown in Fig. 4.73, curve (b). As seen, water was absorbed very quickly. A saturation limit of U about 0.2% gravimetric was achieved in less than 2 days. This maximum degree of water saturation in PU sheets, corresponding to the atmospheric conditions of 50% RH and 25°C, was also confirmed by following the kinetics of desorption on similar PUs sheets which were previously hydrated by immersion in water and then kept under the same atmospheric conditions (Fig. 4.73, curve (a)).

With respect to water absorption process kinetics, a linear dependence between $1/U$ and $1/t$ values was obtained (Fig. 4.74).

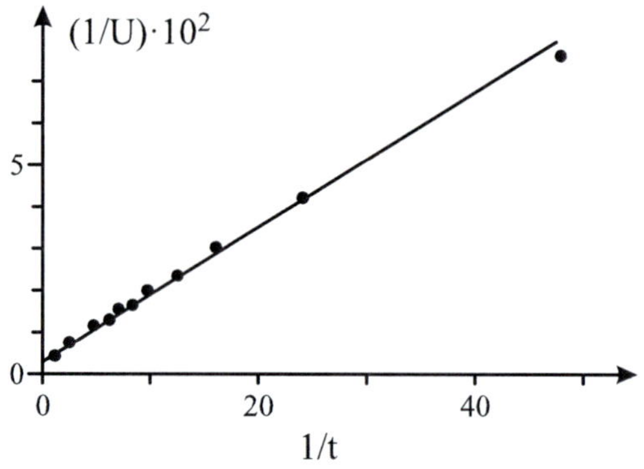

Fig. 4.74 $1/U$ versus $1/t$ dependence during water absorption (U = g water / 100 g PU, t = time days) for $\delta = 2$ mm at 25°C and 50% RH atmosphere [17]

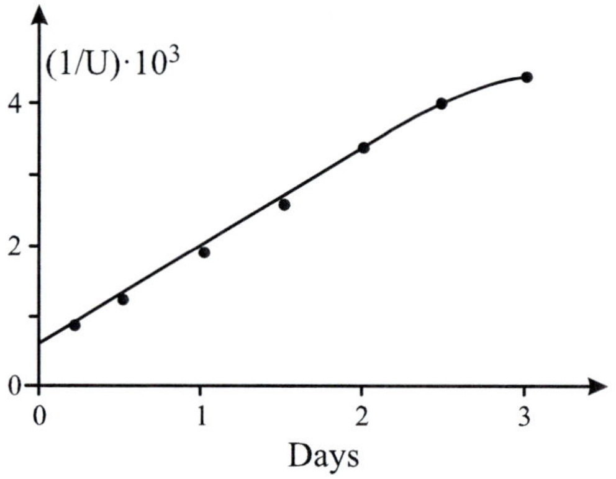

Fig. 4.75 $1/U$ versus time dependence during water desorption (U = g water / 100 g PU) for $\delta = 2$ mm in 25°C and 50% RH atmosphere [17]

Therefore, the hydration process could be represented by equation 4.30:

$$U = \frac{t}{at+b} \tag{4.30}$$

where U = g water / 100 g polymer; a, b = constants.

The constants values resulting from the above graph are $a = 3.89 \cdot 10^{-3}$ and $b = 1.56 \cdot 10^{-3}$.

In the case of the water desorption process, the function which described the decrease of U with time (t) was different. In this case a linear dependence was also obtained when plotting $1/U$ against t, see Fig. 4.75.

As a consequence, the following equation can be written (4.31):

$$U = \frac{c}{t+d} \tag{4.31}$$

where c, d = constants that can be determined from the plot: $c = 0.795$ and $d = 0.4818$.

From the above mentioned data, it has been concluded that for up to 2 mm thin PU sheets, the theoretical quantity of water theoretical necessary to consume the initial 10% NCO excess (i.e., 0.165 g water / 100 g PU) is quickly absorbed in about 1 day, unlike the findings concerning the dynamics of maturation process for these sheets which requires at least 15 days. This fact lead us to conclude that *water diffusion inside the polymer does not control the rate of maturation process and thus, it is necessary to consider also other factors as it is the desorption of* CO_2 resulted from reactions 4.22 (a) or (b).

4.8.5 CO_2 desorption kinetics

The capacity of PUs to absorb appreciable quantities of CO_2 has been relatively recently mentioned [344, 345]. Previous experiments conducted by us have lead to the conclusion that when our PUs samples are kept for 48 hours under liquid CO_2 pressure at room temperature, they may absorb high quantities of CO_2, (up to 14% grav.) [17].

In our experiments, PU sheets of variable thickness were introduced in stainless steel pressure tubes and submitted in every case for two days to different CO_2 pressures (ranging between 0.5 to 6 MPa). During this time, the PU samples absorbed variable quantities of CO_2 depending on the CO_2 pressure to which they were subjected. The samples were withdrawn from the pressure tubes and the CO_2 desorption was followed by the weight decrease in time at atmosphere pressure, 250°C and 50% RH [17].

Unusual phenomena were observed in the case of the reversed phenomenon. The desorption curves were quite different depending on PUs sample thickness and apparently unexpected, on the initial absorbed quantity of CO_2 (Fig. 4.76).

We studied the behaviour of two 2 mm thick PUs sheets with two different initial CO_2 *concentrations* $C_1 = 13.7\%$ *grav. and* $C_2 = 3.8\%$ *grav.*, respectively (Fig. 4.76). After a period of time, the concentration of CO_2 in the first case became equal to the initial concentration of the second case C_2. Surprisingly, the further CO_2 desorption rate for the first sample was not equal to the desorption rate observed in the second case, even though at that time, the CO_2 content was equal in both cases. It was observed that the lower the initial concentration of CO_2, the slower the rate of CO_2 desorption (Fig. 4.76).

This behaviour was explained as owing to a process of supramolecular structure loosening as a result of a polymer plasticization effect determined by CO_2, which allowed quicker removal of CO_2 [17]. The obvious consequence was that the kinetics equations established for every particular case (PUs sheet thickness or CO_2 concentration) depended on the initial CO_2 concentration.

Taking into account our situation when the calculated quantity of CO_2 resulted from reaction is very poor, 9.23×10^{-3} moles of CO_2 / 100 g PUs, (i.e., 0.406 g

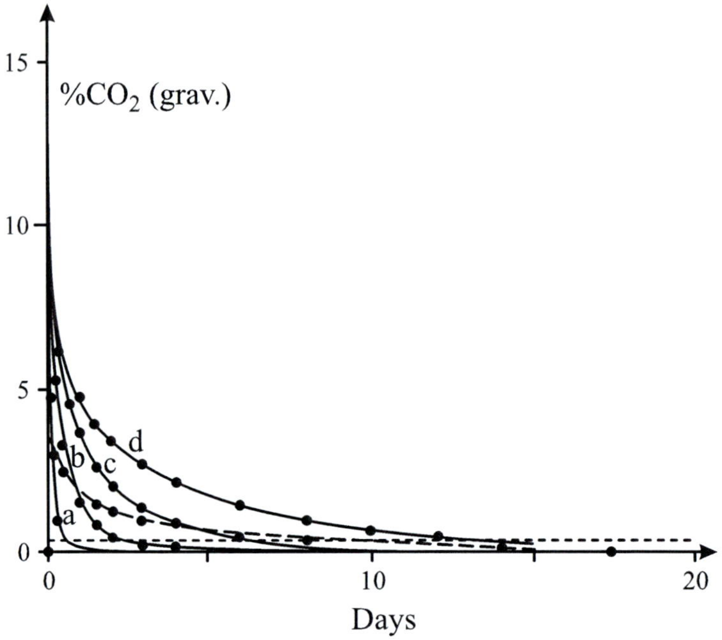

Fig. 4.76 CO_2 desorption rate for different PU sheet thickness and different initial CO_2 content: (——): 13.7% grav. initial CO_2 content: (a) = 1 mm; (b) = 2 mm; (c) = 4 mm; (d) = 6 mm. (– – –): 3.8% grav. initial CO_2 content in 2 mm thick PU sheet; (······): Calculated total stoichiometric CO_2^S content evolved from 10% NCO groups excess reaction with water = 0.406 g CO_2 / 100 g PU [17]

CO_2 / 100 g PUs), we had to focus our attention especially upon the cases of desorptions in samples with a low initial concentration of CO_2, i.e., up to 0.5 g CO_2 / 100 g PUs.

As before we tried to find a correlation between the rate of the CO_2 desorption process and the σ^{300} - time evolution achieved under the same conditions. The experiment was also performed on 2 mm thin PUs sheets.

To enable these different dynamic processes be compared, both of them were expressed in percentages of evolution towards the final state as we proceeded when comparing the cases of σ^{300} and [NCO] group evolution in time. Thus, for each situation we considered the initial and final values of σ^{300} and CO_2 content as seen in Fig. 4.77.

A series of remarks could be made according to equations 4.22 (a) and (b): CO_2 appeared in PUs as time proceeded and consequently, its desorption depended both on the rate of the reactions and on the CO_2 diffusion rate.

As observed from the CO_2 desorption kinetics, the CO_2 diffusion from the polymer was very slow. The quantity of 0.406 g CO_2 / 100 g PUs, equivalent to the whole quantity of CO_2 resulted from the reaction, needed at least two weeks to diffuse totally from the 2 mm PUs sheets. The reaction with water proceeded with time and

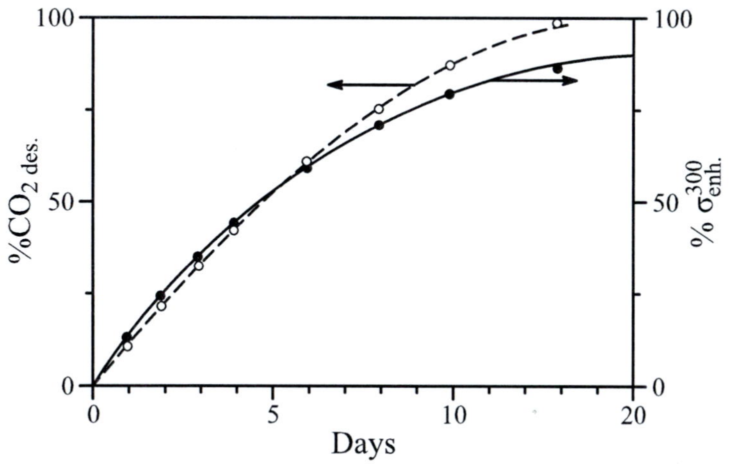

Fig. 4.77 Comparison between percent desorption of CO_2 ($\%CO2_{des}$) and parallel percent σ^{300} enhance ($\%\ \sigma^{300}_{enh}$) in time during the postcure process: $\%CO2_{des} = \frac{\%CO_2{}^S - \%CO_2{}^t}{\%CO_2{}^S}100$, where $\%CO_2{}^S$ = the total stoichiometric quantity of CO_2 resulted from reaction, i.e. 0.406 g CO_2 / 100 g PU; $\%CO_2{}^t$ = the residual CO_2 content in PU at the time t (g CO_2 / 100 g PU); $\sigma^{300}_{enh} = \frac{\sigma^{300}_t - \sigma^{300}_0}{\sigma^{300}_\infty - \sigma^{300}_0}100$, where $\sigma^{300}_0 = 5.15$ MPa, $\sigma^{300}_\infty = 14.1$ MPa and $\sigma^{300}_t = \sigma^{300}$ at time t (days) [17]

CO_2 was formed gradually so that the CO_2 desorption process became even slower and under the circumstances, its rate became comparable with the real rate of the maturation process.

At a first sight, this dependence would seem difficult to be understood due to the fact that the overall reaction of the NCO groups leading to urea groups at room temperature is not a reversible process.

Even in some earlier works [1, 346], it was also assumed that the CO_2 gas evolution is apparently complicated by the stability of a possible intermediary anhydride complex between isocyanate and the carbonic acid (equation 4.22(b)) and this hypothesis could also be emphasized for the PUs solid state conditions.

The discrepancy between the rate of isocyanate group disappearance and the rate of CO_2 evolution was already mentioned even in the case when the reactions were performed in dioxane [346]. On the other hand, interactions between the evolved CO_2 and H_2O could be possible, leading to carbonic acid and other hydrogen bonded mix associates.

The reactivity of water, bounded in such associates towards the NCO groups is expected to decrease so that the maturation process should be prolonged. As a consequence, the diminution in the rate of water –NCO group reaction allowed us to promote, to some extent, the slow competitive reaction of the NCO groups with urethane, leading to allophanate linkages (equation 4.22(d)).

4.8.6 Possible excess –NCO groups consumption reactions performed under water assistance

As shown [17, 331], the postcure process is very intense as the index number I is higher and is strongly influenced by the presence of the atmospheric humidity.

As in other similar cases [337], the main postcure process should be, the appearance of the urea groups according to equations 4.22(a)-(c), together with the slow reaction leading to the appearance of the allophanate group equation 4.22(d).

Taking into account the molecule low mobility in the solid polymer matrix with amine end groups formed as intermediate groups, we should also consider the other possible reaction of amine groups such as that with the urethane or ester groups from the PU chain. The increase of the mechanical properties data would be produced in this case by the substitution of some urethane or ester groups by transreactions leading to the more polar urea or amide groups, respectively [17].

Even though such reactions usually need high temperatures, it should be verified whether or not they would, however, take place to some extent during the long period of the postcure process. The relative higher concentration of urethane and ester groups from the macromolecular chain in the solid polymer could encourage such a reaction.

In order to verify the possibility of the contribution of these reactions to the change in the mechanical properties, some quantities of 4,4'-diamino diphenyl methane (MDA) were incorporated into the polymeric material, so to obtain an equivalent quantity of $-NH_2$ groups equal to 0.02 g / 100 g of a linear PU (equal to the excess initial concentration of –NCO groups corresponding to I = 110). Thin films of PUs with MDA about 0.5 mm thick were obtained by evaporation from solution at 60°C as described elsewhere [17].

The changes in the mechanical properties during 60 days were followed similarly as before. The observed changes in mechanical properties were insignificant both in the case of the films containing MDA, and in the reference films, which lead to the conclusion that the mentioned aminolysis reactions rarely occur at room temperature [17].

4.8.7 Influence of thickness on the final mechanical properties

The final mechanical properties determined on different completely postcured PU sheets after more than two months revealed the interesting fact that even if these samples were made and stored simultaneously under the same conditions, the final mechanical properties were not totally similar (Table 4.30).

As observed, the values of σ^{300} tensile stress and tensile strength were somewhat higher in thinner films probably because in this case the more polar urea group formation was favored to the prejudice of the appearance of allophanate groups. The 100% tensile stress value was similar in all the cases due to the fact that, as

Table 4.30 Influence of thickness on the mechanical properties of a 40 days maturated PU [17]

Thickness, δ (mm)	Hardness (Sh°A)	σ^{100} (MPa)	σ^{300} (MPa)	σ (MPa)	EB (%)	R (%)
1	92	6.0	14.1	73.0	600	10
2	92	5.9	13.1	70.0	550	10
4	92	5.8	12.2	65.0	550	10
6	92	5.8	11.1	60.0	550	10

σ^{100} = stress to produce a 100% strain
σ^{300} = stress to produce a 300% strain
σ = tensile strength
EB = elongation at break
R = residual elongation

shown by IR dichroic and DSC studies [17, 202], up to this elongation value, the main energy in the macromolecule network is consumed for the loosening of the SS tangle. The influence of the HS effect in which the hydrogen bonding is high, becomes decisive at elongations over about 300% when the HS achieve a parallel orientation towards the stretching direction [17, 202].

The results of this research lead to the following overall conclusions: (a) postcuring of PU sheets up to 6 mm thick, synthesized with a small –NCO groups excess, (up to I = 110), proceeds at room temperature mainly with the assistance of atmosphere humidity during a period of about one month; (b) the rate of the process, as well as the final mechanical properties, depend on the thickness of PU sheets; (c) the course of the postcure process depends on the competition between the urea and the allophanate group formation; (d) the most enhanced stress-strain data are achieved for thinner films in which the urea groups formation is favored; (e) the increase of the tensile properties data with time is mainly due to the urea groups, and in a small measure, to the allophanate group formation: this consolidation is produced by the intensification of the internal hydrogen bonding produced by the newly formed urea.

Another aspect of possible postcure reactions, which however, is not yet studied enough, is the role of the atmospheric oxygen on the superficial processes also leading to possible NCO groups consumption both in the thermal cure treatment at 110°C, as well as in the postcure processes.

Although, it is known that the atmospheric oxygen leads to the degradation of PUs in time, it should be possible that oxygen may improve some consolidation of polymer by crosslinking in the first step of the process.

Recently (year 2010), our studies have been continued by Pichon et al. [347], who also investigated the evolution during maturation and the consequences on the PUs elastic properties. The material used in their work was based on the diisocyanate MDI, and on the polyester PCL. Hydroquinone bis(β-hydroxyethyl)ether (HQEE), was preferred to the chain extender BDO, because of its higher incompatibility with the polyester SS [9, 347]. The polymers were cast with a small excess of isocyanate (6%) and kept in time under atmospheric humidity at ambient temperature. DMA and other experimental techniques were performed in order to determine the structural evolutions with time, and their effects on the material mechanical properties.

The evolution of the multiscale morphology of a 2 mm thick PU plate and its consequences on its strain-stress behaviour during a long maturation time at ambient temperature were studied (up to 13 weeks). Pichon observed that most part of the modifications occurred during the first 4 weeks. For 2 mm thick PUs plates, a 15% modulus increase and a decrease in $\tan\delta$ at 120°C were obtained when the final properties were reached, 4 weeks after curing, *which is consistent with our previous observations regarding the length of the post curing processes that occur in solid state in PUs, as a result of the hydrolysis of free NCO groups to produce urea linkages at the border of the hard domains* [17, 347]. The hydrolysis of excess NCO groups to produce urea bonds in the border of amorphous hard nodules increased the average chain length of the HS and therefore the glass transition temperature of the amorphous HS. No modification was observed in the soft phase.

The PU obtained by Pichon et al was also tested to fatigue, by using a compressive fatigue test at constant load [347]. They studied the behaviour of 39 mm diameter PU cylinders under stress (between 0.33 and 3.3 MPa at 27 Hz). Because of the viscoelastic behaviour, a self-heating phenomenon was observed that induced failure when the temperature reached the HS melting temperature. The fatigue behaviour was improved during PU maturation, by a decrease in the damping and a simultaneous increase in the modulus. Unlike 2 mm thick plates, the final properties for 39 mm diameter cylinders were obtained only after a 13-week period of maturation. This was due to a thickness effect which delays the moisture diffusion in PUs. This was also consistent with our observations showing that the rate of the PUs post curing process and the final mechanical properties, depend on the thickness of PU sheets.

Chapter 5
Deformation induced morphological developments

5.1 General considerations on strain induced orientation and crystallization developments

There is nothing so stable as change (Bob Dylan)

The strain induced orientation of the PUs has been studied extensively by using a variety of experimental techniques [5, 21, 161, 162.169, 170, 174, 348–350].

Bonart was amongst the first to identify the orientation of the periodic microphase structure under tensile deformation. Bonart suggested that the hard domains are composed of laterally stacked HS. It was noted that in MDI polyether copolymers, the SS become fully extended along the stretch direction and begin to crystallize at strains of about 150%. In contrast, the HS were found initially to orient transverse to the stretch direction and ultimately to break up to allow orientation of the HS in the stretch direction [14]. Similar observations were made by us on DBDI based materials [67], as shown in the following section.

The tensile alignment of initially randomly oriented SS chains exert a local torque force that orients the hard domains perpendicular to the direction up to 300% strain. As shown by Bonart, above 300% strain, the nature of the HS (crystalline or paracrystalline) strongly influences the deformation behaviour. For paracrystalline MDI ethylene diamine or MDI-hydrazine HS, further elongation resulted in the re-orientation and restructuring of the hard domains along the stretching direction as the stress was transferred from the strain-induced crystalline (PTMO) or paracrystalline (mixed polyesters) SS. In the crystalline MDI-BDO hard domains (40wt% diisocyanate), the HS within the hard domains were found to orient along the equator up to 400% strain, which suggested that crystalline HS behave as inert fillers during deformation. Bonart also proposed that the HS reflections observed in diffraction patterns were due to the sterically-hindered lateral arrangement of hydrogen bonds between HS due to a pseudo crystalline lattice formation.

The PUs mechanical behaviour has been investigated extensively and is intimately linked to the composition of the hard and soft domain and their resulting

C. Prisacariu, *Polyurethane Elastomers*, DOI 10.1007/978-3-7091-0514-6_5,

morphology. The initial response to stress is elastomeric stretching of the SS matrix. As reported, strain-induced crystallization of the SS in the stretch direction occurs up to 200% elongation [4, 214].

A good connection between SS and HS due to covalent bonding can cause the hard domains to rotate as rigid units into the strain direction. At a critical strain, the hard domains shear yield. Restructuring of the hard domains occurs indicted by conservation of HS hydrogen bonding [141]. The irreversible deformation of the hard domains during the first loading is responsible for hysteresis in the stress-strain response and for the strain-softening which is typically observed when a PU is subjected to subsequent load/unload cycles [140, 351]. Upon stretching, crystallization of the SS occurs which is reflected by an increase in the stress response [174, 352]. After removing the stress, the crystals melt as the strain recovers. Strain crystallization can be enhanced by increasing the SS molecular weight [219].

The SS nature (polyester or polyether) also influences the PUs mechanical properties. As reported by Nallicheri [292], higher tensile strength values were observed for polyester PUs as compared to polyether materials, due to stronger intermolecular forces and to a high tendency toward crystallization. Elastomers that crystallized upon stretching generated crystalline domains possibly beginning in highly deformed regions near incipient or actual micro-cracks. Crystallinity reduced the strain energy. As we also reported, under tensile deformation polymers derived from the flexible isocyanate DBDI display a high tendency to crystallization and self-association by hydrogen bonding. At high strain energies, plastic deformation of domains causes dissipation of energy which is reflected in high strength and toughness. *The nature of hard domain restructuring during deformation* also continues to be a subject of research. As detailed in section 4.4.3, recently we also reported research regarding the restructuring of HS obtained with the diisocyanates of variable geometries, crsyallizing or not. As shown in previous chapter, in our study the HS were generated from the diisocyanates MDI (non crystallizing), or DBDI (crystallizing). Physical structure was characterized by X-ray scattering (SAXS and WAXS).

Large differences were found by us in the responses to first loading to a given strain. Tensile modulus and work input increased significantly with degree of hard phase crystallinity, but were independent of degree of phase separation [135]. First cycle hysteresis was found to increase with reduced phase separation and with replacement of MDI by DBDI. As shown in section 4.4.3, the relative first cycle hysteresis depends upon the responses of both the initial structure and of the new soft phase and it is also sensitive to the precise mechanism of flow and break-up of the hard domains. The relative first cycle hysteresis was observed to reduce with increasing phase segregation and with replacing DBDI with MDI as diisocyanate, suggesting that in such cases there is relatively more stripping of segments from hard domains, and less plastic deformation of them. A possible explanation is weaker binding of segments at the hard domain boundaries [135].

In second and subsequent load cycles, however, in which the Mullins effect was observed, a remarkable degree of uniformity of response was discovered. A unique linear relation was obtained by us, between second cycle hysteresis and second cycle work input, for all strain levels, and for PUs (especially those with *higher phase mixing*), while some materials with *high phase separation* showed slightly lower second cycle hysteresis. *The results obtained by us were explained in terms of pull-out of segments from the hard phase on the first cycle, to form a new series-coupled soft phase, whose constitutive response then appears almost independent of chemical and physical structure* [135].

As shown by Qi et al [301, 302], the extent of phase mixing is the predominant molecular mechanism that influences the PUs mechanical deformation in terms of both the *high strain-rate hardening* and fold recovery considerably among model segmented PUs. The polymers investigated by Qi et al, were obtained with the SS macrodiol PTMO and variable aromatic diisocyanates. The chain extenders used included 1,4-butanediol (BDO) and 2,2-dimethyl-1,3-propanediol (DMPD). Better phase mixing between the HS and SS occurred when DMPD was used as a chain extender, as revealed by both DSC and SAXS data. The DMPD based materials had no significant interdomain correlations. The BDO based polymers displayed a broad scattering peak typical of segmented PUs. The greater the phase mixing in these model PUs, the higher the dynamic mechanical relaxation strength at all frequencies.

High levels of long-term residual deformation were commonly observed in PUs after removal of stress [259]. For diol based PUs, Estes et al [140] reported residual HS orientation upon removal of applied tensile stress. This effect was particularly pronounced in materials that totally lack N–H groups to participate in hydrogen bonding [353]. In contrast, in polyurethane ureas (extended with diamines) (PUUs), it observed a greater propensity toward microphase segregation, presumably because of the larger number of possible hydrogen bonds between HS chemical groups [10, 354, 355].

The *rate of sample deformation* is also an important factor in defining the hysteresis response surface. As shown by Gorce [291], the rate of sample deformation strongly influences the extent energy dissipation processes.

Xu et al [21] investigated the morphological and mechanical behaviour of semicrystalline segmented PUs with varying degrees of order within the continuous domain, proposing that the incorporation of SS crystallites contributes to the overall reinforcement of the PUs matrix by absorbing energy through a reformable process during deformation [356]. Also, Sonnenschein et al. obtained improved mechanical properties by incorporating semicrystalline polyesterdiols that possess both a high melting temperature and enthalpy of melting [357]. Similar observations were previously reported by us on polyester PUs systems obtained with melt mixtures of the two diisocyantes (MDI+DBDI), where the two isocyanates are inserted more or less randomly into the prepolymer [60].

Additional SAXS studies of the deformation behaviour of segmented PUs microphase-segregated structure, revealed two dominant response modes of the physically crosslinked network to tensile strain [5, 162]. Segmented PUs obtained with the couple MDI/BDO (37wt%MDI), and PUs based on TDI chain extended with either trimethylene glycol bis(*p*-aminobenzoate) (32wt%TDI) or 1,1,1-trimethylolpropane (47wt%TDI) were investigated by Schneider et al. From this study, a generalized model based on shape and structural rigidity of the hard domain response to deformation was obtained [162]. The HS segments with high degrees of structural integrity and cylindrical geometries (length/diameter $\gg$1) were subjected to a shearing mechanism due to the local torque applied by the aligning of the SS chains, adopting a preferred tilt angle of the hard domains to the applied stress and producing a four point scattering pattern. In the case of hard domains with length/diameter $\sim$ 1 (cubic) and less ordered morphologies, the length of the hard block oriented perpendicular to the elongation, resulting in a two-point scattering pattern along the meridian [162, 284].

The data obtained by Schneider et al offered a detailed description of the effect of the PUs deformation process at the microphase level. Depending on the PUs components, it has been shown that the microstructure may deform by either local shearing or local tensile deformation of the SS regions. The shearing mechanisms, which are dominant for disk-shaped HS particles with good structural integrity, results in tilting of the HS lamellae away from the direction of elongation. Such a behaviour was shown as being typified by a four-point SAXS pattern with maxima at 65° to the stretch direction and has been associated with a lower level of mechanical hysteresis. In contrast, in the tensile mechanism, the HS acted as cylinders of comparable thickness and diameter. In this case the repeating direction of the microstructure, along the cylinder axis, turned toward the stretch direction, resulting in a two-point SAXS pattern with maxima parallel to the stretch direction. The SAXS repeat period was seen to increase in the early stages of deformation although between 100% and 200%, a significant decrease in long period was observed [162, 284]. This indicated a break-up of the HS phases into smaller chunks, which was associated with a higher level of mechanical hysteresis.

Over the years, there has been a continuous interest on the impact of mechanical deformation on the mesophase structure in segmented PUs. Some of these previous studies have used SAXS to study the deformation of the mesophase structures [4, 358, 359]. Moreland and Wilkes [309] identified a two-step orientation-elongation mechanism. The region below 50% elongation was associated with low hysteresis in which lamella-like hard domains aligned along the strain direction in a more or less reversible manner. Above 50% elongation, the hard domains were disrupted, with individual HS being pulled away from the hard domains. In many of these studies, the hard-phase domains often were envisaged as lamella like, although this is by no means necessarily the case for all polyurethane materials. Previous studies also have noted the importance of the distribution of the lengths of the hard and soft blocks [3].

A remarkable study of the deformation behaviour of segmented PUs was conducted by Yeh and coworkers [51]. Yeh et al. concluded that disrupted hard domain lamellae and strain-induced aligned SS contribute to equatorial scattering in the *low hard segmented* (12 wt%) PUs, but the intensity may be decreased due to the lack of electron density contrast between the HS fibrils and the aligned SS chains and/or the decrease in coherent scattering due to the small size of the broken down lamellae [51, 284]. The contribution of various orientations (perpendicular, parallel, and at angle) of hard domain lamellae orientation was also probed by Lee et al, on the initial mode of deformation for segmented PUs containing 20-50 wt% HS [360]. Lamellae initially oriented perpendicular to the stretch direction produced an increase in the interdomain spacing under deformation, while those lamellae aligned parallel to the elongation were subjected to shear under strain, which caused a decrease in the interdomain spacing. At *higher* HS contents, hard domain destruction occurred at lower strains. In addition, Lee et al. investigated the SS conformation using SAXS, concluding that the soft chains are only partially extended at high strains due to the hard domain lamellae break-up [360]. In the materials with *lower percentages of aggregated HS* (less structural rigidity), affine deformation occurred in the PU network, which produced an elliptical pattern with scattering intensity on the meridian, similar to the findings of Schneider [162].

Relatively recently a series of investigations was made on PUs with *aliphatic* diisocyanates. The development of the SS phase in the deformation induced microstructure was studied by Xu [21] by comparing selected segmented PUs with either amorphous or semicrystalline SS, but identical HS contents (33% wt%) based on the aliphatic diisocyanate HDI and the chain extender BDO. The goal has been to observe *in-situ* and understand how the restructuring of the SS phase during deformation affects mechanical behaviour, and to probe the role of ordering within the continuous phase as a toughening mechanism. The morphologies of the materials were investigated both in the *unstrained and deformed states*, going from a completely amorphous SS to one with similar chemistry that displayed a high extent of the soft domain crystallization, therefore enhancing phase segregation [21]. The presence of dispersed semicrystalline regions within the continuous soft domain provided a reinforcing effect when compared to that of a PU derived from a non-crystalline SS. Incorporating a semicrystalline SS (PEO, 1 000 g/mol) improved the overall PU toughness. When higher molecular weight (4 600 g/mol), PEO based SS segments was used, extensibility and, consequently, the material toughness was affected due to an increased continuous domain modulus. Increasing the SS molecular weight and crystallinity contributed to an increased incompatibility between the hard and soft domains [21].

Kimura et al. probed further the deformation mechanisms in *polyurethane ureas* with crystalline hard domains [170]. A model was proposed, for the orientation of the HS crystallites. At low tensile strains, the long axis, taken as the length of the lamellae, contributed to scattering at a tilt angle to the stretching direction. As a result, hydrogen bonding between HS produced alignment at a preferred angle to the deformation. At higher strains, the principal axis was along the HS length, due to the disruption of the crystalline texture, forming nanofibrils and resulting in diffuse scattering along the equator. Nanofibrils were HS aggregates aligned in the stretch direction.

5.2 Thermal investigations of deformation induced morphological developments

If you saw a heat wave, would you wave back? (Stephen Wright)

Not applicable for PUs, but what about us? I would gladly wave back when it is cold outside. ☺ Numerous studies had been made on the influence of temperature on the PUs deformation. A series of model MDI based PUs were characterized by Crawford et al [220] before, and after varying degrees of strain aging to study the effect of strain on the polymer morphology and mechanical properties. The segmented materials were synthesized by using the 'one-shot' technique. The molar ratios of the model polymers were varied.

DSC and DMA thermal analyses were performed on polymers *before and after strain aging*. Strain was induced by stretching the polymers at 100% intervals between 100% and 400% elongations [220]. DSC scans showed that the polymers exhibited three prominent thermal transitions including a glass transition below 0°C and two broad endothermic transitions at higher temperatures. The lower temperature endothermic transition was identified with disruption of SS–HS bonds [198, 232, 361], or disruption of short range order within the HS microdomains [178]. The second higher temperature endothermic transition was related to the break-up of inter-urethane hydrogen bonds. This series of transitions was representative of the two phases that are present in the polymer and reflect the relative amounts of HS and SS. In materials achieved with increasing the molar percentages of HS, as the percent HS increased, the peak temperature of the transition increased also [220].

Tensile properties were also evaluated. The data obtained by Crawford showed that *strain aging* disrupts the domain structure. Such disruption may involve phase mixing with, or without hydrogen bonding, between the hard and soft domains, a breaking up of the hard domain microstructure and dispersion of the HS within the soft domain, or plastic deformation of the hard domains. Similar observations were made by us for materials obtained with MDI and DBDI via the prepolymer synthesis route [60, 135]. However, as shown by us, irrespective of the type of macrodiol (polyester or polyether), first loading of the materials caused significant damage to the initial structure, such that second loading corresponded to deformation of a new structure: subsequent loading inside the envelope of previous deformations, was associated by us with lower stiffness and hysteresis. Less efficient molecular packing of HS, once the original two-phase structure starts to break up, suggested a reduction in hydrogen bonding during deformation.

The DMA data obtained by Crawford et al [220], showed that the interaction between the hard and soft domains as a result of straining is quite complex. This was especially true with the polyester PUs which do not show a consistent trend in $\tan\delta$ or E'' with increasing strain. This maybe due to the lower degree of phase separation present in the polyester materials compared to the polyether PUs, and the greater potential of HS–SS interaction of the polyester based materials due to potential hydrogen bonding between the urethane HS and the SS carbonyl. Data relating to $\tan\delta$ showed an increase in T_G up to a 200% strain elongation for polyester PUs.

Heat aging the of the materials without strain elongation was shown to enhance domain ordering and phase separation and resulted in improved abrasion resistance.

The mechanical hysteresis of a PU investigated by Gorce based on MDI/BDO and PTMO of molar 2 000 g/mol was also found to be sensitive *to the temperature-dependent morphologies* of this material and to the morphological changes induced by the strain level and deformation energy during a deformation cycle [291].

Low mechanical hysteresis occurred at small strains where the morphology of the polymer was not permanently disrupted. In these cases, the energy was dissipated mostly as heat, which was due to the viscoelastic nature of the polymer. Increases in mechanical hysteresis were attributed to processes which lead to permanent deformation and morphological changes. *At low temperatures*, this occurred at small strain levels and involved the disruption of the glassy matrix. *Above the glass transition temperature*, this occurred at higher strain levels and was attributed to the thermal and strain induced break-up of the HS domains acting as physical crosslinks to prevent viscous flow [291]. Also, strain-induced crystallization and the disruption of the crystalline matrix contributed to mechanical hysteresis.

As reported by Blackwell et al [201], the X-ray analyses made on MDI/diol/PBA based PUs prepared by using hexanediol (HDO), butanediol (BDO), EG and propanediol (PDO) revealed the development of a second crystalline structure in the HS *as a result of thermal elongation and annealing*. The HDO polymer crystallized initially in the fully extended conformation, but a second crystal structure, a contracted form, developed with stretching and annealing at 130°*C*. In contrast, the BDO and PDO polymers crystallized initially in contracted conformations, but fully extended forms developed as a result of *elongation and annealing* [201].

The development of the new crystal structures was accompanied by the appearance of new HS melting peaks in the DSC traces. For the HDO polymer, the new peak for the contracted form appeared at a higher temperature than that for the original extended form; for the BDO and PDO polymers the new peaks for the extended forms appeared at lower temperatures than those for the contracted forms. Only a single peak was seen for the polymer extended with EG, for which only one crystal structure was detected for the HS. These results indicated that the development of polymorphic crystal structures have to be considered in the interpretation of the multiple melting phenomena seen for the hard domains in PUs.

DSC measurements were also used by ourselves to investigate the PUs thermal behaviour during the material stretching. Example DSC events are shown in Table 5.1 for two EG/PEA based PUs with similar structures, by using a single diisocyanate (DBDI or MDI). The materials were compared to their corresponding homopolyurethanes [EG-DBDI] and [EG-MDI] based on HS only. It observed that there coexist two fields of temperatures: 40°C up to 62°C when it occurrs the melting of the SS and 168°C up to 277°C characteristic to the HS melting zones.

As shown in Table 5.1 after stretching from a 300% to a 600% strain amplitude, the SS crystallized both in the materials with DBDI and MDI. The SS degree of crystallinity increased with increasing the strain amplitude, whereas, more of less the HS crystallinity was not significantly influenced by the increase of the strain amplitude. As seen in Table 5.1, for the materials studied, the HS crystallinity was

Table 5.1 DSC events for the PU studied initially, after stretching to 300% and 600% elongation [250]

PU #	Structure	Melting Peaks (°C) Soft segment (SS) zone t_{max} / ΔH			Heat of Fusion (J/g) Hard segment (HS) zone t_{max} / ΔH		
		initially[a]	300%	600%	initially[a]	300%	600%
1	DBDI:PEA:EG (4:1:2.64)	–	40°/3.46	54°/5.58	277°/18.21	277°/25.22	277°/16.4
2	MDI:PEA:EG (4:1:2.64)	–	–	62°/0.5	173°/8.3	168°/6.7	152°/2.94
3	DBDI:EG homopolymer	Without soft segments			287°/91.83	–	–
4	MDI:EG homopolymer	Without soft segments			237°/41.7	–	–

not significantly influenced by the degree of elongation. The materials PU_1 and PU_2 (where the SS was present) had lower values of the melting peaks than their corresponding homopolymers (materials 3 and 4 in Table 5.1), made up only by HS. In the case of PU_2 with MDI, the HS–SS interaction within mesophases [60] was preponderant, which affected the position of the PU melting peaks against models (homopolyurethanes). This was reflected by a significant difference between the melting peak of PU4 with MDI and its corresponding MDI homopolymer (4). Similar observations where made by us for other materials based on diisocyanates like MDI, and chain extended with EG.

However, an interesting observation could be made in the case of the DBDI material, PU_1. For this polymer, the melting peak ($t_{max} = 277°C$) was very closed to the melting peak of the corresponding homopolyurethane $[EG–DBDI]_n$ ($t_{max} = 287°C$). Due to the special spatial structure of DBDI, the HS crystallization of the EG-DBDI couple in the material PU_2 performed in a high proportion (89.7%) [250]. This process was not affected by the presence of the SS.

5.3 Deformation induced morphological developments as revealed by SEM

Every beginning is a consequence—every beginning ends some thing (Paul Valery)

For selected materials with MDI or DBDI obtained by us, we examined the polymer surface structure by means of SEM, under three conditions: *(a) initially after synthesis, on original PUs sheets; (b) while stretching the polymers to a 300% strain amplitude; (c) on materials recovered after rupture.* SEM was performed on the outer surface of the original cast sheets as prepared and after stretching up to a 600% level of elongation, as well as after etching, as described elsewhere [60, 204]. The samples subjected to stretching were rotated so that the draw direction lied approximately from bottom left to top right of the rectangular pictures displayed.

Example SEM results are shown in Fig. 5.1 and 5.2 for two materials with analogous structures that differ only in the type of diisocyanate, DBDI or MDI. The two polymers were designed, to reveal the roles of choice of flexible or rigid diisocyanate, and of chain extender (EG) in determining the performance as an elastomer [204]. The SS macrodiol was built using PEA of molar mass 2 000 g/mol. The polymers were synthesized by the prepolymer route. The molar ratio between components was diisocyanate:macrodiol:chain extender = 4:1:2.64 resulting in an isocyanate index I = 110. Both of the materials with DBDI and MDI displayed a relatively coarse structure on 10 μm scale but which varied between the two polymers.

In a first stage the PUs morphology has been studied by SEM *initially after synthesis, on original PUs sheets.* The results were detailed previously in section 2.3. The material with DBDI which displayed X-ray crystallinity gave a rougher surface morphology than did the material with MDI.

In a second stage the PUs morphology has been studied by SEM *while the materials where stretched to a 300% level of elongation.* Fig. 5.1.a and 5.1.b. display the features of the DBDI based PU as compared to the MDI based material. As before, the polymer with DBDI which displays crystallinity due to the dibenzyl structures, tends to give a much rougher surface morphology than the MDI based material. Rotation around the central $-CH_2-CH_2-$ bridge in DBDI allows alignment of aromatic rings and hence crystallization within the PU hard phase, which is not available with MDI in melt-cast polyurethanes [1–5].

Further SEM investigations were done by us on the same two materials, (using different samples) *after rupture occurred.* This is shown in Fig. 5.2.a and 5.2.b. for both polymers. Fig. 5.2.a displays the features of the DBDI material after rupture where it observes the banding parallel to the draw direction running from bottom left to top right. Fig. 5.2.b displays the features of the MDI based PU where the banding parallel to the draw direction is even more prominent. There are different cracking patterns. The elliptical cracks in the material with DBDI are characteristic of where a surface void has opened up and a work hardened ring has formed around it. As seen in Fig. 5.2.a, the polymer with DBDI did not retract much after rupture.

Fig. 5.1 and 5.2 should be compared with Fig. 2.12 and 2.13 in section 2.3, where the morphology of these two materials has been studied by SEM initially after synthesis, on the original PUs sheets.

The material with MDI (Fig. 5.2.b) appears to have retracted almost to its original length after rupture, the cracks opened up seem to be real cracks formed by sideways stresses. This is intimately connected with the chemical hard block variable structure as a function of the type of the adopted isocyanate.

Overall different cracking patterns between the DBDI and MDI based polymers were observed after material rupture when the dibenzyl based polymer did not retract. On materials stretched to a 300% strain amplitude, the presence of the dibenzyl

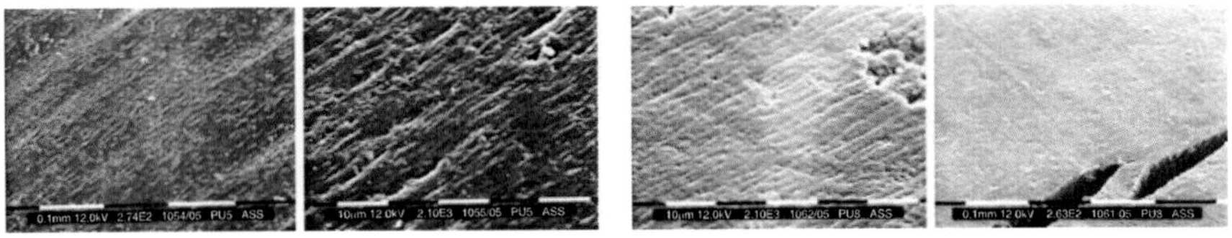

(a) (b)

Fig. 5.1 SEM of a DBDI based material (a) and a MDI based material (b) while stretched to a 300% strain amplitude, at two magnifications [204]

(a) (b)

Fig. 5.2 SEM of a DBDI based material (a) and of a MDI based material (b) after rupture at two magnifications [204]

structures gave a much rougher surface morphology than that observed in the MDI based polymers.

Other aspects regarding the morphology of different DBDI, MDI-based PUs or combination of these, as revealed by chemical etching techniques were reported elsewhere [60].

5.4 Orientation and crystallization of microstructure as revealed by IR dichroic measurements

Not everything that counts can be counted, and not everything that can be counted counts.
(Einstein's office at Princeton)

Everything that counts could be counted by using IR dichroic measurements to study the PUs orientation and crystallization ☺. Numerous studies on the PUs segmental orientation by means of infrared IR dichroic measurements have been reported. Estes et al [140] studied the segmental orientation of model MDI based PUs. The NH, CH and bonded and free C=O peaks were used to determine the degree and direction of orientation of the HS and SS. They found that the mode of HS orientation was strongly dependent on the type of ordering present within the hard domains. For PUs with para-crystalline hard domain order it was shown that HS orient parallel to the stretch direction throughout the entire deformation process. For materials comprised of crystalline hard domains, the orientation of the HS at low strains was shown to be transverse to the direction of stress. Allegrezza et al [353] showed that after elongation, the SS tend towards a random configuration while the overall HS orientation for these compositions does not relax. Hysteresis experiments suggested, however, that orientation of crystalline HS domains is reversible so long as the domains are not disrupted by the strain.

Hun et al. [361] used IR dichroism to study segmental orientation in PUs with varied structures and composition. A significant change in orientation behaviour was seen in MDI/BDO materials, with PTMO soft segment, when the MDI content was changed from 24 to 28%. The lower data of mechanical properties and orientability at 24% (MDI, 33wt% HS) was attributed to the inability of this material to form an interlocking domain structure. Similar conclusions have been drawn by Sung et al [362] based on their IR dichroic data on poly(urethane ureas).

Sung and Schneider predicted the *glass transition temperatures* (T_G) of the SS by using IR measurements to determine the degree of phase mixing along with a modified copolymer equation that treats the hydrogen bonds as crosslinks. The T_G of the SS was used by Sung and Schneider [134], Clough and Schneider [82], Clough et al [6] and Srichatrupimuk and Cooper [153, 154] as evidence of phase separation. The lower the proportion of dissolved HS in the SS domains the lower the T_G of the SS.

Seymour et al. [363] also studied the segmental orientation of MDI based PUs during deformation using differential IR dichroism. The NH, CH and bonded and free C=O peaks were used to determine the degree and direction of orientation of the HS and SS. They found that the mode of the HS orientation was strongly dependent on the type of ordering present within the hard domains. In the case of materials derived from crystalline hard domains, the orientation of the HS at low strains was shown to be transverse to the direction of stress. Similar observations were made by us for materials based on the diisocyanate DBDI crystallizing, especially when obtained with diol chain extenders like EG or BG [67].

Seymour found that at *higher strains* the HS realign to orient in the direction of the stress. This observation was also made by us in for PUs with DBDI [67] It was suggested that at this point the crystalline hard domains are disrupted, and the individual HS order along the direction of strain.

Kimura et al. [170] studied polyether based *polyurethane-ureas* using IR dichroism. They observed that the hard domains were disrupted at higher strains to give way to HS oriented along the stress direction. These remarks lead to the conclusion that, depending on the structure, size and molecular organization of the hard domains, it is possible to observe both transverse and parallel orientation of the hard domains during tensile elongation.

The orientation of the HS during tensile deformation was also investigated by Hammond in *diacetylene containing PUs*, by using dichroism [364]. The materials were obtained with PTMO of molar mass 1 000 g/mol. HS were obtained with MDI and 2,4-hexadiyne-1,6 diol. The diacetylene groups introduced into the HS remained dormant until exposed to electron radiation, UV light or heat [364]. The visible spectra of 4,4'-MDI and PTMO based PUs films containing HS with fully extended polydiacetylene chains were measured in polarized light in the stretched and relaxed state. It observed that the HS were preferentially oriented normal to the direction of stress at strain levels up to about 500%, indicating that the long axis of the hard domains was orienting parallel to the stretch direction during the deformation process. The orientation of these domains was relatively reversible, with moderate degrees of residual orientation. Up to strains of 300%-400%, the hard domains remained structurally intact. At strain levels higher than 500%, there was evidence of hard domain disruption [364].

In the model presented by Hammond, the HS, tied together by a polydiacetylene backbone, are oriented laterally in lamellar-like hard domains oriented in random directions before stress is applied, and possibly forming spherulitic type superstructures. The SS are randomly coiled macromolecules spaced between the hard domains. At low moderate strains, when the material is stretched, stress is transferred to the hard domains. The HS orient perpendicular to the stress direction. Upon removal of the stress, the SS relaxation allows the hard domains to return in a random orientation. This results in a small residual hard domain orientation. Therefore, the two-phase microstructure of the material is not highly interconnected. It consists of discontinuous hard domains dispersed throughout a continuous SS phase [364].

By using IR dichroic measurements, Alegrezza et al [353] investigated a series of PUs containing hard blocks with incorporated piperazine units which *have no possibility for hydrogen bonding*. In such cases, the SS where observed to orient into the direction of stretch at all strains but related towards a random configuration when the load is removed. HS orientation was transverse to the stretch direction at low strains, turning into the direction of stretch at higher strains. At higher strains the HS orientation relaxed very little when the load is removed.

Compared to hydrogen bonding, the PUs investigated in this study showed higher HS orientability and strain hysteresis. This may be partly attributed to the reduced strength of the intersegment bonding and to the more crystalline nature of the HS in the piperazine extended polymers.

A series of studies on the PUs segmental orientation by using IR dichroic measurements were also reported by us [67, 127]. We investigated a series of polyesteric PUs based on PEA, and the isocyanate of variable geometry DBDI (ca 40% HS fraction). To avoid HS crystallinity given by diisocyanate couples such as DBDI/EG or DBDI/BG, the materials were chain extended with DEG. PUs were synthesized by a standard prepolymer procedure as reported elsewhere [67].

As mentioned before, the ability of the two benzene rings in the DBDI unit to rotate in two discrete "syn" and "*anti*" positions around the $-CH_2CH_2-$ ethylene bridge results in a specific hard domain coplanar packing structure in the case of dibenzyl based polymers. This has an important impact on the material properties.

A stretching device described elsewhere [67] was made by us which extended the PU samples uniformly from both ends while it was centered in the IR beam. The samples were placed at an intermediate focus in the infrared source beam at an angle of *45°* to the slits. Wire grid polarizers were placed in the sample and reference beams at a setting of 45°. The entire stretching device rotated around the centre of the IR polarized beam (and the sample) to allow absorption experiments while stretching the sample both parallel and transverse to the plane of the polarized radiation. Testing was done at 23° initially with a fresh sample being used for each individual strain test. Subsequently, the experiments involved extending the PU sample by stepwise increments to 600% strain (and unloading stepwise until the stress equal was zero). During each step the sample was scanned with the polarized IR so that the same profile at different strain levels (up to 600% elongation) was followed for each sample. Then the polarized IR spectra were analyzed quantitatively using selected adsorption bands [67].

The recorded output was the dichroic difference, from which the orientation function (F^D) could be calculated [140, 363], which is unity for perfect orientation in the stretch directions, $-1/2$ for perfect orientation transverse to the stretch direction, and *zero* for random orientation.

The applicability of infrared dichroism to the study of PUs has been determined by the availability of at least one well-characterized absorption band in each type of domain. It has been shown [67, 140, 364] that the soft domains are comprised of prepolymer (polyether or polyester) segments, while the hard domains contain urethane segments. The N–H stretching which is located in the urethane linkage characterizes the orientation of the higher modulus hard domains. For the DBDI polymers, we have shown that it appears at 3 320 cm^{-1} [67]. Similarly, the asymmetric C–H stretching absorption was determined. It appears at 2 940 cm^{-1} and was used to describe the orientation of prepolymer segments which comprise the rubbery soft domains.

Orientation hysteresis studies were performed by us to probe the mechanism of the transverse orientation and orientation inversion. The determination of strain recovery and strain energy recovery on cycling (hysteresis) was made by stretching the PU samples by stepwise increments to a 300% strain (and unloading stepwise until the stress was zero) while recording the polarized IR spectra at each step. The PU samples were subjected to two subsequent load-unloading cycles after which the sample was stretched to 600% extension, as shown in Fig. 5.3 and 5.4.

By monitoring the N–H and C–H infrared stretching absorptions, it observed that the orientation of HS and SS strongly depends upon the strain history. A higher degree of orientation and a greater distinction between the behaviour of the N–H and C–H groups was observed. The explanation resides in the fact that the N–H groups belong to the HS which are strongly associated by reciprocal H bonds, that tend to maintain the initial polymer network state for a longer time during the increase of the stress. Instead, the C–H groups which display a reduced polarity do not form associates and are more susceptible to change their positions in the polymer network [67]. The spatial movement of the N–H and C–H groups while the polymer network was subjected to stress was different. As seen in Fig. 5.3, the maximum of the N–H moiety orientation parallel to the stretching direction occurred at a 50% elongation where the HS became oriented perpendicular to the stretching direction. Then as a consequence of breaking-up of some of the primary hydrogen bonds and as a result of the macromolecular chains tendency to orient towards the stretching direction, the N–H bonding started to orient transverse to the stretching direction, and implicitly the HS tended to orient in parallel to the stretching direction. The higher strain amplitude caused a more pronounced tendency of the HS to align in the stretching

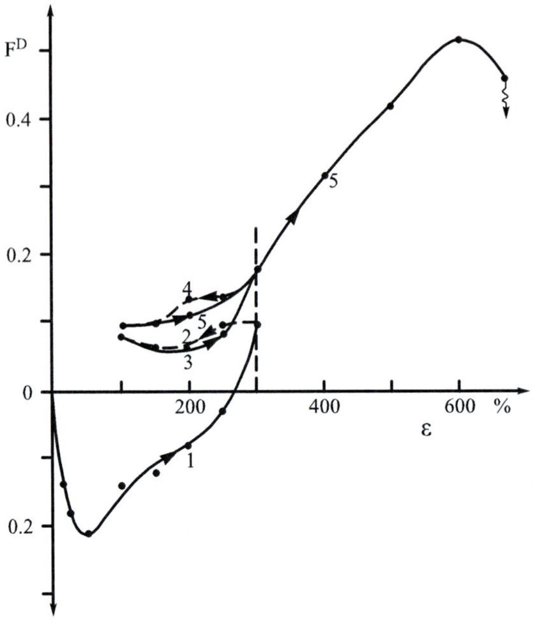

Fig. 5.3 Orientation function (F^D) as a function of strain (ε) for a DBDI based PU (DEG 40-PEA, DBDI, I = 100) at loading-unloading (0% to 600% elongation) for ν_{NH} = 3 320 cm^{-1} [67]

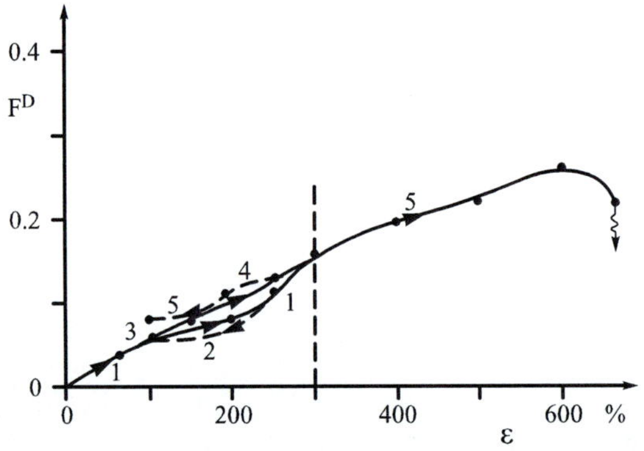

Fig. 5.4 Orientation function (F^D) as a function of strain (ε) for a DBDI based PU (DEG 40-PEA, DBDI, I = 100) at loading-unloading (0% to 600% elongation) for ν_{CH} = 2 940 cm^{-1} [67]

direction. The maximum of the N–H moiety perpendicular orientation was reached at a 600% strain level (Fig. 5.3). This process required an input energy E_C = 24 MPa and a recoverable energy E_R = 15.6% of E_C.

Subsequent stretches by the hysteresis curves showed a higher and higher orientation and ordination of the N–H bonding under load towards the transverse

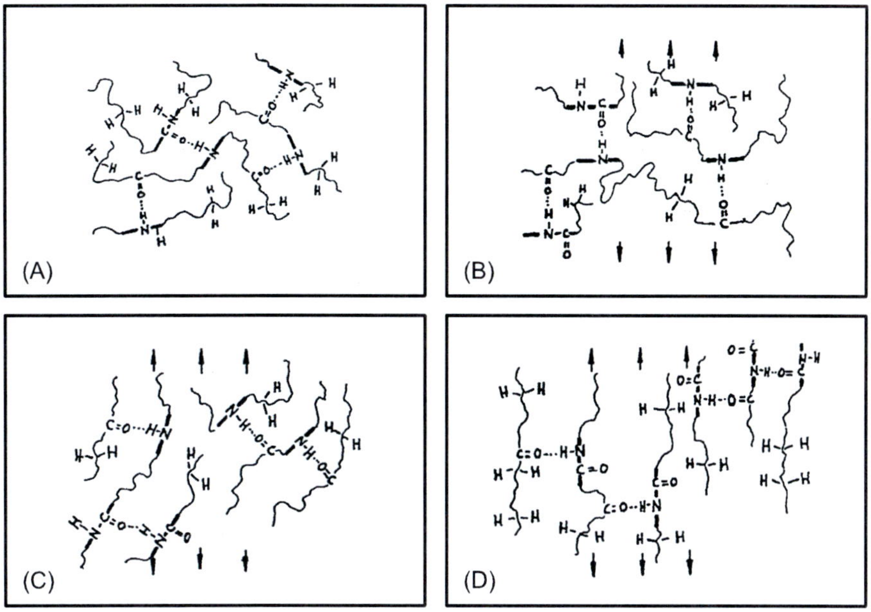

Fig. 5.5 Schematic of the changes in the supramolecular structure of a dibenzyl based PUs (DEG 40-PEA, DBDI, I = 110) at polymer extension [67]

direction. In subsequent load-unload experiments the N–H bonding remained oriented transverse to the stretching direction (see Fig. 5.5). After relaxation, the chains containing the N–H group remained oriented to a substantially higher degree than those containing C–H functional groups. When the sample was stretched again, the N–H containing chains were more highly oriented than during initial extension. The distinct difference between the orientation functions for the two types of chains provided evidence that the different block polymer segments reside in different domains, which react independently to the application and removal of stress.

Gradually decreasing the tensile stress caused the HS to exhibit slow orientation to the stretching direction. The HS did not return to the transverse position observed at the 50% elongation. The SS displayed different degrees of crystallization as evidenced by the DSC melting peaks. However it should be noted that we also investigated other DBDI based PUs with a high elasticity (residual elongation = 0%–5%), where a permanent molecular ordering did not occur. An example was the case of a PUs with a similar structure as before, but with a lower content of DEG, and molar concentrations such as DEG:DBDI:PEA = 3:1:1.73 (I = 110) [67] Depending on the structure, PUs crystallization under tensile deformation was observed.

Chapter 6
Perspectives. Novel crosslinked polyurethanes as shape-memory materials

You never know when you are making a memory. (Rickie Lee Jones)

True, but do we really desire to remember *all* our memories? I do not think so ☺. But this does not apply here to our research in shape memory polyurethanes.

A relatively new and exciting application for polymers is as shape 'memory' materials. Therefore, one of the objectives which we will continue to follow in the immediate future will be to focus on novel crosslinked shape-memory polyurethanes.

Recently we have done a series of preliminary studies on this topic. Here we report only a few particular aspects of our research. More complete details are given in one of our most recent works (2010) [63].

In the recent study made by us [63], the usual diol chain extender was replaced by a triol (TMP), producing crosslinked PU networks without phase segregation. The aim was to ensure high degrees of strain recoverability, to produce candidate thermally-triggered shape-memory polyurethanes.

Again MDI and DBDI were compared as hard segments, but also toluene diisocyanate was used.

Numerous polymers have been proposed as shape-memory polymers (SMPs), and many of them are based on polyurethanes. This is because of the intrinsic versatility of segmented copolyurethane systems. By suitable choice of diisocyanate and macrodiol, a wide variation in properties may be obtained, allowing the possibility of tuning the shape-memory response to suit different applications. Usually they are phase-segregated materials. For example, a dispersed rigid phase (usually based on the diisocyanate) provides physical crosslinks, while the macrodiol provides a soft amorphous phase with low glass transition that provides the trigger temperature for shape recovery [63].

An ideal shape-memory polymer for a given application has a small set of parameters that determine the usefulness, or otherwise, for a given application: (1) The fractional recoverable strain: the maximum fraction of the imposed strain that can be recovered when triggered. (2) The temperature of maximum recovery rate $T_{\max}$ for a given rate of heating: a measure of the temperature required to trigger

C. Prisacariu, *Polyurethane Elastomers*, DOI 10.1007/978-3-7091-0514-6_6,

recovery of shape. (3) The width ΔT of the window of temperature within which shape recovery occurs. (4) The tensile modulus E_R at the temperature of full recovery: this determines the maximum shape-recovery restoring force if shape recovery is resisted [63]. Since different applications make different demands on SMPs, it is particularly desirable to find polymer systems where parameters $T_{\max}$, ΔT, and E_R are variable, so the same basic system can be tuned to suit different applications. Preferably, these will be systems where all the imposed strain is recoverable, that is the fractional recoverable strain is 100% [63].

These attributes are difficult to achieve, all in the same polymer. Further advances require improved understanding of how molecular structure controls these parameters.

The source of "memory" of polymers is their capacity to store preferred molecular orientation on large molecular length scales. The mechanism by which the shape memory is locked-in is usually cooling through the glass transition T_G (and consequent lengthening of mechanical relaxation times) or formation of a hard phase such as crystal domains. Shape recovery is triggered by subsequent heating, causing passage through T_G (and hence shortening of relaxation times) or softening of the hard phase. Lendlein and Kelch [365] provide a comprehensive review of the possibilities and of many of the shape-memory polymers proposed so far.

Previous authors have highlighted the merits of polyurethane polymers as potential SMPs [365]. This arises largely from the chemical versatility of segmented polyurethane copolymer systems. Thus by suitable choice of diisocyanate and macrodiol phase-segregated materials may be produced, for example where one component comprises HS that provide physical crosslinking, to enable the retention of shape memory, while the other component provides a softer phase with a lower softening temperature, either glass transition or melting temperature, to enable the locking-in and release of shape memory at convenient temperatures [63].

The softening temperature, which acts as the shape-memory trigger temperature (to be precise, $T_{\max}$), may be changed systematically by variation of the HS content [366–369] or the chain length of the macrodiol employed [63, 370, 371]. A feature of practical importance for biomedical applications of SMPs is the fact that these systems may be designed with trigger temperatures in the convenient range 20–50°C [63, 366, 369–372].

The width of the trigger window ΔT is also important in applications of SMPs, Published shape-recovery measurements, for example as reported by Lin and Chen [371] for polyurethanes, suggest values in the range 28–46 K, when $T_{\max}$ exceeds room temperature.

The third parameter, E_R, poses the greatest challenge for SMPs relative to shape-memory metal alloys, which are orders of magnitude stiffer. Shape recovery in polymers is usually driven by entropy-elastic recoil of molecular chains, but this is associated with extremely low elastic moduli, for example Young's moduli in the range 1–10 MPa [63]. Thus there has been interest in reinforcing SMPs with hard particles such as chopped glass fibres [373], sub-micron SiC particles [374] and [375] or molecular-level mixing with silica [376]. However, the price for a small increase in E_R (e.g. to 10 MPa) may be a significant deterioration in the recoverability of shape, as observed by Ohki et al. [373]. Even 20% by weight of sub-micron SiC in a crosslinked epoxy gave only a 50% increase in shape restoring force [63, 374].

Our purpose for this study has been to present results for a new family of polyurethanes that have particular promise as SMPs [62, 63, 377]. They are all chemically crosslinked by means of a triol crosslinking agent acting as a chain extender, and hence have the advantage of 100% strain recoverability. Systematic variations in chain flexibility and HS content have been achieved by changing the chemical structures of hard and soft segments, and by changing the length of SS chains, respectively.

All the materials were produced by the prepolymer route followed by melt casting of sheets, using diisocyanate HS and macrodiol SS. The synthesis of the materials has been described elsewhere [62, 63].

Such materials have the usual versatility with respect to varying HS content and macrodiol chain length, allowing the systematic variation of $T_{\max}$ and E_R (without using reinforcing particles). *However, they are also single phase, chemically crosslinked polymers, and hence have the advantage of 100% recoverability of shape.* This is achieved by employing a triol as chain extender, giving a network structure with tri-functional crosslinks. *The crosslinker triol used by us in all cases was 1,1,1-trimethylol propane (TMP).*

We have characterized these polymers with respect to their thermorheological response in tension, in the linear viscoelastic regime. Then their performance in a shape-memory sequence was computed from the theory of temperature-dependent linear viscoelasticity, allowing comparisons to be made between them [63].

To detail, the new family of SMPs was synthesized in our Romanian laboratory, each of the materials comprising three components, in molar ratios 4:1:2 respectively. (1) a macrodiol (MD) of various chain lengths: either polytetrahydrofurane diol (PTHF) or polycaprolactone diol (PCD); (2) a diisocyanate: a classical diisocyanate with a rigid structure 4,4'-diphenyl diisocyanate (MDI), a classical isocyanate containing a single phenyl group and non-identical isocyanic groups (2,4 TDI 80% + 2,6 TDI 20%), or 4,4'-dibenzyldisocyanate (DBDI). The latter diisocyanate (DI) is similar to MDI but has variable geometry because of rotation around the central $-CH_2-CH_2-$ ethylene bridge which links the two benzene rings. (3) the TMP acting as a 3-arm star crosslinker.

The crosslink density was quantified in terms of the number n_c of moles of 3-arm star junction points per 100 g of polymer. The macrodiols were of various lengths (molar mass 650 to 2 000). For PTMO of molar mass 2 000 g/mol polymers, the glass transition was near to or below room temperature.

As shown in previous chapters, the diisocyanate DBDI provides an interesting contrast with MDI, as it has more mobility because of rotation of the two benzene rings around the central $-CH_2-CH_2-$ bridge which links them [60, 61, 127, 135]. This degree of freedom is not available to MDI, where the bridge is merely $-CH_2-$.

A complete list of the polymers produced by us in this study, is given in Table 6.1. It includes a measure of crosslink density n_c for each material, expressed as the number of moles of three-arm stars per 100 g of polymer.

Table 6.1 List of the crosslinked polyurethane polymers synthesized and studied in this work [63]

SMP type	MD 1 mol	DII 4 mol	Moles of stars per 100 g of SMP
1	$PTHF_{650}$	2,4 + 2,6 TDI	0.1239
2	$PTHF_{650}$	MDI	0.1043
3	$PTHF_{650}$	DBDI	0.1013
4	$PTHF_{1000}$	DBDI	0.0861
5	$PTHF_{1000}$	MDI	0.0882
6	$PTHF_{1000}$	2,4 + 2,6 TDI	0.1018
7	$PTHF_{2000}$	DBDI	0.0602
8	$PTHF_{2000}$	MDI	0.0612
9	$PTHF_{2000}$	2,4 + 2,6 TDI	0.0505
10	PCD_{830}	DBDI	0.0928
11	PCD_{830}	MDI	0.0953
12	PCD_{830}	2,4 + 2,6 TDI	0.1115
13	PCD_{1250}	DBDI	0.0777

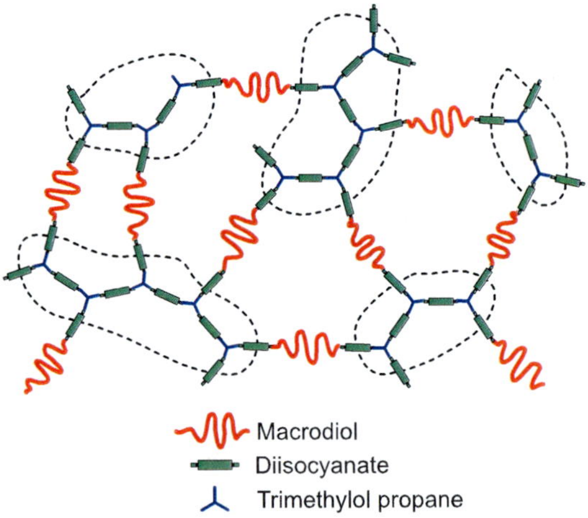

Fig. 6.1 Schematic diagram of the structure of the triol-crosslinked polyurethanes. Dashed lines indicate DI-rich regions of reduced mobility [62, 63]

The result is a network polymer with structure as shown schematically in Fig. 6.1 [63]. Each three-arm star is surrounded by three diisocyanate molecules, and these are linked together partly by the macrodiol. However, given the molar compositions of these polymers, there must also be some connectivity of junction points through just the diisocyanate, producing localities within the network with much reduced mobility — see dotted lines in Fig. 6.1.

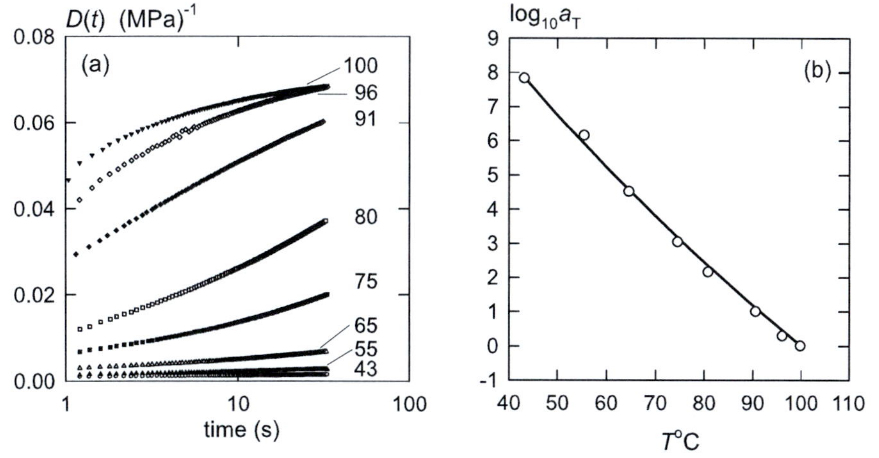

Fig. 6.2 (a) A sample set of creep curves for various temperatures T, and (b) their time-temperature shift factors for reference temperature $T^* = 100°C$ [63] (b), for SMP3 based on DBDI/PTHF$_{650}$ [63]

6.1 Thermorheological characterization

Experience is directly proportional to equipment ruined. (unknown source)

Yes, it sometimes happens to us. The thermorheological response of each polymer in Table 6.1 was characterized by us in the Oxford laboratory. Fortunately after doing this study no equipment was ruined ☺.

The materials were characterized by us, by means of a series of short-term (30 s) tensile creep tests at a range of increasing temperatures above room temperature, using a custom-built, lever-loading, creep machine. Specimens were prismatic rectangular bars of width ca 3 mm, thickness ca 1 mm, and length ca 60 mm between the grips. The length/width ratio (20:1) was sufficient for grip displacement to be used for strain measurement [377]. The creep tests were carried out in the linear viscoelastic range, and a recovery period of at least 300 s was allowed between tests. Strain was found to be fully recoverable. Creep strain for a test temperature T was expressed as tensile creep compliance thus: $D(t,T) = \varepsilon(t)/\sigma$. A typical set of isothermal creep curves is shown for polymer DBDI/PTHF$_{650}$ in Fig. 6.2(a). Creep data such as this were adjusted to a reference temperature T^* by multiplication by T/T^* (to allow for temperature-dependence of relaxed compliance D_R, neglecting the smaller effect of change of density), and combined using time-temperature superposition to obtain the complete creep curve — "master curve". The required shift factor a_T for this polymer is given in Fig. 6.2(b).

In each case, the master curve was then fitted to the expression

$$D^*(t) = D_U^* + (D_R^* - D_U^*) \int_{-\infty}^{\infty} \varphi^*(\ln\tau)[1-\exp(-t/\tau)]\mathrm{d}\ln\tau \qquad (6.1)$$

to obtain the unrelaxed and relaxed compliances and the retardation spectrum: D_U^*, D_R^* and $\varphi^*(\ln\tau)$ respectively, for the reference temperature. Even for the short creep period (30 s) adopted, there was sufficient overlap of the creep curves to generate the master curves with adequate redundancy.

6.2 Linear theory of shape recovery

It is the theory which decides what we can observe. (Albert Einstein)

For comparing the materials, their performances were modelled for a standard shape-recovery sequence, in which a strain ε_0 is imposed at sufficiently high temperature to produce complete relaxation, before quenching to freeze-in the strain. In simulating re-heating to generate shape recovery, the evolution of residual strain was computed from the theory of thermo-rheologically simple linear viscoelasticity, giving the evolution of fractional recovery as follows:

$$\frac{(\varepsilon_0 - \varepsilon(T))}{\varepsilon_0} = 1 - \int_{-\infty}^{\infty} \varphi^*(\ln\tau)\exp\left(-\frac{-\xi}{\tau}\right)\mathrm{d}\ln\tau \qquad (6.2)$$

where $\xi = \int_{20}^{T} \frac{\mathrm{d}T}{\dot{T}a_T}$.

The integral in equation (6.2) was evaluated numerically for the retardation spectra obtained from master curves, together with a_T from the time-temperature superposition. A typical result is demonstrated in Fig. 6.3(a), where recovery is shown for the DBDI/PTHF$_{650}$ polymer, for a simulated constant rate of heating $\dot{T} = 0.1$ K/s. This Figure also illustrates the definitions of two parameters used to compare the responses: the temperature of maximum recovery rate $T_{\max}$ and the width of the recovery window ΔT. These parameters of the SMPs were compared with respect to chemical composition and crosslink density n_c, expressed as the number of moles of 3-point star crosslinks, per 100 g of polymer.

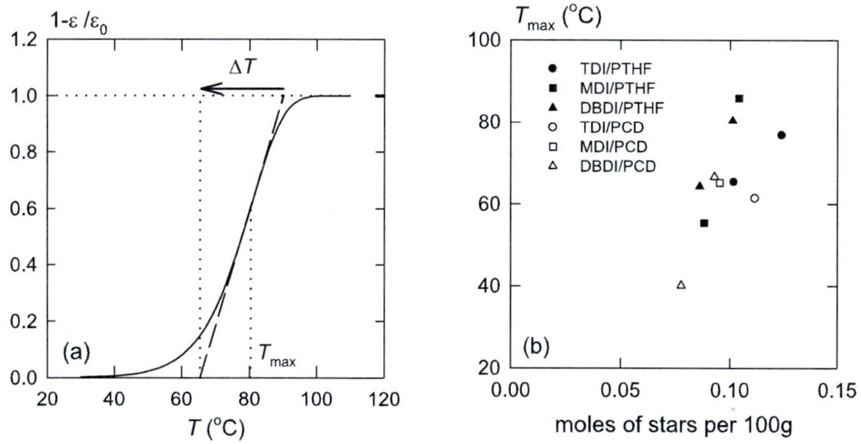

Fig. 6.3 (a) Fractional recovery as calculated from the data in Fig. 6.2 for a heating rate of 0.1 K/s, showing definitions of T_{max} and ΔT; (b) T_{max} for all the polymers studied [63]

6.3 Recovery temperature

The quality of recovery is proportional to the quality of surrender. (unknown source)

Indeed, sometimes surrender may be the only way to recover. Susspicion, apprehension, uselessness and even angerness may lead to the act of surrender. Sometimes this is justified but certainly not here ☺.

Fig. 6.3(b) shows the variation of temperature of maximum recovery rate versus n_c for the various combinations of DI and MD. It may be seen that there is a clear trend of increasing T_{max} with increasing crosslink density. This reflects increasing retardation times — decreasing molecular mobility. It is consistent with reduction in the mass fraction of the relatively mobile MD chains, located between *relatively rigid diisocyanate* groups surrounding each crosslinks. Nevertheless it is interesting to note there is also a variation with chemical composition. The recovery temperature of TDI-based polymers lies clearly below those of the other polymers for both MDs. This reflects the greater mobility of the chains where DI contains only one phenyl ring instead of two.

6.4 Width of recovery window

If the facts don't fit the theory, change the facts. (Albert Einstein)

Clearly this could never apply to us if "facts" refer to "results". We would never change our results to fit theory ☺. Fig. 6.4(a) shows the width of the recovery

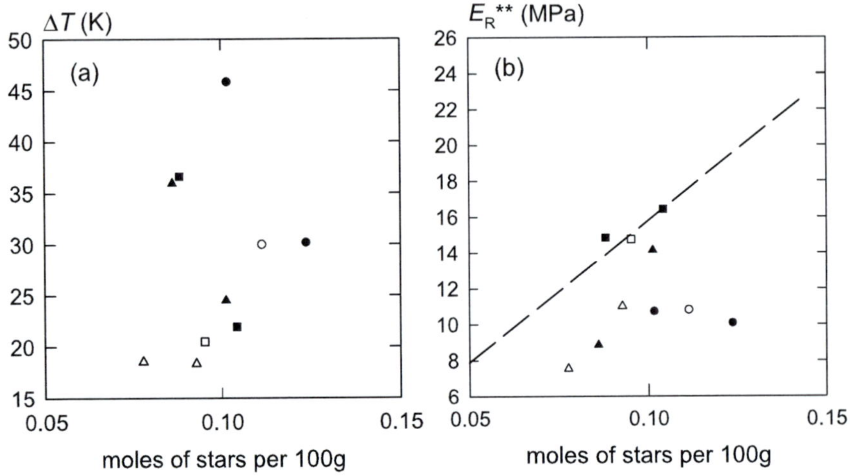

Fig. 6.4 (a) Width of the recovery window [63], and (b) relaxed creep modulus at 92.3°C. Also shown (dashed line) is the relaxed modulus predicted by the affine theory of rubber elasticity. Symbols are as in Fig. 6.3 [63, 377]

window plotted versus n_c. This parameter reflects primarily the width of the retardation spectrum (see [63]). For the PTHF-based polymers there is a clear trend of decreasing ΔT with increasing crosslink density. The explanation for this lies in the range of molecular environments of the MD chains. Close to their terminations at relatively immobile DI groups the chains are constrained, and retardation times are expected to be lengthened relative to the central portions of chains where there is less constraint. With increasing crosslink density the influence of the DI groups extends over a larger portion of the chains, and hence the contribution of the relatively mobile (short relaxation time) fraction diminishes and the retardation spectrum narrows, in addition to shifting to longer retardation times. A further interesting point to emerge from Fig. 6.4(a) is the relatively wide recovery window of the TDI-based polymers (shown as circles). This may be attributed to the fact that TDI comprises two isomers: 2,4-TDI (80%) and 2,6-TDI (20%), which will have different mobilities, and a broader retardation spectrum is expected. Notwithstanding the comments above concerning the PTHF polymers, it is notable from Fig. 6.4(a) that PCD-based polymers appear not to show the same trend. This inconsistency needs further investigation.

6.5 Relaxed modulus

Every time I find the meaning... they change it. (Leonardo da Vinci)

Not here ☺. Also of great interest in relation to shape-memory performance is the relaxed modulus E_R. It determines the recovery stress available to overcome any resistance to shape recovery during heating. It is determined by entropic elasticity of the network and is therefore expected to be proportional to absolute temperature. To compare the various polymers, therefore, relaxed moduli were all reduced to a common temperature T^{**} = 92.3°C: the mean of the reference temperatures. Thus values were obtained from relaxed compliances thus

$$E_R^{**} = \left(\frac{T^{**}}{T^*}\right)\left(\frac{1}{D_R^*}\right). \tag{6.3}$$

Fig. 6.4(b) shows the results of these calculations plotted versus crosslink density, together with the classical prediction from the affine theory of rubber elasticity: $E_R^{**} = 3N_c k_B T^{**}$ where N_c is the number density of chains. It is clear there is approximate agreement with the theory (to within a factor 2), but there is also significant influence of chemical composition. In particular it is interesting to note the order of stiffness for polymers based on the three DIs: MDI>DBDI>TDI. This reflects the relative intrinsic stiffness of the DI groups.

Results have shown systematic trends with crosslinked density and chemical composition. The recovery temperature primarily increases with crosslink density, but is reduced by use of TDI and the DI. The width of the recovery window generally decreases with crosslink density and is least for PCD-based polymers. The relaxed modulus generally increases with crosslink density, and is highest for polymers based on MDI and lowest for polymers based on TDI.

As detailed elsewhere [63], *the results obtained by us have shown that the new triol-crosslinked polyurethanes appear to have great potential as shape-memory polymers.* Complete recovery is ensured by chemical crosslinking through a triol, while adjustment of the shape-memory trigger temperature is achieved by variation of the chemical composition of the hard and soft segments, and the chain length of the soft segments [63]. By appropriate choice of the hard segment, the trigger temperature has been varied from below room temperature up to 86°C, illustrating the versatility of the system. The shape-memory performance of these materials was predicted, by first measuring their thermorheological response in the linear viscoelastic region, and then predicting their shape recovery during the heating phase in a standard shape-memory test [63].

The predicted temperature of peak recovery rate $T_{\max}$ increased with crosslink density for any given combination of diisocyanate and macrodiol, as a result of decreased molecular mobility, revealed through shifting of the retardation spectra to longer times [63]. The greater flexibility of TDI was also apparent, as compared to MDI and DBDI, producing a lower $T_{\max}$ for given crosslink density. The width of the shape-recovery temperature window ΔT was found to decrease with decreasing macrodiol chain length. The retardation spectra revealed that this was caused primarily by a suppression of the shortest retardation times [63]. The shorter chain length reduced the mobility of the portions of macrodiol chains most remote from the influence of the hard segments at the network junctions.

As shown in more detail elsewhere, the rubbery-state modulus E_R showed a trend of increasing with crosslink density, with measured values lying close to or between the predictions of the affine and phantom chain theories of rubber elasticity [63]. However, we also observe an influence from the chemical composition, with the actual values between these two limits reflecting the intrinsic stiffnesses of the three diisocyanates and hence the molecular mobility at the network junctions [63].

These results illustrate the potential of this family of polyurethanes to act as shape-memory polymers, and to be tailored chemically to suit particular practical applications.

Our further research in this field will include also other systems of PUs that are based on various diol chain extenders and diisocyanates, crystallizing or non-crystallizing. Thermally induced shape-memory effects, including thermodynamic aspects significant for the shape memory effect will be followed. The molecular mechanism of the shape memory effect will be investigated for larger systems of materials based on hard segments crystallizing or not. Macroscopic shape memory effects, will be studied by quantifying the shape memory effect in terms of cyclic thermomechanical investigations.

Another promising field which we will follow will be to do research in the area of biomedicine where candidate shape memory polyurethane materials could enable technology for future applications.

Other perspectives

Polyurethanes elastomers are a highly versatile class of materials in their own right. For their potential to be exploited more fully, a better understanding of their behaviour is still needed. Understanding the properties of the polyurethanes in terms of their chemistry and molecular architecture raises many new scientific challenges.

Polyurethane structures can be varied systematically, so they are an ideal vehicle for providing information of generic interest in the *modelling of hard phase reinforced elastomers*. Modelling the mechanical properties of polymers phase-segregated on the nm length scale is of profound significance for polymer science, with far-reaching implications in the design and synthesis of future materials.

The perspectives to advance require a multidisciplinary approach combining synthetic macromolecular chemistry and microstructural studies, with the study and quantitative modelling of mechanical properties.

Such a strategy will enable us to continue doing progress on the characterization of these materials, with respect to their physical and mechanical response of large practical interest and to generate understanding in terms of a physically-based finite-strain constitutive model; and to use microstructural information to assist in the interpretation.

This will offer us the opportunity to find out more performant methods to select suitable raw materials in order to achieve high performance polyurethanes. Thus there is great scope to achieve further new fundamental science and to assist the development of new polyurethanes.

Closing remarks

A conclusion is the place where you got tired of thinking. (Martin Fischer)

...A well known quote... but it does not apply to the topic of polyurethanes ☺.

Conclusions on polyurethane research are like circles as they end where they begin. The topic is inexhaustible. Progress in polyurethane development is being made continuously. Polyurethane research is as organized as a spider silk, where the most difficult part, the construction of the first thread (on which the rest of the web is hanging) was made decades ago. But the wheel-shaped webs, their bridges, are built continuously by every novel valuable contribution to polyurethane research while the whole web is extended by progress of research continuously expanding.

The world of polyurethanes is enormous, and so it is the literature that has been dedicated to these materials over the past seventy years. Academia and industry have equally brought contributions of very high merit to the development of polyurethanes. Thus, obviously it is impossible to write a book about polyurethanes where to include all the valuable works that have been reported in this field.

Therefore in this book I only referred to *particular* aspects, regarding the synthesis and characterization model and novel polyurethane elastomers.

The objective of this book was to review, evaluate and improve understanding of the relationship between molecular/supramolecular architecture, and nm-scale and macroscopic mechanical properties of this uniquely versatile family of polyurethanes categorized in engineering elastomers. The study included numerous conventional block polyurethane elastomers, based on several diisocyanates, macrodiols and chain extenders.

The book reviews aspects from the up-to-date literature focus on these topics. In addition, in order to widen the range of structures achievable beyond those normally available, in this book I also included recent developments that have been made by us, by producing polymers based on the diisocyanate of variable geometry, 4,4'-dibenzyl diisocyanate (DBDI) (only available from Romania), that allows the variation of hard domain crystallinity as a key structural variable. The conformational mobility of DBDI causes an unusually wide range of mechanical, physical and chemical properties, associated with the possibility of pronounced phase separation

C. Prisacariu, *Polyurethane Elastomers*, DOI 10.1007/978-3-7091-0514-6, into a domain - matrix morphology, and with a high tendency to crystallization and self-association by hydrogen bonding, which is not available with the conventional diisocyanates in traditional melt-cast polyurethanes. The mechanical performance of such elastomers has been shown to be strongly affected by the higher-ordered structure of hard segments on the macromolecular chain.

In the present book, materials derived from the diisocyanate DBDI were mainly compared with the conventional diisocyanate 4,4'-diphenylmethane diisocyanate (MDI), as one of the most representative and commonly used aromatic diisocyanates.

A key feature of the work described in this book, has been to focus on materials synthesized in the Romanian laboratory, under known and controlled conditions with known reactants. I consider this is a much more promising route to doing good science than working with commercial materials where the exact synthesis conditions and composition are likely to be unknown. Science is feasible when the variables and their combinations are distinct and clear. We are tending toward the condition of science and aspiring to do it.

What it has not been done in this book:

(1) – the block copolyurethanes, consisting of isocyanate hard segments, macrodiol soft segments and chain extenders, represent an extraordinarily versatile family of polymers, where a wide variety of physical properties may be achieved via variations in chemical composition and synthesis route. However *our* research described in this book had to be limited mainly to the comparison between the novel DBDI materials and conventional materials obtained with hard segments of MDI. Further developments by also including other diisocyanates, macrodiols and chain extenders will be reported in another book to be published in year 2013;

(2) As shown in the present book, in recent years, we have made a series of studies of different aspects of the mechanical behaviour of these materials, aimed at improving our understanding of the way macroscopic properties of these materials depend on chemical structure and nm-scale physical structure. However, in view of the difficulty of knowing in detail the physical processes active on the nm scale of elastomers, there still is an urgent need for more comprehensive information on the sensitivity of the inelasticity to structural detail on the nm scale. There have been many attempts to explain the inelastic features and to capture them in constitutive models, with some success at the phenomenological level. But, somewhat inevitably in view of the difficulty of knowing in detail the physical processes active on the nm scale, much of previous work in this area has been speculative in terms of providing a physical interpretation. For progress to be possible, there is an urgent need for more comprehensive information on the sensitivity of the inelasticity to structural detail on the nm scale.

(3) Further work is planned, to exploit a series of novel partially or fully deuterated polyurethane elastomers recently made by us, in obtaining a better understanding of the relation between mechanical properties and the nature and degree of phase separation that occurs in these polymers. The ongoing research in this area will be included in our further book.

(4) Due to the limited length of the book, particular aspects like thermal investigation as revealed by thermogravimetry or thermomechanical analyses had to be limited to only a few examples of DBDI based polymers.

(5) Similarly, due to the same reason, in the present book I had to refer only to few aspects regarding the synthesis and characterization of polyurethane ureas. I have mainly focused on polyurethanes extended with diols because our most recent works are dedicated to the investigation of polyurethanes with diols. Further works on the chemistry, morphology and mechanical performance of polyurethane ureas based on isocyanates of variable geometries will be included in our second book.

(6) Some people write beautifully and effortlessly while others feel like they are sweating out each word. But over time authors with both writing styles make successful contributions to science. For myself it has not been quite easy to write this book as a non-native English speaker. I often felt that *inspiration* is what is the tail for a cat. You cannot catch it, but it follows you if you ignore it ☺.

As always, a conclusion is a chance to have a final say, to summarize thoughts, to demonstrate the importance of the ideas developed in the book so to give the readers an overall view. This is what I attempted to do here in the hope readers have enjoyed this book, but while I am also aware that, as Leonardo da Vinci said, for readers *"it is easier to resist at the beginning than at the end."* ☺

I hope this book will inspire you to take action as well and continue to embark on further study in the so challenging field of polyurethanes. Our mobility is such that we can ask questions and answering them. Our curiosity and willingness are insatiable. Our mind is in constant turmoil, solving problems, accumulating knowledge within a short range of time. Our chances and our ideas turn our daily life into a mosaic of intense moments, where the joy of doing valuable research plays a key role. The future of polyurethanes fundamental research and industry will be driven by the continued innovation in both the chemistry and the polymer physics of these highly versatile materials.

However, as Paul Valery said, *"the trouble with our times is that the future is not what it used to be"*.

References

[1] J.H. Saunders, K.C. Frisch, *Chemistry and Technology, Part I and II* (Interscience Publishers., New York, 1962)
[2] C. Hepburn, *Polyurethane elastomers* (Applied Science, London, 1982)
[3] G. Oertel, *Polyurethane Handbook* (Hanser Publishers, Munich, 1985)
[4] Z.S. Petrović, *Polyurethanes* (Marcel Dekker, Inc., New York, 2005), 2nd edn. Handbook of Polymer Synthesis
[5] R. Bonart, Journal of Macromolecular Science, Part B: Physics **B2**(1), 115 (1968)
[6] S.B. Clough, N.S. Schneider, A.O. King, Journal of Macromolecular Science, Part B: Physics **B2**(4), 641 (1968)
[7] K. Nakayama, T. Ino, I. Matsubara, Journal of Macromolecular Science, Part A: Chemistry **A3**(5), 1005 (1969)
[8] A. Tager, *Physical Chemistry of Polymers* (Mir Publishers, Moscow, 1978)
[9] S.A. Visser, G. Pruckmayr, S.L. Cooper, Macromolecules **24**(25), 6769 (1991)
[10] C.S. Paik Sung, C.B. Hu, C.S. Wu, Macromolecules **13**(1), 111 (1980)
[11] O.E.A. Bolduan, R.S. Bear, J. Polym. Sci. **6**, 271 (1951)
[12] J. Brandrup, E.H. Immergut (eds.), *Polymer Handbook* (Interscience Publishers, New York, 1975)
[13] J. Blackwell, M. Ross, Journal of Polymer Science: Polymer Letters Edition **17**, 447 (1979)
[14] R.J. Bonart, L. Morbitzer, G.J. Hentze, Journal of Macromolecular Science, Part B: Physics **B3**, 337 (1969)
[15] R.J. Bonart, L. Morbitzer, E.H. Muller, Journal of Macromolecular Science, Part B: Physics **B9**(3), 447 (1974)
[16] N.R. Legge, G. Holden, H.E. Schroeder (eds.), *Thermoplastic elastomers: a comprehensive review* (Hanser, Munich, 1987)
[17] C. Prisacariu, I. Agherghinei, Journal of Macromolecular Science, Part A: Pure and Applied Chemistry **37**(7), 785 (2000)
[18] L.E. Nielsen, *Mechanical Properties of Polymers* (Van Nostrand Reinhold, Princeton, New Jersey, 1962)

[19] N.P. Apukhtina (ed.), *Sintezi Svoystva Uretanovykh Elastomerov* (Khimiya, Leningrad, 1976)
[20] H.K. Frensdorff, Macromolecules **4**(4) (1971)
[21] M. Xu, W.J. MacKnight, C.H.Y. Chen-Tsai, E.L. Thomas, Polymer **24**(10), 1327 (1983)
[22] C.H.Y. Chen-Tsai, R.M. Briber, E.L. Thomas, M. Xu, W.J. MacKnight, Polymer **24**(10), 1333 (1983)
[23] R.J. Cella, Journal of Polymer Science: Polymer Symposia **42**, 727 (1973)
[24] K.W. Rausch, A.A.R. Sayigh, I&EC Product Research and Development **4**(2), 92 (1965)
[25] C. Prisacariu, A.A. Caraculacu, in *13th International Conference on Deformation, Yield and Fracture of Polymers, Book of Abstracts* (DYFP, 2006), pp. 429–432
[26] J. L. L. Harrell, Macromolecules **2**, 607 (1969)
[27] J.M. Buist, *Development in Polyurethanes – 1* (Applied Science Publishers, London, 1978)
[28] E.F.T. White, British Polymer Journal **18**(6), 403 (1986)
[29] C. Prisacariu, A.A. Caraculacu, in *Book of Abstracts* (The 6th International Conference on Residual Stresses, 2000)
[30] K.C. Frisch, Rubber Chemistry and Technology **45**(5), 1442 (1972)
[31] M. Szycher, V. Poirier, D. Dempsey, W. Robinson, Trans. Soc. Biomater. **6**, 49 (1983)
[32] C.S. Schollenberger, F.D. Stewart, J. Elastoplastics **4**, 294 (1972)
[33] S. Brauman, G. Mayorga, J. Heller, Annals of Biomedical Engineering **9**, 45 (1981)
[34] C.S. Schollenberger, H. Scott, G.R. Moore, Rubber World (137), 549 (1958)
[35] R. Adams, J.L. Anderson, Journal of the American Chemical Society **72**(11), 5154 (1950)
[36] O.G. Takakanov, L.V. Nevskji, V.K. Beljakov, J. Polym. Sci. Part C **23**, 193 (1968)
[37] K. Weissermel, H.J. Arpe, *Industrielle Organische Chemie*, 2nd edn. (Verlag Chemie, Weinheim, 1978)
[38] D.H. Chadwick, E.E. Hardy, *Encyclopedia of Chem. Technology*, vol. 12, 2nd edn. (Interscience, New York, 1967)
[39] R. Vieweg, A. Hochtlen (eds.), *Kunststoff-Handbuch (Band VII).* Polyurethane (Carl Hanser Verlag, Munchen, 1966)
[40] D.K. Chattopadhyay, K.V.S.N. Raju, Progress in Polymer Science **32**(3), 352 (2007)
[41] Y. Li, J. Liu, H. Yang, D. Ma, B. Chu, Journal of Polymer Science Part B: Polymer Physics **31**(7), 853 (1993)
[42] R.V. Honeychuck, T. Ho, K.J. Wynne, R.A. Nissan, Chemistry of Materials **5**(9), 1299 (1993)
[43] C.G. Seefried, J.V. Koleske, F.E. Cfitchfield, Journal of Applied Polymer Science **19**, 3185 (1975)

[44] A.A. Caraculacu, I. Agherghinei, M. Gaspar, C. Prisacariu, J. Chem. Soc., Perkin Trans. 2 pp. 1343–1348 (1990)
[45] A.A. Caraculacu, I. Agherghinei, C. Prisacariu, V. Cozan, Journal of macromolecular science. Chemistry **27**(12), 1547 (1990)
[46] I. Agherghinei, C. Prisacariu, A.A. Caraculacu, Rev. Roumaine de Chimie **36**(9-10), 1135 (1991)
[47] M.V. Pandya, D.D. Deshpande, D.G. Hundiwale, Journal of Applied Polymer Science **32**(5), 4959 (1986)
[48] I. Javni, W. Zhang, Z.S. Petrović, Journal of Applied Polymer Science **88**(13), 2912 (2003)
[49] K. Wiegel, *Polyurethane* (Lacke, Holz-Verlag GmbH, Mering Germany, 1966)
[50] P.G. Forcier, J. Blackwell, Acta Crystallographica Section B **37**(1), 286 (1981)
[51] F. Yeh, B.S. Hsiao, B.B. Sauer, S. Michel, H.W. Siesler, Macromolecules **36**(6), 1940 (2003)
[52] P. Knaub, Y. Camberlin, Journal of Applied Polymer Science **32**(6), 5627 (1986)
[53] S.A. Guelcher, K.M. Gallagher, J.E. Didier, D.B. Klinedinst, J.S. Doctor, A.S. Goldstein, G.L. Wilkes, E.J. Beckman, J.O. Hollinger, Acta Biomater. **1**(4), 471 (2005)
[54] Y. Liu, C.Y. Pan, European Polymer Journal **34**(5-6), 621 (1998)
[55] M. Szycher, *Handbook of polyurethanes* (CRC Press, Boca Raton, FL (US), 1999)
[56] S. Gogolewski, Colloid & Polymer Science **267**, 757 (1989)
[57] O. Bayer, E. Muller, S. Peterser, H. Dieperbrink, E. Windemuth, Angewandte Chemie **62**(57) (1950)
[58] D.J. Lyman, J. Heller, M. Barlow, Die Makromolekulare Chemie **84**(1), 64 (1965)
[59] A.A. Caraculacu, G. Caraculacu, Macromol. Sci. Chem. **A22**, 631 (1995)
[60] C. Prisacariu, R.H. Olley, A.A. Caraculacu, D.C. Bassett, C. Martin, Polymer **44**(18), 5407 (2003)
[61] C. Prisacariu, C.P. Buckley, A.A. Caraculacu, Polymer **46**(11), 3884 (2005)
[62] C. Prisacariu, C.P. Buckley, A.A. Caraculacu, PU Deutsche Magazin **4**, 253 (2006)
[63] C.P. Buckley, C. Prisacariu, A.A. Caraculacu, Polymer **48**(5), 1388 (2007)
[64] E. Scortanu, C. Prisacariu, A.A. Caraculacu, High Performance Polymers **16**(1), 113 (2004)
[65] E. Scortanu, C. Prisacariu, A.A. Caraculacu, M. Bruma, N. Sulitanu, High Performance Polymers **18**(2), 127 (2006)
[66] E. Scortanu, C. Prisacariu, E.G. Hitruc, A.A. Caraculacu, High Performance Polymers **18**(6), 877 (2006). URL http://hip.sagepub.com/content/18/6/877.abstract
[67] C. Prisacariu, E. Scortanu, High Performance Polymers **20**(2), 117 (2008). URL http://hip.sagepub.com/content/20/2/117.abstract

[68] C. Prisacariu, E. Scortanu, High Performance Polymers **in press** (2011)
[69] C. Prisacariu, E. Scortanu, C.P. Buckley, International Journal of Polymer Analysis and Characterization **14**(13), 527 (2009)
[70] T.A. Speckhard, et al., Polymer **26**, 55 (1955)
[71] B. Bengtson, C. Feger, W.J. MacKnight, N.S. Schneider, Polymer **26**(6), 895 (1985)
[72] L.M. Sergeyeva, I.B. Belov, Y.S. Lipatov, T.T. Todosiichuk, Z.Y. Kogan, A.Y. Kalaus, Polymer Science U.S.S.R. **12**(9), 2339 (1970)
[73] Z.S. Petrović, Z. Zavargo, J.H. Flyn, W.J. MacKnight, Journal of Applied Polymer Science **51**(6), 1087 (1994)
[74] R.S. Waletzko, L.T.J. Korley, B.D. Pate, E.L. Thomas, P.T. Hammond, Macromolecules **42**(6), 2041 (2009)
[75] J. Kloss, M. Munaro, G.P. De Souza, J.V. Gulmine, S.H. Wang, S. Zawadzki, L. Akcelrud, Journal of Polymer Science Part A: Polymer Chemistry **40**(23), 4117 (2002)
[76] B. Fernández d'Arlas, L. Rueda, P.M. Stefani, K. de la Caba, I. Mondragon, A. Eceiza, Thermochimica Acta **459**(1-2), 94 (2007)
[77] A. Eceiza, M.D. Martin, K. de la Caba, G. Kortaberria, N. Gabilondo, M.A. Corcuera, I. Mondragon, Polymer Engineering & Science **48**(2), 297 (2008)
[78] B. Fernández d'Arlas, L. Rueda, K. de la Caba, I. Mondragon, A. Eceiza, Polymer Engineering & Science **48**(3), 519 (2008)
[79] L. Rueda-Larraz, B. Fernández d'Arlas, A. Tercjak, A. Ribes, I. Mondragon, A. Eceiza, European Polymer Journal **45**(7), 2096 (2009)
[80] C. Prisacariu, E. Scortanu, International Journal of Polymer Analysis and Characterization **15**(5), 277 (2010)
[81] C.S. Paik Sung, N.S. Schneider, Macromolecules **8**, 68 (1976)
[82] S.B. Clough, N.S. Schneider, Journal of Macromolecular Science, Part B: Physics **B2**, 533 (1968)
[83] K. Shibayama, M. Kodanm, J. Polym. Sci. **A1**(4), 83 (1966)
[84] K.C. Frisch, J.A. Dieter, Polymer-Plastics Technology and Engineering **4**(1), 1 (1975)
[85] Y.K. Godovsky, N.P. Bessonova, Thermochimica Acta **247**(1), 19 (1994)
[86] M. Rogulska, A. Kultys, S. Pikus, Journal of Applied Polymer Science **110**(3), 1677 (2008)
[87] C.S. Schollenberger, (Wiley (Interscience), New York, 1969), p. 197. Polymer Engineering and Technology
[88] J. Furukava, S. Yamashita, M. Murahashi, K. Harada, Makromol Chem. **85**, 80 (1965)
[89] A.A. Caraculacu, S. Coseri, Progress in Polymer Science **26**(5), 799 (2001)
[90] N.V. Kozak, Y.N. Nizel'skii, Theoretical and Experimental Chemistry **24**(2), 148 (1988)
[91] Y.N. Nizel'skii, N.V. Kozak, T.E. Lipatova, Ukr. Khim. Zh. **53**(7), 772 (1987)
[92] V.N. Nicolaev, M.M. Izeeva, I. Semcikov, Vysokomol Soedin **A22**(4), 857 (1980)
[93] S. Ramesh, G. Radhakrishnan, I Indian Chem Soc **74**(4), 347 (1997)

[94] J.W. Peruchin, W.P. Archirejew, E.W. Kusnezow, Plaste Kautsch **22**, 394 (1975)
[95] T.E. Lipatova, N.V. Kozak, Y.N. Nizel'skii, N.E. Kruleak, Theor Exp Khim **20**, 86 (1984)
[96] A.M. Nesterenko, V.G. Maslov, Journal of Structural Chemistry **20**(5), 817 (1979)
[97] S. Siggia, J.G. Hanna, Anal. Chem. **20**, 1084 (1948)
[98] W. Reichen, Chemical Reviews **78**(5), 569 (1978)
[99] C.D. Nenitescu, V. Ioan, *Manualul Inginerului Chimist*, vol. II (Editura Tehnica, Bucuresti, 1952)
[100] J. Baker, J. Gaunt, J. Chem. Soc. pp. 19–24 (1949)
[101] D.P.N. Stachell, R.S. Satchell, Chem Soc. Rev. **4**, 231 (1975)
[102] E. Sacher, Journal of Macromolecular Science, Part B: Physics **16**(4), 525 (1979)
[103] A.E. Oberth, R.S. Bruenner, The Journal of Physical Chemistry **72**(3), 845 (1968)
[104] M. Orozco, F.J. Luque, J. Comput. Chem, **2**, 165 (1991)
[105] S. Scheiner, *Calculating the Properties of Hydrogen Bonds by Ab Initio Methods* (VCH Publishers, New York, 1991), *Reviews in Computational Chemistry*, vol. 2, pp. 165–218
[106] P.P. Schmidt, The Journal of Physical Chemistry **97**(17), 4249 (1993)
[107] N. Shida, J. Almlof, P.F. Barbara, The Journal of Physical Chemistry **95**(25), 10457 (1991)
[108] C.G. Park, M. Tasumi, The Journal of Physical Chemistry **95**(7), 2757 (1991)
[109] C. Panayiotou, I.C. Sanchez, The Journal of Physical Chemistry **95**(24), 10090 (1991)
[110] W.A.P. Luck, Angewandte Chemie **92**(1), 73 (1980)
[111] N. Sokolov, M.V. Vener, V.A. Savellev, Fiz Mnogochastichnykh Sist **19**, 51 (1991)
[112] I.V. Stasyuk, A.L. Ivankiv, Mod. Phys. Lett. B **6**(2), 85 (1992)
[113] H. Graener, T.Q. Ye, A. Laubereau, in *Ultrafast phenomena in spectroscopy: proceedings of the sixth international symposium, Neubrandenburg, German Democratic Republic, August 23-27, 1989* (Springer Verlag, 1990), p. 252
[114] B.A. Veytsman, The Journal of Physical Chemistry **94**(23), 8499 (1990)
[115] W.G.P. Robertson, J.E. Stutchbury, J. Chem. Soc. pp. 4000–4003 (1964)
[116] L.P. Kuhn, Journal of the American Chemical Society **74**(10), 2492 (1952)
[117] L.P. Kuhn, P. Von R. Schleyer, W.F. Baitinger, L. Eberson, Journal of the American Chemical Society **86**(4), 650 (1964)
[118] E.F. Fishman, T.L. Chen, Spectrochim Acta **A**(25), 1231 (1969)
[119] A.A. Caraculacu, I. Agherghinei, P. Baron, S. Coseri, Revue Roumaine de Chimie **41**, 539 (1996)
[120] P.A. Kollman, L.C. Allen, Chemical Reviews **72**(3), 283 (1972)
[121] T. Dziembowska, Pol. J. Chem. **68**, 1455 (1994)
[122] A.J. Benigno, E. Ahmed, M. Berg, The Journal of Chemical Physics **104**(19), 7382 (1996)

[123] R.P. Tiger, L.S. Behli, B. SP, S.G. Entelis, Zh. Org. Khim. **9**, 1563 (1973)
[124] A.A. Caraculacu, S. Coseri, Rev. Roumaine de Chimie **45**(2), 139 (2000)
[125] C. Prisacariu, E. Scortanu, C.P. Buckley, in *AIP Conference Proceedings*, vol. 1255, ed. by A. D'Amore, D. Acierno, L. Grassia (AIP, 2010), vol. 1255, pp. 4–6
[126] D. Dietrich, H. Hespe, (Hanser Publishers, Munich, 1985), pp. 37–53
[127] C. Prisacariu, E. Scortanu, *Structural Studies and the Correlation with the Stress–Strain Response in Polyurethanes* (John Wiley & Sons, Ltd, 2006), pp. 1–24
[128] G. Woods, *The ICI Polyurethane Book*, 2nd edn. (John Wiley & Sons, 1990)
[129] A. Noshay, J.E. McGrath, *Block Copolymers: Overview and Critical Survey* (Academic Press, New York, 1977)
[130] M.D. Lelah, S.L. Cooper, CRC Press, Inc. p. 225 (1986)
[131] K. Kojio, S. Kugumiya, Y. Uchiba, Y. Nishino, M. Furukawa, Polym. J **41**(2), 118 (2009)
[132] D.S. Tiffan, J.F. Terenzi, J. Polym. Sci. **28**, 443 (1958)
[133] Y.I. Boyarchuk, L.Y. Rappoport, V.N. Nikitin, N.P. Apukhtina, Vysokomol. Soyed. **7**, 778 (1965)
[134] C.S. Paik Sung, N.S. Schneider, Journal of Materials Science **13**, 1689 (1978). 10.1007/BF00548732
[135] C.P. Buckley, C. Prisacariu, C. Martin, Polymer **51**(14), 3213 (2010)
[136] J.T. Koberstein, R.S. Stein, Journal of Polymer Science: Polymer Physics Edition **21**(8), 1439 (1983)
[137] L.M. Leung, J.T. Koberstein, Journal of Polymer Science: Polymer Physics Edition **23**(9), 1883 (1985)
[138] R.L. McKiernan, S.P. Gido, J. Penelle, Polymer **43**(10), 3007 (2002)
[139] R.L. McKiernan, A.M. Heintz, S.L. Hsu, E.D.T. Atkins, J. Penelle, S.P. Gido, Macromolecules **35**(18), 6970 (2002)
[140] G.M. Estes, R.W. Seymour, S.L. Cooper, Macromolecules **4**(4), 452 (1971)
[141] R.W. Seymour, G.M. Estes, S.L. Cooper, Macromolecules **3**(5), 579 (1970)
[142] C.S.P. Sung, N.S. Schneider, Macromolecules **8**(1), 68 (1975)
[143] T. Yamamoto, M. Shibayama, S. Nomura, Polymer Journal **21**(11), 895 (1989)
[144] Y.S. Lipatov, V.P. Privalko, Vysokomol. Soyed. **A13**, 103 (1971)
[145] L. Ning, W. De-Ning, Y. Sheng-Kang, Polymer **37**(14), 3045 (1996)
[146] C.S. Paik Sung, N.S. Schneider, Macromolecules **10**(2), 452 (1977)
[147] C.M. Brunette, S.L. Hsu, W.J. MacKnight, Macromolecules **15**(1), 71 (1982)
[148] M.M. Coleman, K.H. Lee, D.J. Skrovanek, P.C. Painter, Macromolecules **19**(8), 2149 (1986)
[149] H.S. Lee, Y.K. Wang, S.L. Hsu, Macromolecules **20**(9), 2089 (1987)
[150] C.D. Eisenbach, W. Gronski, Die Makromolekulare Chemie, Rapid Communications **4**(11), 707 (1983)
[151] J.T. Koberstein, T.P. Russell, Macromolecules **19**(3), 714 (1986)
[152] W. Neumüller, R. Bonart, Journal of Macromolecular Science, Part B: Physics **23**(1), 1 (1984)

[153] V.W. Srichatrapimuk, S.L. Cooper, Journal of Macromolecular Science, Part B: Physics **15**(2), 267 (1978)
[154] S.L. Cooper, R.W. Seymour, J.C. West, *Polyurethane block polymers* (Wiley-Interscience, New York, 1976), vol. 1, p. 521
[155] K.A. Pigott, Kirk-Othmer Encyclopedia of Chemical Technology **21**, 56 (1970)
[156] R. Bonart, Die Angewandte Makromolekulare Chemie **58**(1), 259 (1977)
[157] J.A. Koutsky, N.V. Hien, S.L. Cooper, Journal of Polymer Science Part B: Polymer Letters **8**(5), 353 (1970)
[158] M. Xu, W.J. MacKnight, C.H.Y. Chen-Tsai, E.L. Thomas, Polymer **28**(13), 2183 (1987)
[159] C.H.Y. Chen-Tsai, E.L. Thomas, W.J. MacKnight, N.S. Schneider, Polymer **27**(5), 659 (1986)
[160] M. Serrano, W.J. MacKnight, E.L. Thomas, J.M. Ottino, Polymer **28**(10), 1667 (1987)
[161] I.D. Fridman, E.L. Thomas, Polymer **21**(4), 388 (1980)
[162] N.S. Schneider, C.R. Desper, J.L. Illinger, A.O. King, D. Barr, Journal of Macromolecular Science, Part B: Physics **11**(4), 527 (1975)
[163] Y. Li, Z. Ren, M. Zhao, H. Yang, B. Chu, Macromolecules **26**(4), 612 (1993)
[164] M.A. Dezhu, M. Ruhseng, L. Xiaolie, Chinese Journal of Polymer Science **8**(4) (1990)
[165] M. Serrano, W.J. MacKnight, E.L. Thomas, J.M. Ottino, Polymer **28**(10), 1674 (1987)
[166] C. Li, S.L. Goodman, R.M. Albrecht, S.L. Cooper, Macromolecules **21**(8), 2367 (1988)
[167] A. Wasiak, D. Peiffer, R.S. Stein, Journal of Polymer Science: Polymer Letters Edition **14**(7), 381 (1976)
[168] T.P. Russell, J. Koberstein, R. Prud'Homme, A. Misra, R.S. Stein, J.W. Parsons, R.L. Rowell, Journal of Polymer Science: Polymer Physics Edition **16**(10), 1879 (1978)
[169] S.L. Samuels, G.L. Wilkes, Journal of Polymer Science: Polymer Symposia **43**(1), 149 (1973)
[170] I. Kimura, H. Ishihara, H. Ono, N. Yoshihara, S. Nomura, H. Kawai, Macromolecules **7**(3), 355 (1974)
[171] A.L. Chang, E.L. Thomas, Journal of the American Chemical Society **31**, 176 (1979)
[172] M. Xu, M.L. Ye, L.H. Shi, C. Zhu, D.Z. Ma, X.L. Luo, Polym. Commun. (China) (1), 63 (1985)
[173] P.R. Laity, J.E. Taylor, S.S. Wong, P. Khunkamchoo, M. Cable, G.T. Andrews, A.F. Johnson, R.E. Cameron, Macromolecular Materials and Engineering **291**(4), 301 (2006)
[174] C.E. Wilkes, C.S. Yusek, Journal of Macromolecular Science, Part B: Physics **7**(1), 157 (1973)
[175] C.P. Buckley, C. Prisacariu, A.A. Caraculacu, C. Martin, *Constitutive Models for Rubber IV* (Taylor and Francis, London, 2005), pp. 465–470

[176] S. Abouzahr, G.L. Wilkes, Z. Ophir, Polymer **23**(7), 1077 (1982)
[177] I.M. Pereira, R.L. Oréfice, Macromolecular Symposia **299-300**(1), 190 (2011)
[178] L.M. Leung, J.T. Koberstein, Macromolecules **19**(3), 706 (1986)
[179] S. M., *Szychers handbook of polyurethanes* (CRC Press LLC, Boca Raton, 1999)
[180] J.T. Koberstein, A.F. Galambos, L.M. Leung, Macromolecules **25**(23), 6195 (1992)
[181] S.H. Chen, Annual Review of Physical Chemistry **37**(1), 351 (1986)
[182] V.V. Shilov, E.I. Oranskaya, Y.S. Lipatov, Vysokomol. Soyed. Set A **26**(12), 2508 (1984)
[183] Y.S. Sun, U.S. Jeng, Y.S. Huang, K.S. Liang, T.L. Lin, C.S. Tsao, Physica B: Condensed Matter **385-386**(Part 1), 650 (2006). Proceedings of the Eighth International Conference on Neutron Scattering
[184] L.Y. Chiang, L.Y. Wang, C.S. Kuo, Macromolecules **28**(22), 7574 (1995)
[185] C. Prisacariu, E. Scortanu, High Performance Polymers **22**(7), 876 (2010)
[186] A. Saiani, W.A. Daunch, H. Verbeke, J.W. Leenslag, J.S. Higgins, Macromolecules **34**(26), 9059 (2001)
[187] A. Saiani, C. Rochas, G. Eeckhaut, W.A. Daunch, J.W. Leenslag, J.S. Higgins, Macromolecules **37**(4), 1411 (2004)
[188] B. Chu, T. Gao, Y. Li, J. Wang, C.R. Desper, C.A. Byrne, Macromolecules **25**(21), 5724 (1992)
[189] R.A. Assink, Journal of Polymer Science: Polymer Physics Edition **15**(1), 59 (1977)
[190] L.C. Dickinson, J.F. Shi, J.C.W. Chien, Macromolecules **25**(4), 1224 (1992)
[191] J.J. Dumais, L.W. Jelinski, L.M. Leung, I. Gancarz, A.F. Galambos, J.T. Koberstein, Macromolecules **18**(1), 116 (1985)
[192] M.D. Meadows, C.P. Christenson, W.L. Howard, M.A. Harthcock, R.E. Guerra, R.B. Turner, Macromolecules **23**(9), 2440 (1990)
[193] J.A. Kornfield, H.W. Spiess, H. Nefzger, H. Hayen, C.D. Eisenbach, Macromolecules **24**(17), 4787 (1991)
[194] H.W. Spiess, Angew. Makromol. Chem. **202/203**, 331 (1992)
[195] L.C. Sawyer, D.T. Grubb, G.F. Meyers, *Polymer microscopy* (Springer Verlag, 2008)
[196] A. Aneja, G.L. Wilkes, Polymer **44**(23), 7221 (2003)
[197] C. Harrison, P.M. Chaikin, D.A. Huse, R.A. Register, D.H. Adamson, A. Daniel, E. Huang, P. Mansky, T.P. Russell, C.J. Hawker, D.A. Egolf, I.V. Melnikov, E. Bodenschatz, Macromolecules **33**(3), 857 (2000)
[198] R.W. Seymour, S.L. Cooper, Macromolecules **6**(1), 48 (1973)
[199] S. Yamasaki, D. Nishiguchi, K. Kojio, M. Furukawa, Polymer **48**(16), 4793 (2007)
[200] H.S. Lee, Y.K. Wang, W.J. MacKnight, S.L. Hsu, Macromolecules **21**(1), 270 (1988)
[201] J. Blackwell, M.R. Nagarajan, T.B. Hoitink, Polymer **22**(11), 1534 (1981)

[202] L.A. Gower, T.L.D. Wang, D.J. Lyman, Journal of Biomaterials Science, Polymer Edition **6**, 761 (1995)
[203] L. Leibler, Macromolecules **13**(6), 1602 (1980)
[204] C. Prisacariu, E. Scortanu, Revue Roumaine de Chimie **53**(9), 821 (2008)
[205] L.E. Alexander, *X-ray Diffraction Methods in Polymer Science*, 1st edn. (John Wiley & Sons Inc., 1969)
[206] C. Prisacariu, E. Scortanu, Macromolecular Symposia **254**(1), 153 (2007)
[207] C. Prisacariu, E. Scortanu, I. Stoica, B. Agapie, V. Barboiu, Polymer Journal (2011)
[208] E.M. Christenson, J.M. Anderson, A. Hiltner, E. Baer, Polymer **46**(25), 11744 (2005)
[209] S.L. Cooper, A.V. Tobolsky, Journal of Applied Polymer Science **10**(12), 1837 (1966)
[210] B. Hartmann, J.V. Duffy, G.F. Lee, E. Balizer, Journal of Applied Polymer Science **35**(7), 1829 (1988)
[211] J. Ferguson, D.J. Hourston, R. Meredith, D. Patsavoudis, European Polymer Journal **8**(3), 369 (1972)
[212] H. Jacobs, E. Jenckel, Makromol. Chem. **43, 47**, 132, 71 (1961)
[213] N.G. McCrumb, B.E. Read, G. Williams, *Anelastic and Dielectric Effects in Polymeric Solids* (Wiley, New York, 1967)
[214] L.T.J. Korley, B.D. Pate, E.L. Thomas, P.T. Hammond, Polymer **47**(9), 3073 (2006)
[215] D.J. Martin, G.F. Meijs, P.A. Gunatillake, S.J. McCarthy, G.M. Renwick, Journal of Applied Polymer Science **64**(4), 803 (1997)
[216] R.F. Harris, M.D. Joseph, C. Davidson, C.D. Deporter, V.A. Dais, Journal of Applied Polymer Science **41**(3-4), 487 (1990)
[217] J.I. Mardel, A.J. Hill, K.R. Chynoweth, M.E. Smith, C.H.J. Johnson, T.J. Bastow, Wear **162-164**(Part 2), 645 (1993)
[218] M. Renier, Y.K. Wu, J.M. Anderson, A. Hiltner, G.A. Lodoen, C.R. Payet, Journal of Biomaterials Science, Polymer Edition **5**(6), 511 (1994)
[219] M.C.E.J. Niesten, J.W.T. Brinke, R.J. Gaymans, Polymer **42**(4), 1461 (2001)
[220] D.M. Crawford, R.G. Bass, T.W. Haas, Thermochimica Acta **323**(1-2), 53 (1998)
[221] T. Murayama, *Dynamic Analysis of Polymeric Material* (Elsevier Scientific, New York, 1978)
[222] K. Bagdi, K. Molnár, B.P. Jr., Journal of Thermal Analysis and Calorimetry **98**(3), 825 (2009)
[223] B. Pukánszky, K. Bagdi, Z. Tóvölgyi, J. Varga, L. Botz, S. Hudak, T. Dóczi, European Polymer Journal **44**(8), 2431 (2008)
[224] A. Marcos, A. Rodríguez, L. González, Journal of Non-Crystalline Solids **172-174**(Part 2), 1125 (1994)
[225] T. Hashimoto, Y. Tsukahara, K. Tachi, H. Kawai, Macromolecules **16**(4), 648 (1983)
[226] Y. Li, T. Gao, J. Liu, K. Linliu, C.R. Desper, B. Chu, Macromolecules **25**(26), 7365 (1992)

[227] M.A. Hood, B. Wang, J.M. Sands, J.J.L. Scala, F.L. Beyer, C.Y. Li, Polymer **51**(10), 2191 (2010)
[228] C. Prisacariu, E. Scortanu, B. Agapie, Proceedia Engineering (2011)
[229] G.L. Wilkes, R. Wildnauer, Journal of Applied Physics **46**(10), 4148 (1975)
[230] J. Blackwell, C.D. Lee, Journal of Polymer Science: Polymer Physics Edition **22**(4), 759 (1984)
[231] P.R. Couchman, Macromolecules **11**(6), 1156 (1978)
[232] R.W. Seymour, S.L. Cooper, Journal of Polymer Science Part B: Polymer Letters **9**(9), 689 (1971)
[233] C.P. Christenson, M.A. Harthcock, M.D. Meadows, H.L. Spell, W.L. Howard, M.W. Creswick, R.E. Guerra, R.B. Turner, Journal of Polymer Science Part B: Polymer Physics **24**(7), 1401 (1986)
[234] T.R. Hesketh, J.W.C. Van Bogart, S.L. Cooper, Polymer Engineering & Science **20**(3), 190 (1980)
[235] G. Oertel (ed.), *Polyurethane Handbook* (Hanser, Munich, Germany, 1994)
[236] T.K. Chen, T.S. Shieh, J.Y. Chui, Macromolecules **31**(4), 1312 (1998)
[237] J.W.C. Van Bogart, D.A. Bluemke, S.L. Cooper, Polymer **22**(10), 1428 (1981)
[238] Y.S. Lipatov, V.P. Privalko, Vysokomol. Soyed. Set A **13**, 103 (1971)
[239] Y. Li, Z. Ren, M. Zhao, H. Yang, B. Chu, Macromolecules **26**, 612 (1993)
[240] J.J. Aklonis, W.J. MacKnight, *Introduction to Polymer Viscoelasticity* (John Wiley & Sons, New York, 1992)
[241] L.H. Sperling, *Introduction to Physical Polymer Science* (John Wiley & Sons, New York, 1992)
[242] M. Jaffe, J.D. Menczel, W.E. Bessey, *Films* (Academic Press, San Diego, 1977), pp. 1955–2089
[243] M. Barikani, M. Barmar, Iranian Polymer Journal **5**(4), 231 (1996)
[244] P. Penkzek, E. Rudnik, B. Arczewska, R. Ostrysz, Polimery **40**, 464 (1995)
[245] M. Rutkowska, Polimery **26**, 360 (1981)
[246] A. Koscielecka, Polimery **26**, 349 (1981)
[247] A. Koscielecka, Polimery **30**, 187 (1985)
[248] P. Król, Progress in Materials Science **52**(6), 915 (2007)
[249] C.J. Paul, M.R.G. Nair, N.R. Neelakantan, P. Koshy, B.B. Idage, A.A. Bhelhekar, Polymer **39**(26), 6861 (1998)
[250] C. Prisacariu, E. Scortanu, Revue Roumaine de Chimie **54**(11-12), 963 (2009)
[251] C. Prisacariu, E. Scortanu, B. Agapie, in *Conference Proceedings 27th Meeting* (Polymer Processing Society, Marakech, 2011)
[252] C. Prisacariu, *Thermal Analysis of Novel Nolyurethane Block Polymers Derived From Flexible Hard Segments* (2001), vol. XLVII (LI), chap. Section I. Mathematics. Theoretical Mechanics. Physics, pp. 143–156
[253] C. Prisacariu, E. Scortanu, V.A. Prisacariu, Revue Roumaine de Chimie **in press** (2011)

[254] C. Prisacariu, C.P. Buckley, *Synthesis and Characterization of Some Novel Polyurethanes Based on Rigid or Flexible Aromatic Diisocyanates and Mixtures of Them* (Cincinnati, OH, SUA, 2007), pp. 992–995

[255] C.W. Bunn, E.V. Garuer, Proc.R.Soc. London **189**(39) (1947)

[256] D.R. Holmes, C.W. Bunn, D.J. Smith, Journal of Polymer Science **17**(84), 159 (1955)

[257] C. Prisacariu, E. Scortanu, B. Agapie, Proceedia Engineering (2) (2011)

[258] ASTM D412-06a, *Standard Test Methods for Vulcanized Rubber and Thermoplastic Elastomers-Tension*

[259] C. Hepburn, *Polyurethane elastomers 2nd Edition* (Elsevier Science Pulbishers LTD, 1992)

[260] E. Muller, F.W. Schmidt, E.W. Weinbrenner, H.F. Piepenbrink. U.S. Patent 2.729.618 (1956)

[261] C. Prisacariu, C.P. Buckley, *Influence of the type of chain extender and urethane group content in the mechanical properties of polyurethanes based on 4,4– dibenzyl diisocyanate* (2006), pp. 456 – 459. 13th International Conference on Deformation, Yield and Fracture of Polymers, Conf. Proceedings

[262] L. Born, H. Hespe, J. Crone, K.H. Wolf, *The physical crosslinking of polyurethane elastomers studied by X-ray investigation of model urethanes* (Springer Berlin / Heidelberg, 1982), vol. 260, pp. 819–828

[263] R. Bonart, L. Morbitzer, H. Rinke, *Beiträge zum physikalischen Verständnis der Erweichungstemperatur von nebenvalenzvernetzten Urethan-Elastomeren* (Springer Berlin / Heidelberg, 1970), vol. 240, pp. 807–819

[264] G.L. Wilkes, S.L. Samuels, R. Crystal, Journal of Macromolecular Science, Part B: Physics **10**(2), 203 (1974)

[265] S.L. Samuels, G.L. Wilkes, Journal of Polymer Science Part B: Polymer Letters **9**(10), 761 (1971)

[266] M.A. Vallance, J.L. Castles, S.L. Cooper, Polymer **25**(12), 1734 (1984)

[267] Z.S. Petrović, J. Budinski-Simendic, Rubber chemistry and technology **58**(4), 685 (1985)

[268] Y.J.P. Chang, G.L. Wilkes, Journal of Polymer Science: Polymer Physics Edition **13**(3), 455 (1975)

[269] M. Shen, U. Mehra, M. Niinomi, J.T. Koberstein, S.L. Cooper, Journal of Applied Physics **45**(10), 4182 (1974)

[270] R.W. Seymour, J.R. Overton, L.S. Corley, Macromolecules **8**(3), 331 (1975)

[271] A. Lilaonitkul, S.L. Cooper, Rubber Chemistry and Technology **50**(1), 1 (1977)

[272] P. Wright, A.P.C. Cumming, *Solid Polyurethane Elastomers*, 1st edn. (MacLaren and Sons, London, 1969)

[273] I. Yilgor, E. Yilgor, Polymer Reviews **47**(4), 487 (2007)

[274] S.L. Huang, J.Y. Lai, European Polymer Journal **33**(10-12), 1563 (1997)

[275] M. Song, H. Chen, C. Jiang, B. Zhao, X. Li, Macromolecular Theory and Simulations **11**(8), 845 (2002)

[276] J.W. Rosthauser, K.W. Haider, C. Steinlein, C.D. Eisenbach, Journal of Applied Polymer Science **64**(5), 957 (1997)
[277] J.L. Stanford, R.H. Still, A.N. Wilkinson, Polymer **44**(14), 3985 (2003)
[278] A.J. Ryan, J.L. Stanford, R.H. Still, Polymer **32**(8), 1426 (1991)
[279] D.W. van Krevelen, P.J. Hoftyzer, *Properties of polymers: correlations with chemical structure* (Elsevier Science Publishers LTD, Amsterdam, New York, 1972)
[280] C.P. Buckley, C. Prisacariu, A.A. Caraculacu, *Constitutive response of model segmented copolyurethane elastomers* (2002), pp. 96–99. Euromech Colloquium and Workshop
[281] C.P. Buckley, C. Prisacariu, A.A. Caraculacu, *Effects of hard segment crystallinity on constitutive response of copolyurethane elastomers* (Cambridge, UK, 2003), pp. 349–352. 12th International Conference on Deformation, Yield and Fracture of Polymers
[282] G. Abouzahr, G.L. Wilkes, Polymer Preprints **21**, 193 (1980)
[283] C.P. Buckley, *Experimental methods for rubberlike solids* (Springer, Vienna, 2004), pp. 1–62. CISM Courses and Lectures No. 452
[284] D.J. Blundell, G. Eeckhaut, W. Fuller, A. Mahendrasingam, C. Martin, Polymer **43**(19), 5197 (2002)
[285] J.A.C. Harwood, L. Mullins, A.R. Payne, Rubber Chemistry and Technology **39**(4), 814 (1966)
[286] J. Diani, B. Fayolle, P. Gilormini, European Polymer Journal **45**(3), 601 (2009)
[287] G. Muller-Riederer, R. Bonart, Prog. Colloid Polym. Sci. **62**, 99 (1977)
[288] R. Bonart, K. Koffman, Colloid & Polymer Science **260**, 268 (1982)
[289] H. Bouasse, Z. Carrière, Annales de la faculté des sciences de Toulouse Sér. 2 **5**(3), 257 (1903)
[290] J.J. Alkonis, W.J. MacKnight, *Introduction to polymer viscoelasticity* (John Wiley and Sons, New York, 1983)
[291] J.N. Gorce, J.W. Hellgeth, T.C. Ward, Polymer Engineering & Science **33**(18), 1170 (1993)
[292] R.A. Nallicheri, M.F. Rubner, Macromolecules **24**(2), 526 (1991)
[293] B. Bengtson, C. Feger, W.J. MacKnight, N.S. Schneider, Polymer **26**(6), 895 (1985)
[294] Y.K. Godovsky, N.P. Bessonova, N.N. Mironova, Colloid & Polymer Science **264**(3), 224 (1986)
[295] W. Li, A.J. Ryan, I.K. Meier, Macromolecules **35**(13), 5034 (2002)
[296] R.W. Ogden, D.G. Roxburgh, Proceedings of the Royal Society of London. Series A: Mathematical, Physical and Engineering Sciences **455**(1988), 2861 (1999)
[297] T.L. Smith, Journal of Polymer Science: Polymer Physics Edition **12**(9), 1825 (1974)
[298] Y.K. Godovsky, N.P. Bessonova, N.N. Mironova, Colloid & Polymer Science **267**(5), 414 (1989)

[299] N.P. Bessonova, A.R. Korigodskii, D.F. Kutepov, Y.K. Godovsky, Polymer Science U.S.S.R. **26**(9), 2098 (1984)

[300] R. Russo, E.L. Thomas, Journal of Macromolecular Science, Part B: Physics **22**(4), 553 (1983)

[301] H.J. Qi, M.C. Boyce, Mechanics of Materials **37**(8), 817 (2005)

[302] J. Yi, M.C. Boyce, G.F. Lee, E. Balizer, Polymer **47**(1), 319 (2006)

[303] J.A. Miller, S.B. Lin, K.K.S. Hwang, K.S. Wu, P.E. Gibson, S.L. Cooper, Macromolecules **18**(1), 32 (1985)

[304] E. Scortanu, C. Prisacariu, Journal of optoelectronics and advanced materials **9**(4), 1025 (2007)

[305] E. Scortanu, C. Prisacariu, A.A. Caraculacu, *Contributions towards the field of polymers with thermal and photo-oxidative stability / Contribuţii în domeniul polimerilor cu stabilitate termică şi foto-oxidativă. Aplicaţii* (Fides, Iaşi, România, 2006)

[306] Z. Hashin, S. Shtrikman, Journal of the Mechanics and Physics of Solids **11**(2), 127 (1963)

[307] R.M. Christensen, *Mechanics of Composite Materials* (Wiley, New York, 1979)

[308] R.W. Gray, N.G. McCrumb, Journal of Polymer Science Part A-2: Polymer Physics **7**(8), 1329 (1969)

[309] J.C. Moreland, G.L. Wilkes, R.B. Turner, Journal of Applied Polymer Science **43**(4), 801 (1991)

[310] C.R. Desper, N.S. Schneider, J.P. Jasinski, J.S. Lin, Macromolecules **18**(12), 2755 (1985)

[311] R. Bonart, Polymer **20**(11), 1389 (1979). Jabtonna Conference on Polymer Networks

[312] H.F. Enderle, H.G. Kilian, B. Heise, J. Mayer, H. Hespe, Colloid & Polymer Science **264**(4), 305 (1986)

[313] C.P. Buckley, D.C. Jones, Polymer **36**(17), 3301 (1995)

[314] C.P. Buckley, J.J. Wu, in *19th Annual Meeting of the Polymer Processing Society* (Melbourne, 2003)

[315] E. Scortanu, C. Prisacariu, J. Optoelectron. Adv. Mater. **9**(4), 1025 (2007)

[316] E. Scortanu, A. Taranu, C. Prisacariu, A.A. Caraculacu, Annals of Science of University "Al. I. Cuza" Iasi **XIV**(1) (2006)

[317] R. Lappin, *1971 Encyclopedia of Polymer Science and Technology*, vol. 14 (Interscience Publishers, Inc., New York, 1971)

[318] H.C. Beachell, I.L. Chang, Journal of Polymer Science Part A-1: Polymer Chemistry **10**(2), 503 (1972)

[319] B. Ranby, J.E. Rabek, *Photodegradation, Photo-oxidation and Photostabilization of Polymers, Principle and Applications* (Wiley (London Interscience), 1971)

[320] L. Geanău, A.A. Caraculacu, V. Taranu, S. Seiler, V. Popescu, *RO Patent 78981* (1982)

[321] A.M. Somlai, D. Creed, F.A. Landis, S. Mahaderan, C.E. Hoyle, A.C. Griffin, Polymer Preprints **41**(1), 371 (2000)

[322] E. Scortanu, I. Beştiuc, A.A. Caraculacu, Iranian Polymer Journal **4**(4), 284 (1995)

[323] S. Michitaka, *JP Patent 63, 56, 557*, vol. 109, p. 74612

[324] D. Munteanu, C. Bolcu, V. Ostafe, Polymer & Polymer Composites **8**(2), 87 (2000)

[325] E. Scortanu, I. Beştiuc, A.A. Caraculacu, *RO Patent 103, 442, 1999* (RO Patent 114252, 1991)

[326] D. Dieterich, K. Uhlig, (Weinheim: Verlag, 2001), pp. 419–493

[327] C. Prisacariu, E. Scortanu, V.A. Prisacariu, in *Lecture Notes in Engineering and Computer Science*, vol. 2 (2010), vol. 2, pp. 1434–1437

[328] C. Prisacariu, E. Scortanu, Journal of Applied Polymer Science **in press** (2010)

[329] C.H. Bamford, C.F.H. Tipper, *Comprehensive Chemical Kinetics*, vol. 14 (Elsevier Scientific Publishing Co., New York, 1975)

[330] R.K. Eby, *Durability of Macromolecular Materials* (ACS Symp, 1979)

[331] C. Prisacariu, A.A. Caraculacu, B. Williams, Polytechnic Bull.Iasi **XLII (XLVI)**(3-4 sect. 1), 123 (1996)

[332] C. Hepburn, Iranian Journal of Polymer Sci. and Technology **1**(2), 84 (1992)

[333] A.M. Kaminsky, M.W. Urban, J. Coat. Technol **69**(873), 113 (1997)

[334] S. Hamed, M. Eslami, S.M. Nabi, Iran. J. Chem, Chem. Eng. **15**(2), 87 (1996)

[335] A. Awater, (Hanser Publishers, Munich, 1985), p. 387

[336] K. Piggot, *Encyclopedia of Polymer Science and Technology*, vol. XI (John Wiley & Sons, Inc., New York, 1969)

[337] D.W. Duff, G.E. Maciel, Macromolecules **24**(2), 387 (1991)

[338] L. Tidhzert, J. Pascault, K. Dusek, E. Querat, Angew. Makromol. Chem. pp. 1–36 (1996)

[339] A.A. Caraculacu, I. Agherghinei, P. Baron, G. Caraculacu, S. Coseri, European Polymer Journal **32**(10), 1235 (1996)

[340] M. Komiya, T. Yokoyama, M. Furukawa, Angew. Makromol. Chem. **240**, 205 (1996)

[341] A. Toffey, N.G. Glasser, Holzforschung **51**(1), 71 (1997)

[342] J.H. Saunders, Rubber Chemistry and Technology **32**, 334 (1959)

[343] P. Pissis, L. Apekis, C. Christodoulides, M. Niaumakis, A. Kyritsis, J. Nedbal, Journal of Polymer Science Part B: Polymer Physics **34**(9), 1529 (1996)

[344] B. Briscoe, C. Kelly, Polymer **37**(15), 3405 (1996)

[345] L.S. Teo, J.F. Kuo, C.Y. Chen, Journal of Applied Polymer Science **59**(10), 1627 (1996)

[346] G. Shkapenko, G.T. Gmitter, E.E. Gruber, Industrial & Engineering Chemistry Research **52**, 605 (1960)

[347] P.G. Pichon, L. David, F. Méchin, H. Sautereau, Macromolecules **43**(4), 1888 (2010)

[348] R.R. Lagasse, Journal of Applied Polymer Science **21**(9), 2489 (1977)

[349] J.W.C. Van Bogart, P.E. Gibson, S.L. Cooper, Journal of Polymer Science: Polymer Physics Edition **21**(1), 65 (1983)

[350] C. Chen, R. Briber, E. Thomas, M. Xu, W. MacKnight, Polymer **24**(10), 1333 (1983)

[351] D. Puett, Journal of Polymer Science Part A-2: Polymer Physics **5**(5), 839 (1967)

[352] N. Reynolds, H.W. Spiess, H. Hayen, H. Nefzger, C.D. Eisenbach, Macromolecular Chemistry and Physics **195**(8), 2855 (1994)

[353] A. Allegrezza, R. Seymour, H. Ng, S. Cooper, Polymer **15**(7), 433 (1974)

[354] G.L. Wilkes, S. Abouzahr, Macromolecules **14**(2), 456 (1981)

[355] N.S. Desper, C.R. Schneider, (Plenum Press, New York, 1983), pp. 233–251

[356] R.W. Hendricks, Journal of Applied Crystallography **11**(1), 15 (1978)

[357] P. Debye, A.M. Bueche, Journal of Applied Physics **20**(6), 518 (1949)

[358] R.A. J., N. Steven, K. Bernd, B. Wim, M.G. R., D.G. E., *Polymer Structure Determination Using Simultaneous Small- and Wide-Angle X-ray Scattering and Differential Scanning Calorimetry* (American Chemical Society, 1994), vol. 581, chap. 13, pp. 162–180

[359] R.W. Seymour, S.L. Cooper, Journal of Polymer Science: Polymer Symposia **46**(1), 69 (1974)

[360] H. Sup Lee, S. Ra Yoo, S. Won Seo, Journal of Polymer Science Part B: Polymer Physics **37**(22), 3233 (1999)

[361] D.S. Huh, S.L. Cooper, Polymer Engineering & Science **11**(5), 369 (1971)

[362] C.S.P. Sung, *Polymer Alloys II* (Plenum Publishing Co., 1980)

[363] R.W. Seymour, A.E. Allegrezza, S.L. Cooper, Macromolecules **6**(6), 896 (1973)

[364] P. Hammond, R. Nallicheri, M. Rubner, Materials Science and Engineering: A **126**(1-2), 281 (1990). Structural Materials: Properties, Microstructure and Processing

[365] A. Lendlein, S. Kelch, *Shape-memory polymers*, vol. 41 (Angewandte Chemie International Edition, 2002)

[366] J.R. Lin, L.W. Chen, Journal of Applied Polymer Science **69**(8), 1563 (1998)

[367] B.S. Lee, B.C. Chun, Y.C. Chung, K.I. Sul, J.W. Cho, Macromolecules **34**(18), 6431 (2001)

[368] J.H. Yang, B.C. Chun, Y.C. Chung, J.H. Cho, Polymer **44**(11), 3251 (2003)

[369] W. Wang, P. Ping, X. Chen, X. Jing, European Polymer Journal **42**(6), 1240 (2006)

[370] T. Takahashi, N. Hayashi, S. Hayashi, Journal of Applied Polymer Science **60**(7), 1061 (1996)

[371] J. Lin, L. Chen, Journal of applied polymer science **69**(8), 1575 (1998)

[372] B.K. Kim, S.Y. Lee, M. Xu, Polymer **37**(26), 5781 (1996)

[373] T. Ohki, Q.Q. Ni, N. Ohsako, M. Iwamoto, Composites Part A: Applied Science and Manufacturing **35**(9), 1065 (2004)

[374] K. Gall, M.L. Dunn, Y. Liu, D. Finch, M. Lake, N.A. Munshi, Acta Materialia **50**(20), 5115 (2002)

[375] Y. Liu, K. Gall, M.L. Dunn, P. McCluskey, Mechanics of Materials **36**(10), 929 (2004). Active Materials

[376] J.W. Cho, S.H. Lee, European Polymer Journal **40**(7), 1343 (2004)

[377] C.P. Buckley, C. Prisacariu, A.A. Caraculacu, in *13th International Conference on Deformation, Yield and Fracture of Polymers, Conf. Proceedings* (DYFP 2006, 2006), pp. 284–287

Index